MW00564355

Springer Transactions in Civil and Environmental Engineering

Editor-in-Chief

T. G. Sitharam, Department of Civil Engineering, Indian Institute of Science, Bengaluru, Karnataka, India

Springer Transactions in Civil and Environmental Engineering (STICEE) publishes the latest developments in Civil and Environmental Engineering. The intent is to cover all the main branches of Civil and Environmental Engineering, both theoretical and applied, including, but not limited to: Structural Mechanics, Steel Structures, Concrete Structures, Reinforced Cement Concrete, Civil Engineering Materials, Soil Mechanics, Ground Improvement, Geotechnical Engineering, Foundation Engineering, Earthquake Engineering, Structural Health and Monitoring, Water Resources Engineering, Engineering Hydrology, Solid Waste Engineering, Environmental Engineering, Wastewater Management, Transportation Engineering, Sustainable Civil Infrastructure, Fluid Mechanics, Pavement Engineering, Soil Dynamics, Rock Mechanics, Timber Engineering, Hazardous Waste Disposal Instrumentation and Monitoring, Construction Management, Civil Engineering Construction, Surveying and GIS Strength of Materials (Mechanics of Materials), Environmental Geotechnics, Concrete Engineering, Timber Structures.

Within the scopes of the series are monographs, professional books, graduate and undergraduate textbooks, edited volumes and handbooks devoted to the above subject areas.

More information about this series at http://www.springer.com/series/13593

Manish Kumar · Francisco Munoz-Arriola ·
Hiroaki Furumai · Tushara Chaminda
Editors

Resilience, Response, and Risk in Water Systems

Shifting Management and Natural Forcings Paradigms

 Springer

Editors
Manish Kumar
Discipline of Earth Sciences
Indian Institute of Technology Gandhinagar
Gandhinagar, Gujarat, India

Hiroaki Furumai
Research Center for Water Environment
Technology, School of Engineering
University of Tokyo
Tokyo, Japan

Francisco Munoz-Arriola
Department of Biological Systems
Engineering
University of Nebraska–Lincoln
Lincoln, NE, USA

Tushara Chaminda
Department of Civil and Environmental
Engineering, Faculty of Engineering
University of Ruhuna
Galle, Sri Lanka

ISSN 2363-7633 ISSN 2363-7641 (electronic)
Springer Transactions in Civil and Environmental Engineering
ISBN 978-981-15-4667-9 ISBN 978-981-15-4668-6 (eBook)
https://doi.org/10.1007/978-981-15-4668-6

© Springer Nature Singapore Pte Ltd. 2020
This work is subject to copyright. All rights are reserved by the Publisher, whether the whole or part of the material is concerned, specifically the rights of translation, reprinting, reuse of illustrations, recitation, broadcasting, reproduction on microfilms or in any other physical way, and transmission or information storage and retrieval, electronic adaptation, computer software, or by similar or dissimilar methodology now known or hereafter developed.
The use of general descriptive names, registered names, trademarks, service marks, etc. in this publication does not imply, even in the absence of a specific statement, that such names are exempt from the relevant protective laws and regulations and therefore free for general use.
The publisher, the authors and the editors are safe to assume that the advice and information in this book are believed to be true and accurate at the date of publication. Neither the publisher nor the authors or the editors give a warranty, expressed or implied, with respect to the material contained herein or for any errors or omissions that may have been made. The publisher remains neutral with regard to jurisdictional claims in published maps and institutional affiliations.

This Springer imprint is published by the registered company Springer Nature Singapore Pte Ltd.
The registered company address is: 152 Beach Road, #21-01/04 Gateway East, Singapore 189721, Singapore

Dedicated to the inspirational life of Padma shri Prof. Sudhir K. Jain, FINAE, Director IIT-Gandhinagar and a sacrificing soul to the world of academics

Foreword by Dr. Virendra M. Tiwari

The identification, transport and fate of pollutants (geogenic and anthropogenic) in the water system are an intricate and challenging issues. I am extremely delighted to note the publication of this contemporary and relevant volume on water systems. The chapters in this book have focused on the water quality monitoring, management and methods for mitigation of associated risks of water contaminations in South Asia and highlighted the most recent advances on hydrogeography, hydrology and water resource management.

There is scanty information available on the surface water–groundwater contaminant interactions under different environmental conditions across the world, as it is considered to be an important driver for prediction of risk and management of water cycle through hydro-geochemical events. Twenty-two contributed chapters are embedded under three sections in this book by the learned experts of their respective fields. The book presents problems, perspectives and challenges of water ecosystem in its first section, whereas in the second section, an average estimation of certain parameters (e.g. precipitation, soil moisture and snow water content) has been highlighted for early prediction of risk and management purposes. On the other hand, the third/last section explores the leading approaches in the field of science and engineering of water management along with the human behaviour and decision making based on life cycle as well as cost-benefit analyses.

Integrating the knowledge of water quality, quantity, monitoring, and engineering of water resource systems are the uniqueness of all the contributed chapters for a better understanding and progress of indispensable freshwater issues. The book is designed to bring together and integrate the subject matter that deals with data science and engineering, remote sensing, modelling, analytics, synthesis and indices, disruptive innovations and its utilization in water management, policy making and mitigation strategies in a single text.

Thus, this book offers a comprehensive text for students and professionals in the Water and Environmental Sciences.

Dr. Virendra M. Tiwari
Director
CSIR-National Geophysical Research Institute
Hyderabad, India

Acknowledgements

This book project is a successful outcome of the joint international collaborative projects namely, IJCSP (India-Japan Cooperative Science Program) and DST-UKIERI (UK-India Education and Research Initiative) under which all the contributed chapters have been put together for the present reference book on the theme of resilient water supply and sustainable water quality management in the urban environment. We also acknowledge the international collaboration and support of Prof. Manish Kumar under the Water Advanced Research and Innovation (WARI) Fellowship programme. We thankfully appreciate the ardent support and continuous cooperation of Indian Institute Under the WARI e of Technology IIT Gandhinagar, the University of Nebraska–Lincoln, USA, the University of Tokyo, Japan, and the University of Ruhuna, Srilanka. Though it is impossible to mention all the names, we are thankful to Dr. Arbind Kumar Patel, Dr. Santanu Mukherjee and WET lab members without whom the editorial responsibilities would not have been possible to complete it in a timely manner. Lastly, our book project would not be accomplished without the heartfelt blessings from our well wishers, family members, peers and friends.

Gandhinagar, India	Dr. Manish Kumar
Lincoln, USA	Dr. Francisco Munoz-Arriola
Tokyo, Japan	Prof. Hiroaki Furumai
Galle, Sri Lanka	Dr. Tushara Chaminda

Contents

Editors and Contributors

About the Editors

Dr. Manish Kumar is a Fellow of the Royal Society of Chemistry (FRSC) and faculty at Discipline of Earth Sciences at the Indian Institute of Technology Gandhinagar, Gujarat, India. He earned his Ph.D. in Environmental Engineering from the University of Tokyo, Japan. He has multifacet research domains like Groundwater Pollution and Remediation, Water Systems and Climate Change, Isotope Hydrology, Emerging Contaminants: Occurrence, Fate and Transport, Arsenic and Metal Toxicity, and Bioremediation. He has been the recipient of Water Advanced Research and Innovation (WARI) Fellowship, Japan Society for the Promotion of Science (JSPS) foreign research fellowship, Brain Korea (BK)-21 post-doctoral fellowship, Monbukagakusho scholarship, Linnaeus-Palme stipend from SIDA, Sweden, and Research Fellowship from CSIR, India and others. He supervised 6 Ph.D. thesis and >20 master dissertations. He published >80 international peer-reviewed journal papers, >120 other scholarly works and has 17 years' research/teaching experience with H-index =25, i10-index=50 with total citation ($\sim$2500)-Google Scholar. He is the core committee member of the International Water Association (IWA)-India Chapter. He is the one of the illutrious members of global collaboration on wastewater based epidemiology of COVID-19.

Dr. Francisco Munoz-Arriola is a faculty member in the Department of Biological Systems Engineering and the School of Natural Resources at the University of Nebraska-Lincoln. His area of interest and expertise includes coupled natural-human systems, hydroinformatics, integrated hydroclimate and water quality and quantity, resilient of complex landscapes, sub-seasonal to seasonal predictability of hydrometeorological and climate extremes and phenotypes, remote sensing applications, and the nexus food-energy-water-ecosystem services. He is fellow of the Robert B. Daugherty Water for food Global Institute, the National Science Foundation Enabling the Next Generation of Hazards and Disasters, the American Meteorological Society Summer Policy Colloquium, and the University of Nebraska Public Policy Center. He has published his research findings in journals of both national and international repute.

Prof. Hiroaki Furumai is a professor of the Research Center for Water Environment Technology, the University of Tokyo with a joint appointment at the Department of Urban Engineering. His research group has conducted research on 1) Sustainable urban drainage management focusing on urban flood modeling and non-point source pollution, 2) Fate and control of pathogenic microorganisms in urban water cycles, 3) Bioremediation of contaminated soil and groundwater, 4) Water quality characterization and its evaluation for sustainable urban water use. He was a member of Board of Director of International Water Association (IWA) since 2012 and then organize the World Water Congress and Exhibition at Tokyo in 2018. He is IWA Distinguished Fellow after working as IWA fellow 2010. He was a vice president (2014–2017), the president (2016–2017) and is now an advisor (2017-) of the Japan Society on Water Environment. He is accredited with 250 research publications.

Dr. Tushara Chaminda is an academic and a professional engineer in Sri Lanka, specialised in the field of civil and environmental engineering. Currently, he is working as a Senior Lecturer in the Department of Civil and Environmental Engineering, Faculty of Engineering, University of Ruhuna, Sri Lanka. His area of expertise includes but not limited to emerging micro-pollutants in the urban water environment, characterisation of DOM, heavy metal speciation, water and wastewater treatment, sustainable water management, etc. He received his PhD from The University of Tokyo, Japan, in the year 2008, in the field of Urban Environmental Engineering. Upon completing his PhD, Dr. Tushara Chaminda worked as a researcher at The University of Tokyo for 4 years. He has been engaged in several administration activities, projects and collaboration work with different international and local stakeholders of various academic and professional institutes. Dr. Tushara Chaminda has around 100 publications in international journal papers, book chapters and international/local conferences. Dr. Tushara Chaminda has also won several international and local awards for his research achievements for the last couple of years. He is also working as an editor and a reviewer in many international journals and international conferences.

Contributors

Bhairo Prasad Ahirvar Department of Environmental Science, Indira Gandhi National Tribal University, Amarkantak, Madhya Pradesh, India

Savita Ahlawat Department of Environmental Science, Maharshi Dayanand University, Rohtak, India

Dr Amjad Aliewi Water Research Center, Kuwait Institute for Scientific Research, Kuwait City, Kuwait

Dr Husam Alomirah Food and Nutrition Program, Environment and Life Sciences Center, Kuwait Institute for Scientific Research, Kuwait City, Kuwait

Chandrasekhar Bhagat Department of Civil Engineering, Indian Institute of Technology Gandhinagar, Gandhinagar, Gujarat, India

Akansha Bhatia Department of Civil Engineering, Indian Institute of Technology, Roorkee, Roorkee, India

T. R. S. B. Bokalamulla Department of Civil and Environmental Engineering, Faculty of Engineering, University of Ruhuna, Galle, Sri Lanka

Dinesh Borse Indian Institute of Technology Gandhinagar, Ahmedabad, India

G. G. T. Chaminda Department of Civil and Environmental Engineering, Faculty of Engineering, University of Ruhuna, Galle, Sri Lanka

Laya Das Indian Institute of Technology Gandhinagar, Ahmedabad, India

Pallavi Das Department of Environmental Science, Indira Gandhi National Tribal University, Amarkantak, Madhya Pradesh, India

K. C. Ellawala Department of Civil and Environmental Engineering, University of Ruhuna, Galle, Sri Lanka

Tejaswini Eregowda IHE Delft Institute of Water Education, Delft, The Netherlands

W. B. Gunawardena Department of Civil Engineering, University of Moratuwa, Moratuwa, Sri Lanka

Kuldeep Gupta Department of Molecular Biology and Biotechnology, Tezpur University, Napaam, Tezpur, Assam, India

Seiya Hanamoto Environment Preservation Center, Kanazawa University, Kanazawa, Ishikawa, Japan

Hiroe Hara-Yamamura Faculty of Geosciences and Civil Engineering, Institute of Science and Engineering, Kanazawa University, Kanazawa, Japan

A. K. Haritash Department of Environmental Engineering, Delhi Technological University, Delhi, India

Pawan Kumar Jha Centre of Environmental Science, University of Allahabad, Prayagraj, India

Tanuj Joshi Faculty, Department of Pharmaceutical Sciences, Bhimtal, Uttarakhand, India

Y. B. P. Kahatagahawatte Faculty of Geosciences and Civil Engineering, Institute of Science and Engineering, Kanazawa University, Kanazawa, Japan

A. A. Kazmi Department of Civil Engineering, Indian Institute of Technology, Roorkee, Roorkee, India

Ashwini Khandekar Discipline of Earth Sciences, Indian Institute of Technology Gandhinagar, Gandhinagar, Gujarat, India

Manish Kumar Discipline of Earth Sciences, Indian Institute of Technology Gandhinagar, Gandhinagar, Gujarat, India

Manabendra Mandal Department of Molecular Biology and Biotechnology, Tezpur University, Napaam, Tezpur, Assam, India

Ajay Kumar Manhar Department of Molecular Biology and Biotechnology, Tezpur University, Napaam, Tezpur, Assam, India

Jaivik Mankad Indian Institute of Technology Gandhinagar, Ahmedabad, India

Payal Mazumder Centre for the Environment, Indian Institute of Technology Guwahati, North Guwahati, Assam, India

Michael T. Meyer U.S. Geological Survey, Kansas Water Sciences Center, Lawrence, KS, USA

Sanjeeb Mohapatra Environmental Science and Engineering Department, Indian Institute of Technology Bombay, Mumbai, India

Meenakshi Nandal Department of Environmental Science, Maharshi Dayanand University, Rohtak, India

Dhrubajyoti Nath Department of Molecular Biology and Biotechnology, Tezpur University, Napaam, Tezpur, Assam, India

M. Otaki Department of Human Environmental Sciences, Ochanomizu University, Tokyo, Japan

Y. Otaki Faculty of Sociology, Hitotsubashi University Graduate School of Sociology, Tokyo, Japan

Nitin Padhiyar Indian Institute of Technology Gandhinagar, Ahmedabad, India

Shivani Panday Department of Environmental Science, Indira Gandhi National Tribal University, Amarkantak, Madhya Pradesh, India

Chitra Pande Kumaun University, Nainital, India

Efthimia Papastavros Department of Chemistry, University of Nebraska Lincoln, Lincoln, NE, USA

Arbind Kumar Patel Discipline of Earth Sciences, Indian Institute of Technology Gandhinagar, Gandhinagar, Gujarat, India

Kiran Patni Graphic Era Hill University, Bhimtal, Uttarakhand, India

M. S. Priyanka Department of Civil Engineering, National Institute of Technology Calicut, Kozhikode, India

Ankur Rajpal Department of Civil Engineering, Indian Institute of Technology, Roorkee, Roorkee, India

Nirav Raval Discipline of Earth Sciences, Indian Institute of Technology Gandhinagar, Gandhinagar, Gujarat, India

Fernando Rubio Eurofins-Abraxis Inc, Warminster, PA, USA

Veerendra Sahoo Department of Civil Engineering, Indian Institute of Technology, Roorkee, Roorkee, India

Devabrata Saikia Department of Molecular Biology and Biotechnology, Tezpur University, Napaam, Tezpur, Assam, India

Venkata Sandeep Discipline of Earth Sciences, Indian Institute of Technology Gandhinagar, Gandhinagar, Gujarat, India

Vandana Shan Department of Environmental Engineering, Delhi Technological University, Delhi, India

M. N. M. Shayan Department of Civil and Environmental Engineering, University of Ruhuna, Galle, Sri Lanka

S. K. Singh Department of Environmental Engineering, Delhi Technological University, Delhi, India

Ashwin Singh Discipline of Civil Engineering, Indian Institute of Technology, Gandhinagar, Gandhinagar, India

Anjali Singhal Department of Botany, University of Allahabad, Prayagraj, India

Daniel D. Snow Nebraska Water Center, University of Nebraska Lincoln, Lincoln, NE, USA

Babji Srinivasan Indian Institute of Technology Gandhinagar, Ahmedabad, India

Atul Srivastava Centre of Environmental Science, University of Allahabad, Prayagraj, India

Medhavi Srivastava Discipline of Earth Sciences, Indian Institute of Technology Gandhinagar, Gandhinagar, Gujarat, India

Kaling Taki Department of Civil Engineering, Indian Institute of Technology Gandhinagar, Gandhinagar, Gujarat, India

G. G. Tushara Chaminda Department of Civil and Environmental Engineering, University of Ruhuna, Galle, Sri Lanka

Vinay Kumar Tyagi Department of Civil Engineering, Indian Institute of Technology, Roorkee, Roorkee, India

Part I
Risk Management and Data Science (Engineering) for Water Supply

Chapter 1
History, Evolution, and Future of Rapid Environmental Assays Used to Evaluate Water Quality and Ecosystem Health

Daniel D. Snow, Michael T. Meyer, Fernando Rubio, and Efthimia Papastavros

1.1 Introduction

In general, an ideal analytical method should have a high recovery rate, a low detection limit, high selectivity and sensitivity, and good reproducibility. Currently, a number of techniques that meet these criteria are available for identifying and quantifying the presence of analytes of interest in environmental samples. Selecting the appropriate technique depends on variables such as budget, location, availability of technical personnel, throughput, and project goals. Instrumental analytical techniques such as gas and liquid chromatography are very sensitive and reliable, but are simply not practical for high throughput or for field use. These techniques are time-consuming, expensive, and must be performed by highly trained operators. Rapid or field tests can detect or quantify the presence or absence of analytes through the determination of the appearance or lack of a distinct color or signal (Hottenstein et al. 1996; Rubio et al. 2005; Sanchis et al. 2012). A number of commercial rapid test methods, using various formats, are available for quickly screening and quantifying the presence of contaminants in environmental samples (Table 1.1).

D. D. Snow (✉)
Nebraska Water Center, University of Nebraska Lincoln, Lincoln, NE 68583, USA
e-mail: dsnow1@unl.edu

M. T. Meyer
U.S. Geological Survey, Kansas Water Sciences Center, 1217 Biltmore Drive, Lawrence, KS 66046, USA

F. Rubio
Eurofins-Abraxis Inc, 124 Railroad Drive, Warminster, PA 18974, USA

E. Papastavros
Department of Chemistry, University of Nebraska Lincoln, Lincoln, NE 68588, USA

© Springer Nature Singapore Pte Ltd. 2020
M. Kumar et al. (eds.), *Resilience, Response, and Risk in Water Systems*,
Springer Transactions in Civil and Environmental Engineering,
https://doi.org/10.1007/978-981-15-4668-6_1

Table 1.1 Analytes for which there are commercially available environmental immunoassays

2,4-D	Chlordane	Imidacloprid	Penoxsulam
Acetochlor	Coplanar PCBs	Isoproturon	Progesterone
Acrylamide	Cotinine	Lasalocid	Total Pyrethroids
Alachlor	Cyclodienes	Maduramicin	Salinomycin
Alkyl Ethoxylate	Cylindrospermopsin	Malachite Green	Saxitoxin
Alkylphenol	DDT	Metolachlor	Spinosyn
Alkylphenol Ethoxylate	Dioxin	Microcystins	Streptomycin
Anabaenopeptins	Diuron	Monensin	Sulfamethazine
Anatoxin-a	Domoic Acid	Narasin	Sulfamethoxazole
Atrazine/Triazines	Estradiol	Nodularins	Testosterone
Benzo(a)pyrene	Estrone	Okadaic Acid	Tetracycline
Bisphenol A	Ethinyl Estradiol	Organophosphates	Total Estrogens
Brevetoxin	Fluridone	Total PAH	Toxaphene
Total BTEX	Fluoroquinolones	Paraquat	Triazine Metabolites
Caffeine	Gentamicin	Total PBDE	Triclopyr
Carbamates	Glyphosate	Total PCB	Triclosan
Carbamazepine			

Here, we briefly review the typical rapid test methods that are used to assess the presence of contaminant residues in environmental samples. An overview is presented of the development and use of rapid, economical, quantitative testing methods for measuring pesticides, pharmaceuticals, natural toxins, and synthetic chemicals in environmental matrices.

1.2   Background and Theory of Immunoassays

Immunoassay is a bio-analytical technique used in many fields to detect and quantify the presence of an analyte in a sample by an antibody. The principle of immunoassays is based on the reaction of an antigen or analyte (Ag) with a specific antibody (Ab) to give a product (Ag–Ab). The product is measured by using a detectable label; the label can involve an isotope (radioimmunoassay, or RIA), enzyme (ELISA), fluorescence (FIA), chemiluminescence (CLIA), latex, gold, and other options. As with any chemical reaction, the reaction between an antibody and antigen may be described by the law of mass action:

$$[Ag] + [Ab] \rightarrow [Ag\text{–}Ab]$$

Since the invention of immunoassays in 1959, there has been constant innovation of the technique as well as an immense range of new applications. The technique was created by R. Yalow and S. Berson at the Bronx, N.Y. VA Hospital to study insulin levels in diabetes mellitus (Yalow and Berson 1959) by radioimmunoassay. They were later awarded the Nobel Prize in Medicine, in 1977 (Plaza et al. 2000). Since RIA uses radioactive markers, researchers sought a safer alternative, and in 1972, Engvall and Perlman (1972) published a paper describing methods for the use of enzymes in combination with a substrate/chromogen for the performance of enzyme immunoassays (EIA, ELISA). Enzymes are usually obtained from microorganisms or plants such as *Escherichia coli, Rhizopus niveus, Aspergillus niger, Bacillus pasteurii, and Armoracia rusticana*. Examples of commonly used enzymes include alkaline phosphatase, horseradish peroxidase, β-galactosidase, glucose oxidase and urease (Plaza et al. 2000).

Lateral flow sticks or immunochromatographic strips or devices are another variation of an immunoassay, where the sample flows along a solid substrate via capillary action. The sample is added directly to the device where it encounters a colored reagent which mixes with the sample as it migrates up the substrate. The solutions encounter lines or zones which have been pretreated with a capturing reagent (antibody or analyte protein). After a brief period of time, the color between a control and a test line is compared and the variation in intensity determines the analyte concentration. The first reported lateral flow device used an enzyme as the label (Pappas et al. 1983). Current lateral flow sticks use simpler and more stable labels such as colloidal gold.

Lateral flow devices dramatically reduce the test time from hours to minutes and are true field tests requiring no instrumentation or operator training, allowing almost anyone to use them. These tests can also be used as a qualitative, semi-quantitative, and even quantitative detection of analytes and can be configured so that several analytes can be tested simultaneously on the same strip. In the environmental field, the sample may be water, soil, or vegetation extract. Since the first commercial strip test in 1988, a pregnancy test, this technique has become a very popular platform in clinical, veterinary, agricultural, food safety, biowarfare, and environmental applications. There are many different formats, and the choice of format depends on the target molecule and application. Environmental immunoassays are usually competitive assays employing coated plates, tubes, magnetic particles, or lateral flow sticks.

Since its inception, immunoassay and its various formats have been extensively used, from over the counter pregnancy tests to proteomic research. Immunoassay has been applied to detect and quantify well over 1000 substances including hormones, vitamins, drugs of abuse, viruses, environmental pollutants, mycotoxins, marine and freshwater toxins, allergens, metals, and genetically modified proteins. It is estimated that over one billion tests are run worldwide every year. The sensitivity and specificity of immunoassays are remarkable; some immunoassays have detected concentrations in the fg/mL range (Vashist and Luong 2018).

1.3 History

Pesticide (2,4-D, triazine, acetamide herbicides) immunoassay and enzymatic test kits have been developed. Enzymatic test kits are used to detect the presence of a wide range of organophosphate and carbamate pesticides in water, soil, vegetable/fruit extracts, and other environmental matrices. These kits are qualitative, colorimetric assays (modification of the Ellman method) based on the inhibition of the enzyme acetylcholinesterase (AChE). AChE in the absence of an inhibitor (organophosphate, carbamate) hydrolyzes acetylcholine (ATC), which in turn reacts with 5, 5'-dithiobis-(2-nitrobenzoic acid) to produce a yellow color that can be read visually or at 405 nm with a colorimeter. If organophosphate or carbamate compounds are present in a sample, they will inhibit AChE, and therefore, color formation will be less intense or absent, depending on the concentration present in the sample. Enzyme activity, before and after exposure to pesticides, can be measured using amperometry, potentiometry, spectrophotometry, fluorimetry, and thermometry.

Since the early work on the development of a radioimmunoassay for aldrin and dieldrin (Langone and Van Vunakis 1975), the development of an RIA for 2,4-D and 2,4,5-T in surface waters (Rinder and Fleeker 1981), the development of an RIA for parathion (Ercegovich et al. 1981), followed by the prolific work of Bruce Hammock's group at U.C. Davis, several commercial environmentally focused immunoassay companies were founded in the late1980s, mainly in the USA (Agri-Diagnostics, Ensys, Immunosystems, Ohmicron). The number of commercial environmental immunoassays currently in the market has expanded to 65 + kits, offered mainly by three US-based companies: Abraxis, Beacon, Envirologix, and a UK-based company: Modern Water.

There are several commercially available enzyme-based kits for pesticide detection. Among them are an organophosphate/carbamate screening kit (Abraxis, Warminster, PA, USA), the Neuro-IQ Tox Test kit (Aqua Survey, Flemington, NJ, USA), Eclox pesticide strips (Severn Trent Services, Fort Washington, PA, USA), and a test kit for pesticides in water (PRO-LAB, Weston, FL, USA). Early field studies using commercial kits for triazine and chloroacetanilide herbicides conducted in the early 1990s by Thurman et al. (1992) where the ELISA kits were used inside a vessel navigating down the Mississippi River, showed that immunoassays were a rapid, reliable, and low-cost analytical technique that could be used for the screening of herbicides in water.

1.4 Accreditation/Validation of Environmental Rapid/Field Tests

The EPA Office of Solid Waste was receptive to the use of immunoassay for regulatory purposes, and in 1992, the first commercial immunoassay, for pentachlorophenol. was approved for inclusion into the compendium of Test Methods for Evaluating

Solid Waste or SW-846. Between 1993 and 1995, ten other ELISA-based screening methods for various environmental analyte classes were included. All of these methods were officially approved by the US EPA in 1997. The US EPA under the Superfund Innovative Technology Evaluation (SITE) has conducted performance verification studies; examples of verified technologies are rapid tests for dioxins and coplanar PCBs (http://www.epa.gov/esd/cmb/site/index.htm).

The US Environmental Technology Verification (ETV) Program was established by the US EPA in 1995. This program developed test protocols and verified the performance of innovative technologies that had the potential to improve the protection of human health and the environment. The ETV program aimed to provide users of environmental technologies with a greater degree of confidence in the technology that they were buying. It also provided the vendors of the technology with accreditation that would give confidence to prospective buyers. The US EPA did not endorse, certify, or approve technologies. Verification reports were published on the ETV Web site (www.EPA.gov/etv). Since 1995, over 250 environmental technologies were verified; some examples of field test technologies verified under the ETV program were various test kits for atrazine, organophosphates/carbamates, and microcystins. This program concluded operations in 2014.

Canada, South Korea, Bangladesh, Japan, China, Malaysia, Philippines, Cambodia, and the EU have worked toward establishing individual ETV Programs. The ETV International Working Group (IWG) worked until 2010 toward international recognition to ensure that a technology verified in a member program would be accepted as verified in other member programs: "verify once, accept everywhere" (http://www.epa.gov/etv/inter-partic.html). IWG eventually made a request to the International Standards Organization to develop an ETV standard. The EPA then worked with ISO to develop an environmental management—ETV standard, ISO 14034:2016.

More recently, the US EPA validated and promulgated Abraxis Microcystins-ADDA ELISA as US EPA Method 546 (USEPA Method 546). Method 546 is a procedure for the determination of "total" microcystins (MC) and nodularins (NOD) in finished drinking water and in ambient water using enzyme-linked immunosorbent assay (ELISA). The term "total microcystins and nodularins" is defined as the sum of the congener-independent, intracellular and extracellular microcystin and nodularin that is measurable in a sample. Method 546 measures the total concentration based on detection of a characteristic feature common to microcystin and nodularin congeners (structural variants), specifically, the Adda amino acid side chain: $(4E,6E)$-3-amino-9-methoxy-2,6,8-trimethyl-10-phenyldeca-4,6-dienoic acid.

For over 120 years, AOAC has been meeting the needs of analytical scientists for confidence in analytical results through the development and validation of AOAC Official Methods[SM]. These validated methods have been primarily related to food analysis, as well as bacterial contamination. In September 2010, the AOAC Stakeholder Panel on Endocrine Disruptors (SPED) approved standard method performance requirements (SMPRs) for the quantitative measurement of endocrine-disrupting compounds (EDCs) in freshwater. These are voluntary standards that can be used to evaluate test kit methods for their appropriateness in meeting AOAC's needs for quick analytical solutions that are reliable and accurate in environmental matrices (http://www.aoac.org).

1.5   Current and Future Directions in Rapid Screening Methods

Environmental contaminants are often analyzed using LC/MS, LC/MS/MS, and GC/MS. While these methods provide high accuracy, sensitivity, and reliability, they are expensive, time-consuming and require highly trained personnel, as well as extensive sample pretreatment. Alternative, rapid, field-portable and easy-to-use screening methods, such as the immunoassay, at first proved to be reliable, sensitive and selective in clinical applications (Plaza et al. 2000). The first environmental application of immunoassays was in 1971 when Ercegovich used this technique for pesticide analysis (Ercegovich 1971). The first commercially available immunoassay kit to detect pesticides became available in 1988 and was used for atrazine (Andrews 1992). Immunoassays are ideal for large-scale screening of pollutants and have been developed for a wide range of analytes in water, soil, and sediment samples. These include pesticides, metals, and organic compounds such as polyaromatic hydrocarbons (PAHs), polychlorinated biphenyls (PCBs), microbial toxins, and trinitrotoluene (TNT) (Plaza et al. 2000).

There are some disadvantages to using immunoassays, and these include development costs as well as the risk of cross-reacting compounds and non-specific interferences (Plaza et al. 2000). In addition, they need to be validated by another method (Plaza et al. 2000). Immunoassays are best used for qualitative assessment of environmental contamination (Plaza et al. 2000). Farré et al. (2006) compared the performance of four commercially available ELISA kits for the determination of estrogens in water to results obtained using HPLC-MS/MS with a quadrupole mass analyzer. The kits, based on monoclonal antibodies, were aimed at 17-β-estradiol, estrone, 17-α-ethynyl estradiol and estrogens in general. Their design involved a direct competitive format with 96 microtiter plates. The samples analyzed included unaltered and spiked river water, groundwater, as well as urban and industrial wastewater. There was very good agreement between the two methods. However, up to 20% overestimation occurred using the ELISA kits. This inaccuracy was greater in wastewater samples and likely resulted from the presence of interfering compounds, such as estrone sulfate or other conjugated estrogens, residual concentrations of other progestogens and phytoestrogens, as well as the complexity of the sample matrix (Farré et al. 2006). The lowest limit of detection for the ELISA analyses was 0.05 μg/L. Despite overestimation of estrogens with use of the ELISA kits, the analysis was very rapid; 2.5 h for 40 samples, as opposed to 90 h using LC-MS/MS. In addition, the cost of analysis using ELISA was 50 times less expensive than LC-MS/MS. The authors concluded that the simple and rapid prescreening that ELISA kits allow make them valuable for preselection of samples to undergo further analysis by instrumental methods (Farré et al. 2006).

In the last few years, nanoparticle (NP)-based optical, electrochemical, and magnetic relaxation environmental sensors have been developed. They have especially high selectivity, sensitivity, and stability and are inexpensive and amenable to high-throughput analyses. NP-based environmental sensors have been used to detect toxins, heavy metals, and organic pollutants in water, soil, and air (Wang et al. 2010). For example, Wang et al. (2009) used filter paper impregnated with single-walled carbon nanotubes (SWNTs) and antibodies to detect microcystin-LR (MC-LR) in water from Tai Lake in Wuxi, China. Detection is electrochemical, with the SWNT-paper strip serving as the working electrode, Pt wire serving as the counter electrode and saturated Hg_2Cl_2 serving as the reference electrode. As the analyte spreads through the SWNT layers, it interacts with the antibodies within the paper and forms an Ag-Ab complex, driving SWNT layers further apart and decreasing the current passing through them. Hence, there is a decrease in current with increasing analyte concentration. This method is simple, portable, sensitive, specific, and inexpensive. It has a linear range of 0–10 μmol/L and a limit of detection of 0.6 ng/ml, comparable to those of ELISA, but is at least 28 times faster and does not require the training and reagents that ELISA requires.

Zhang et al. (2010b) also used a NP-based immunoassay with electrochemical detection to analyze MC-LR in water samples obtained from Tai Lake. Their device was prepared by immobilizing MC-LR on oxidized single-walled carbon nanohorns (SWNHs). The SWNHs coated a glassy carbon electrode (GCE). There was no pretreatment of samples, other than mixing the water with MC-LR antibodies. The linear range was 0.05–20 μg/L, and the limit of detection was 0.03 μg/L. There was good correlation between results obtained using the immunoassay and those obtained via HPLC for the lake samples.

NPs are suitable for sensing for several reasons. They have a large surface-to-volume ratio, their electrons or holes are confined at the nanoscale, and their chemistries can be modified. Materials used as sensors include gold NPs (GNPs), quantum dots (QDs), and magnetic NPs (MNPs). Table 1.2 summarizes nanoparticle-based environmental sensors (Wang et al. 2010).

There are several types of biosensors available for pesticide detection. These include sensors based on enzymes, antibodies, artificial macromolecules, and whole cells (Liu et al. 2013). Biosensors offer several attractive characteristics for use in pesticide detection. These include rapidity and simplicity of analysis, selectivity, sensitivity, low cost, and portability (Liu et al. 2013). Enzyme-inhibition-based pesticide biosensors make use of the fact that organophosphorus and carbamate pesticides inhibit cholinesterase (ChE) by blocking the serine in the active site. By measuring the decrease in enzyme activity after exposure to the analyte, pesticide toxicity is determined. Pesticide concentration can be measured by determining the percent inhibition of enzyme activity resulting from pesticide exposure. Enzyme activity can be measured by detecting substrates or products of enzymatic reactions using amperometry, potentiometry, spectrometry, fluorimetry, or thermometry (Liu et al.

Table 1.2 Summary of nanoparticle-based environmental sensors

Type of sensor	Particular feature	Nanoparticle used	Detection mechanism	Analytes evaluated
Optical	High signal-to-noise ratio, GNPs used in immunochromatographic strip assays	GNPs	Color change during aggregation or dispersion of aggregates	Toxins, heavy metals, nitrite, small molecules, TNT, bacteria (gold nanorods)
		QDs	Photoluminescence changes	
Electrochemical	NP labels can also be used for spectroscopic detection	GNPs, modified gold colloid surface, SWNTs	Oxidation of metal NPs	Copper ions, microcystin-LR
Magnetic relaxation	Radiofrequency-based, unaffected by light-based interferences (scattering, etc.)	Biocompatible magnetic NPs	Switching between dispersed and clustered states produces spin–spin relaxation times of water T_2 signals	Toxins, bacteria

Compiled from Wang et al. (2010)

2013). Enzyme-based biosensors are being improved through changes in immobilization methods and the use of nanomaterials (Liu et al. 2013). Whole-cell biosensors are based on the inhibition of photosynthesis or phosphorylation by herbicides. This effect can be detected by an oxygen electrode, amperometrically, or optically. Whole-cell sensors based on organophosphorus hydrolase (OPH) have been prepared using microalgae and bacteria. OPH expressed on the cell surface, or in whole cells, catalyzes hydrolysis of organophosphorus pesticides such as paraoxon, parathion, and methyl parathion. The ensuing release of p-nitrophenol is detected electrochemically or colorimetrically.

Molecularly imprinted polymers (MIPs) provide a synthetic alternative to enzymes and antibodies, and avoid potential problems such as difficulty in finding suitable antibodies, instability, binding that is too strong and high cost. MIPs are stable under a wide range of temperatures and chemical environments. Polymerization to prepare the MIP can take place on the surface of a transducer, allowing for binding of the target analyte to create a measurable optical (e.g., fluorescence, chemiluminescence, surface plasmon resonance), electrochemical or mass-based signal. Polymerization takes place in the presence of a template molecule representing the target analyte. Upon removal of the template, the polymer has recognition sites to which the analyte will bind. Possible disadvantages of MIPs include low binding capacity and slow binding kinetics. The latter problem is a result of binding sites being located inside a rigid polymer matrix. A possible way to overcome poor site accessibility is to use nanomaterials as the support for preparation of MIPs (Liu et al. 2013).

Aptamers provide another option for artificial-receptor-based biosensors. Aptamers are single-stranded DNA, RNA, or peptide sequences that bind target analytes with affinities and specificities comparable to those of antibodies. As with MIPs, aptamers are very stable. A disadvantage of aptamers is the time-consuming process of selecting them. This procedure is known as Systematic Evolution of Ligands by Exponential Enrichment (SELEX). It involves repeated rounds of refining selections of appropriate random combinations of nucleic acid or peptide sequences. Electrochemical, optical, and mass-based transducers may be used with aptamer biosensors (Liu et al. 2013).

Biosensors for pesticide determinations include optical transducers used in fluorescence, Surface plasmon resonance (SPR) and chemiluminescence assays, electrochemical transducers, and mass-based transducers, such as a quartz crystal microbalance (QCM). Optical transducers are based on a change in the intensity or wavelength of absorption or fluorescence. Although optical biosensors have the advantages of quick response times, resistance of the signals to electrical or magnetic interference and the ability to process multiple signals, instrumentation costs can be rather high. Fiber-optic biosensors are more amenable to miniaturization, remote sensing and in situ monitoring. Some examples include fiber-optic biosensors based on sol–gel immobilized ChE to detect dichlorvos; the enzyme glutathione S-transferase I to detect atrazine; a disposable microbial membrane for detection of methyl parathion; the immunological detection of 2,4-D and the MIP-based detection with europium as the signal transducer (Liu et al. 2013). Detection may also depend on the conversion of a substrate to a product, in which case the biosensor is reusable. Potentiometric, optical, and amperometric transducers may be used in this type of biosensor (Liu et al. 2013).

Surface plasmon resonance (SPR) immunosensors have been used to detect atrazine (Minunni and Mascini 1993; Farré et al. 2007), 2,4-D (Kim et al. 2007; Revoltella et al. 1998; Gobi et al. 2007), chlorpyrifos (Mauriz et al. 2006a, 2007a), carbaryl (Mauriz et al. 2006b), DDT (Mauriz et al. 2007b), isoproturon (Gouzy et al. 2009), and other pesticides (Mauriz et al. 2006c; Chegel et al. 1998; Mouvet et al. 1997; Tanaka et al. 2007; Nakamura et al. 2003). SPR is an attractive detection option because it does not require the use of labels and provides high sensitivity. SPR immunoassays, including those in portable formats, have been used to detect 2,4-D and other pesticides at nanogram per liter levels (Mauriz et al. 2007a; Svitel et al. 2000; Yang and Kang 2008). It was possible to reuse the same sensor surface for over 200 assays. AChE coupled to gold nanoparticles has also been used with SPR for pesticide detection (Lin et al. 2006; Huang et al. 2009).

A lateral flow immunochromatographic assay, in the form of a strip, has been developed as a portable, easy-to-use, inexpensive and rapid screening method for pesticides (Liu et al. 2013; Su et al. 2010; Lisa et al. 2009; Blazkova et al. 2009; Guo et al. 2009; Kaur et al. 2007; Lan et al. 2020). Immunochromatographic assays using gold nanoparticle–antibody conjugates have been used to detect the toxins ochratoxin A (OTA), zearalenone (ZEA) and aflatoxin B1 (Liu et al. 2008; Shim et al. 2009; Shim et al. 2007), and for the cyanobacterial toxins anatoxin-a, cylindrospermopsin, and microcystin (Eurofins-Abraxis). Antibiotics and pesticides have also been detected, using a microimmunoassay on a disk (Morais et al. 2009). An immunochromatographic fluorescent biosensor was also developed for the detection of 3,5,6-trichloropyridinol (TCP) (Zou et al. 2010). Colorimetric sticks or strips, based on ChE, to detect organophosphorus and carbamate pesticides have also been developed (Fu et al. 2019; Pohanka et al. 2010; No et al. 2007).

While most biosensors have been evaluated using standard solutions, there are several examples of some that have been employed with real environmental samples (Liu et al. 2013). These include several enzyme-based biosensors for pesticides in lake, river and tap water (Andreou and Clonis 2002; Joshi et al. 2005; Halamek et al. 2005; Walker and Asher 2005; Arduini et al. 2006; Viswanathan et al. 2009; Wang et al. 2011), as well as immunoassays for antibiotics and pesticides in natural waters (Morais et al. 2009). There are many examples of fluorescent labels being used for detection in immunological methods in order to improve sensitivity. However, high background signals can be problematic. Time-resolved fluorescence makes use of europium and terbium chelates that have narrow and strong emission bands around 600 nm, as well as very long decay times. The use of these types of labels eliminates high background signals.

Majima et al. (2002) used a fluorescent europium chelate label in a time-resolved fluoroimmunoassay (TRFIA) to detect 17β-estradiol and estriol in river water. This method was sensitive and selective and enabled the authors to attain a detection limit of 2.3 pg/ml for 17β-estradiol, which is on the same order of magnitude as that obtained by ELISA. For estriol, the detection limit using the TRFIA was 4.3 pg/ml, which is 1–2 orders of magnitude better than ELISA. This TRFIA method was used to measure levels in river water of 32 pg/ml for 17β-estradiol and 5.5 pg/ml for estriol. A volume of 20 ml of river water was required for extraction and concentration. Time-resolved fluoroimmunoassay (TRFIA) was tested by Zhang et al. (2010) for the routine screening and quantification of sulfonamide antibiotics in environmental waters. This method was chosen in order to avoid sensitivity issues, matrix interferences, and other inadequacies of ELISA methods (Zhang et al. 2010). Limits of detection ranged from 5.4 ng/L for sulfadiazine to 9.8 ng/L for sulfamethazine. These were in the same range as the limits of detection for LC-MS/MS, which were 1–10 ng/L. In contrast to LC-MS/MS, TRFIA did not require sample preconcentration. Limits of quantitation were 18.0 ng/L for sulfamethazine, 61.5 ng/L for sulfamethoxazole, and 14.7 ng/L for sulfadiazine.

As with ELISA methods, TRFIA has the advantages of high sample throughput, low cost, simple operation, and minimal sample preparation, i.e., filtration was the only step required prior to analysis. In addition, TRFIA had superior sensitivity and suffered less from matrix effects. Surface water and wastewater samples were tested (Zhang et al. 2010a). Bacigalupo and Meroni (2007) also developed a rapid and sensitive TRFIA screening method for the herbicide diuron in surface and groundwater. This inexpensive method required no sample preconcentration or cleanup and offers the possibility of analyzing 100 water samples simultaneously in 2 h. Detection relied on antibodies conjugated with Eu^{3+} labeled chelates, and detection at levels 20 ng/L below the European Community limits was achieved.

A rapid TRFIA screening method involving liposomes trapping a terbium/citrate complex was used by Bacigalupo et al. (2003) to monitor atrazine in surface water. A detection limit of 0.1 ng/ml was achieved with this method, which allows many samples to be analyzed simultaneously. The cytolytic agent consisted of atrazine covalently linked to the polypeptide mastoparan and allowed the release of terbium citrate from the liposomes after a short incubation time. Mastoparan is easily prepared as it is composed of 14 amino acids. Its lysine residues make it possible to conjugate to a variety of target analyte molecules. An excess of dipicolinic acid (DPA) was added because of the higher stability constant of the Tb/DPA complex than the Tb/citrate complex and the enhanced fluorescence that results (Bacigalupo et al. 2003). A rapid and sensitive competitive TRFIA method was developed by Zhao et al. (2009) to measure the endocrine disruptor diethylstilbestrol (DES) in water samples. This method made use of europium chelate labeled antibodies and offered a detection limit of 0.595 pg/mL (Zhao et al. 2009; Kumar et al. 2019).

Underwater mass spectrometry (MS) systems were developed by Kibelka et al. (2004) and used in situ, to monitor municipal wastewater in St. Petersburg, FL, USA (Kibelka et al. 2004). A modular design was employed, in order to facilitate the anticipated need for repeated reconfiguration. The three subsystems were each housed in a separate pressure vessel, one for the sample intake component, one for the mass analyzer component, and one for the roughing pump component. A linear quadrupole mass spectrometer system was used in the unit that was deployed within the wastewater treatment plant, although the group also developed a system employing an ion trap mass spectrometer for other applications.

A polydimethylsiloxane (PDMS) membrane was used to introduce analytes into the mass spectrometry system via membrane diffusion. The MS system was placed in the treatment tank, after coarse filtration of reclaimed water and raw sewage. The sampling system automatically injected 1 mL samples of wastewater into a continuously flowing stream of charcoal-filtered wastewater in contact with the membrane interface, at 12-min intervals. The system monitored chloroform, a by-product of the chlorination of drinking water (National Research Council 1987), as well as toluene, in the range of 10 to 50 ppb, peaking daily at approximately 5 pm local time. After 30 hours, there was clogging of the stainless steel intake filter and the membrane capillary, and data acquisition was stopped. However, all other components of the underwater quadrupole MS system continued to function properly during the entire deployment period. A replaceable filter for the MS inlet line could potentially allow uninterrupted use of the sampling component.

1.6 Summary and Conclusion

Since the beginning of immunoassays in 1959, they have been continuously developed for countless analytes, including those of environmental concern. Commercially available, field-portable test kits exist for the rapid analysis of environmental samples containing hormones, pharmaceuticals, pesticides, and other classes of contaminants. Immunoassays have provided the foundation for research that has produced environmental sensors, nanoparticle-based sensors, as well as sensors based on surface plasmon resonance. These techniques have been used to detect a very wide range of environmentally relevant compounds at low concentrations.

Although research is being done to develop a range of rapid techniques with low detection limits, work with real environmental samples, such as lake and river water, has been relatively limited. There has been considerable success in obtaining results comparable to instrumental methods of analysis, and field methods can also be useful for prescreening samples before the decision is make to spend time analyzing them in the laboratory. Aptamers and molecularly imprinted polymers are synthetic alternatives to enzymes and antibodies. Although they have disadvantages such as low binding capacity, slow binding kinetics, and time-consuming production, their high stability and potential use with nanomaterials, for example, warrant investment in future work.

References

Andreou VG, Clonis YD (2002) Novel fiber-optic biosensor based on immobilized glutathione S-transferase and sol-gel entrapped bromocresol green for the determination of atrazine. Anal Chim Acta 460:151–161

Andrews R (1992) Immunoassay advances as tool for environmental testing. The Scientist, March 1992

Arduini F, Ricci F, Tuta CS, Moscone D, Amine A, Palleschi G (2006) Detection of carbamic and organophosphorus pesticides in water samples using a cholinesterase biosensor based on Prussian Blue-modified screen-printed electrode. Anal Chim Acta 580:155–162

Bacigalupo MA, Meroni G (2007) Quantitative determination of diuron in ground and surface water by time-resolved fluoroimmunoassay: seasonal variations of diuron, carbofuran, and paraquat in an agricultural area. J Agric Food Chem 55:3823–3828

Bacigalupo MA, Ius A, Longhi R, Meroni G (2003) Homogeneous immunoassay of atrazine in water by terbium-entrapping liposomes as fluorescent markers. Talanta 61:539–545

Blazkova M, Mickova-Holubova B, Rauch P, Fukal L (2009) Immunochromatographic colloidal carbon-based assay for detection of methiocarb in surface water. Biosens Bioelectron 25:753–758

Chegel VI, Shirshov YM, Piletskaya EV, Piletsky SA (1998) Surface plasmon resonance sensor for pesticide detection. Sens Actuators B 48:456–460

Engvall E, Perlman P (1972) Enzyme-linked immunosorbent assay,ELISA. J Immunol 109:129–135

Ercegovich CD (1971) Analysis of pesticide residues: immunological techniques. In: Gould RF (ed) Pesticides identification at the residue level. American Chemical Society, Washington, D.C., 162

Ercegovich CD, Vallejo RP, Gettig RR, Woods L, Bogus ER, Mumma RO (1981) Development of a radioimmunoassay for parathion. J Food Ag Chem 29:559–563

Eurofins-Abraxis. https://abraxis.eurofins-technologies.com

Farré M, Birx R, Kuster M, Rubio F, Goda Y, López de Alda MJ, Barceló D (2006) Evaluation of commercial immunoassays for the detection of estrogens in water by comparison with high-performance liquid chromatography tandem mass spectrometry-HPLC-MS/MS (QqQ). Anal Bioanal Chem 385:1001–1011

Farré M, Martinez E, Ramon J, Navarro A, Radjenovic J, Mauriz E, Lechuga L, Marco MP, Barcelo D (2007) Part per trillion determination of atrazine in natural water samples by a surface plasmon resonance immunosensor. Anal Bioanal Chem 388:207–214

Fu Q, Zhang C, Xie J, Li Z, Qu L, Cai X, Ouyang H, Song Y, Du D, Lin Y, Tang Y (2019) Ambient light sensor based colorimetric dipstick reader for rapid monitoring organophosphate pesticides on a smart phone. Anal Chim Acta 1092:126–131

Gobi KV, Kim SJ, Tanaka H, Shoyama Y, Miura N (2007) Novel surface plasmon resonance (SPR) immunosensor based on monomolecular layer of physically-adsorbed ovalbumin conjugate for detection of 2,4-dichlorophenoxyacetic acid and atomic force microscopy study. Sens Actuators B 123:583–593

Gouzy M-F, Kess M, Kraemer PM (2009) A SPR-based immunosensor for the detection of isoproturon. Biosens Bioelectron 24:1563–1568

Guo Y-R, Liu S-Y, Gui W-J, Zhu G-N (2009) Gold immunochromatographic assay for simultaneous detection of carbofuran and triazophos in water samples. Anal Biochem 389:32–39

Halamek J, Pribyl J, Makower A, Skladal P, Scheller FW (2005) Sensitive detection of organophosphates in river water by means of a piezoelectric biosensor. Anal Bioanal Chem 382:1904–1911

Hottenstein CS, Rubio FM, Herzog DP, Fleeker JR, Lawruk TS (1996) Determination of trace atrazine levels in water by a sensitive magnetic particle-based enzyme immunoassay. J Agric Food Chem 44:3576–3581

Huang X, Tu H, Zhu D, Du D, Zhang A (2009) A gold nanoparticle labeling strategy for the sensitive kinetic assay of the carbamate-acetylcholinesterase interaction by surface plasmon resonance. Talanta 78:1036–1042

Joshi KA, Tang J, Haddon R, Wang J, Chen W, Mulchandani A (2005) A disposable biosensor for organophosphorus nerve agents based on carbon nanotubes modified thick film strip electrode. Electroanalysis 17:54–58

Kaur J, Singh KV, Boro R, Thampi KR, Raje M, Varshney GC, Suri CR (2007) Immunochromatographic dipstick assay format using gold nanoparticles labeled protein-hapten conjugate for the detection of atrazine. Environ Sci Technol 41:5028–5036

Kibelka GPG, Short RT, Toler SK, Edkins JE, Byrne RH (2004) Field-deployed underwater mass spectrometers for investigations of transient chemical systems. Talanta 961–969

Kim SJ, Gobi KV, Iwasaka H, Tanaka H, Miura N (2007) Evaluation of a portable SPR immunosensor equipped with multichannel flow-cell for detecting 2,4-D. Chem Sens 23 Issue Suppl A:10–12

Kumar M, Ram B, Honda R, Poopipattana C, Canh VD, Chaminda T, Furumai H (2019) Concurrence of Antibiotic Resistant Bacteria (ARB), Viruses, Pharmaceuticals and Personal Care Products (PPCPs) in ambient waters of Guwahati. India: urban vulnerability and resilience perspective. Sci Total Environ 693:133640

Lan J, Sun W, Chen L, Zhou H, Fan Y, Diao X, Wang B, Zhao H (2020) Simultaneous and rapid detection of carbofuran and 3-hydroxy-carbofuran in water samples and pesticide preparations using lateral-flow immunochromatographic assay. Food Agric Immunol 31:165–175

Langone JJ, Van Vunakis H (1975) Radioimmunoassay for dieldrin and aldrin. Res Commun Chem Pathol Pharmacol 10:163

Lin T-J, Huang K-T, Liu C-Y (2006) Determination of organophosphorus pesticides by a novel biosensor based on localized surface plasmon resonance. Biosens Bioelectron 22:513–518

Lisa M, Chouhan RS, Vinayaka AC, Manonmani HK, Thakur MS (2009) Gold nanoparticles based dipstick immunoassay for the rapid detection of dichlorodiphenyltrichloroethane: an organochlorine pesticide. Biosens Bioelectron 25:224–227

Liu B-H, Tsao Z-J, Wang J-J, Yu F-Y (2008) Development of a monoclonal antibody against ochratoxin A and its application in enzyme-linked immunosorbent assay and gold nanoparticle immunochromatographic strip. Anal Chem 80:7029–7035

Liu S, Zheng Z, Li X (2013) Advances in pesticide biosensors: current status, challenges, and future perspectives. Anal Bioanal Chem 405:63–90

Majima K, Fukui T, Yuan J, Wang G, Matsumoto K (2002) Quantitative measurement of 17β-estradiol and estriol in river water by time-resolved fluoroimmunoassay. Anal Sci 18:869–874

Mauriz E, Calle A, Lechuga LM, Quintana J, Montoya A, Manclus JJ (2006a) Real-time detection of chlorpyrifos at part per trillion levels in ground, surface and drinking water samples by a portable surface plasmon resonance immunosensor. Anal Chim Acta 561:40–47

Mauriz E, Calle A, Abad A, Montoya A, Hildebrandt A, Barcelo D, Lechuga LM (2006b) Determination of carbaryl in natural water samples by a surface plasmon resonance flow-through immunosensor. Biosens Bioelectron 21:2129–2136

Mauriz E, Calle A, Manclus JJ, Montoya A, Escuela AM, Sendra JR, Lechuga LM (2006c) Single and multi-analyte surface plasmon resonance assays for simultaneous detection of cholinesterase inhibiting pesticides. Sens Actuators B 118:399–407

Mauriz E, Calle A, Manclus JJ, Montoya A, Lechuga LM (2007a) Multi-analyte SPR immunoassays for environmental biosensing of pesticides. Anal Bioanal Chem 387:1449–1458

Mauriz E, Calle A, Manclus JJ, Montoya A, Hildebrandt A, Barcelo D, Lechuga LM (2007b) Optical immunosensor for fast and sensitive detection of DDT and related compounds in river water samples. Biosens Bioelectron 22:1410–1418

Minunni M, Mascini M (1993) Detection of pesticide in drinking water using real-time biospecific interaction analysis. Anal Lett 26:1441–1460

Morais S, Tortajada-Genaro LA, Arnandis-Chover T, Puchades R, Maquieira A (2009) Multiplexed microimmunoassays on a digital versatile disk. Anal Chem 81:5646–5654

Mouvet C, Harris RD, Maciag C, Luff BJ, Wilkinson JS, Piehler J, Brecht A, Gauglitz G, Abuknesha R, Ismail G (1997) Determination of simazine in water samples by waveguide surface plasmon resonance. Anal Chim Acta 338:109–117

Nakamura C, Hasegawa M, Nakamura N, Miyake J (2003) Rapid and specific detection of herbicides using a self-assembled photosynthetic reaction center from purple bacterium on an SPR chip. Biosens Bioelectron 18:599–603

National Research Council (1987) Drinking water and health. In: Disinfectants and disinfectant by-products, vol 7. The National Academies Press, Washington, DC. https://doi.org/10.17226/1008

No H-Y, Kim YA, Lee YT, Lee H-S (2007) Cholinesterase-based dipstick assay for the detection of organophosphate and carbamate pesticides. Anal Chim Acta 594:37–43

Pappas MG, Hajkowski R, Hockmeyer WT (1983) Dot-enzyme linked immunosorbent assay (Dot:ELISA): a micro technique for the rapid diagnosis of visceral leishmaniasis. J Immunol Methods 64:205–214

Plaza G, Ulfig K, Tien AJ (2000) Immunoassays and environmental studies. Pol J Environ Stud 9:231–236

Pohanka M, Karasova JZ, Kuca K, Pikula J, Holas O, Korabecny J, Cabal J (2010) Colorimetric dipstick for assay of organophosphate pesticides and nerve agents represented by paraoxon, sarin and VX. Talanta 81:621–624

Revoltella RP, Robbio LL, Liedberg B (1998) Comparison of conventional immunoassays (RIA, ELISA) with surface plasmon resonance for pesticide detection and monitoring. Biotherapy (Dordrecht, Neth) 11:135–145

Rinder DF, Fleeker JR (1981) A radioimmunoassay to screen for 2,4-D-diclorophenoxyacetic acid and 2,4,5-trichlorophenoxyacetic acid in surface water. Bull Environ Contamin Tox 26:375–380

Rubio F, Parrotta CC, Li QX, Shelver WL (2005) Development of a sensitive magnetic particle immunoassay for polybrominated diphenyl ethers. Organohalogen Compd 67:27–30

Sanchis J, Kantiani L, Llorca M, Rubio F, Ginebreda A, Fraile J, Garrido T, Farré M (2012) Determination of glyphosate in groundwater samples using an ultrasensitive immunoassay and confirmation by on-line solid-phase extraction followed by liquid chromatography coupled to tandem mass spectrometry. Anal Bional Chem 402:2335–2345

Shim W-B, Yang Z-Y, Kim J-S, Kim J-Y, Kang S-J, Woo G-J, Chung Y-C, Eremin SA, Chung D-H (2007) Development of immunochromatography strip-test using nanocolloidal gold-antibody probe for the rapid detection of aflatoxin B1 in grain and feed samples. J Microbiol Biotechnol 17:1629–1637

Shim W-B, Kim K-Y, Chung D-H (2009) Development and validation of gold nanoparticle immunochromatographic assay (ICG) for the detection of zearalenone. J Agric Food Chem 57:4035–4041

Su J, Yang H, Chen J, Yin H, Tang R, Xie Y, Song K, Huyan T, Wang H, Wang W, Xue X (2010) Development of a class-specific immunochromatographic strip test for the rapid detection of organophosphorus pesticides with a thiophosphate group. Hybridoma (29):291–299

Svitel J, Dzgoev A, Ramanathan K, Danielsson B (2000) Surface plasmon resonance based pesticide assay on a renewable biosensing surface using the reversible concanavalin A monosaccharide interaction. Biosens Bioelectron 15:411–415

Tanaka M, Sakamoto K, Nakajima H, Soh N, Koji N, Chung D-H, Imato T (2007) Development of surface plasmon resonance immunosensor for the determination of methyl parathion. Bunseki Kagaku 56:705–712

Thurman EM, Goolsby DA, Meyer MT, Mills MS, Pomes ML, Kolpin DW (1992) A reconnaissance study of herbicides and their metabolites in surface water of the midwestern United States using immunoassay and gas chromatography/mass spectrometry. Environ Sci Technol 26:2440–2447

USEPA Method 546: Determination of total microcystins and nodularins in drinking water and ambient water by adda enzyme-linked immunosorbent assay

Vashist SK, Luong JH (2018) Immunoassays: future prospects and possibilities. In: Vashist SK, Luong JH (eds) Handbook of immunoassay technologies: approaches, performances, and applications. Academic Press, Cambridge, MA, p 455

Viswanathan S, Radecka H, Radecki J (2009) Electrochemical biosensor for pesticides based on acetylcholinesterase immobilized on polyaniline deposited on vertically assembled carbon nanotubes wrapped with ssDNA. Biosens Bioelectron 24:2772–2777

Walker JP, Asher SA (2005) Acetylcholinesterase-based organophosphate nerve agent sensing photonic crystal. Anal Chem 77:1596–1600

Wang L, Chen W, Xu D, Shim B-S, Zhu Y, Sun F, Liu L, Peng C, Jin Z, Xu C, Kotov NA (2009) Simple, rapid, sensitive, and versatile SWNT: paper sensor for environmental toxin detection competitive with ELISA. Nano Lett 9:4147–4152

Wang L, Ma W, Xu L, Chen W, Zhu Y, Xu C, Kotov NA (2010) Nanoparticle-based environmental sensors. Mater Sci Eng R 70:265–274

Wang K, Wang L, Jiang W, Hu J (2011) A sensitive enzymatic method for paraoxon detection based on enzyme inhibition and fluorescence quenching. Talanta 84:400–405

Yalow RS, Berson SA (1959) Assay of plasma insulin in human subjects by immunological methods. Nature 184:1648–1649

Yang G, Kang S (2008) SPR-based antibody-antigen interaction for real time analysis of carbamate pesticide residues. Food Sci Biotechnol 17:15–19

Zhang J, Lei J, Xu C, Ding L, Ju H (2010a) Carbon nanohorn sensitized electrochemical immunosensor for rapid detection of microcystin-LR. Anal Chem 82:1117–1122

Zhang Z, Liu J-F, Shao B, Jiang G-B (2010b) Time-resolved fluoroimmunoassay as an advantageous approach for highly efficient determination of sulfonamides in environmental waters. Environ Sci Technol 44:1030–1035

Zhao Y, Liang Y, Qian J, Li L, Wang S (2009) Determination of diethylstilbestrol by time-resolve fluoroimmunoassay. Anal Lett 42:216–227

Zou Z, Du D, Wang J, Smith JN, Timchalk C, Li Y, Lin Y (2010) Quantum dot-based immunochromatographic fluorescent biosensor for biomonitoring trichloropyridinol, a biomarker of exposure to chlorpyrifos. Anal Chem 82:5125–5133

Chapter 2
Development of Operational Resilience Metrics for Water Distribution Systems

Jaivik Mankad, Dinesh Borse, Laya Das, Nitin Padhiyar,
and Babji Srinivasan

2.1 Water Distribution Networks

Water distribution networks (WDNs) have prevailed in civilization since ancient time. The Romans built the aqueducts, which later got upgraded by closed conduits. In the modern era, sophisticated WDNs are developed to fulfil the water requirements of the society. They are an integral component of town planning due to its economic and social significance. Apart from domestic, commercial, agriculture and energy generation sectors are the major consumers of water. Each year most of the nations make a sizable amount of investment in their water systems (Van Ginneken et al. 2019). This makes WDNs critical below-ground infrastructure just like gas pipelines, metro transportation and sewerage networks.

Pipes, pumps, tanks, reservoirs, valves and sensors are the major elements of a WDN. The design layout of WDN depends on factors like the nodal demands, topological parameters and presence of active elements. They are designed with the primary objective to supply the required quantity of water to a population at an adequate pressure. Since decades, least cost has been the primary factor considered for the design of WDN (Geem 2006). However, with the increasing stresses on the WDN, multi-objective factors are being deployed for its design (Farmani et al. 2005, 2006; Babayan et al. 2007).

Urbanization has stirred the focus towards a sustainable framework for the operation, monitoring, maintenance and security of the WDN. Intelligent networks are being developed with an aim to uphold the performance and safeguard the services provided by the WDNs. The task of cementing the communication between the physical systems and the cyber modules with necessary features is performed by cyber-physical systems (CPS). The water cyber-physical systems consist of sensing, networking, computing and control technologies (Wang et al. 2015; Lin et al. 2009).

J. Mankad · D. Borse · L. Das · N. Padhiyar · B. Srinivasan (✉)
Indian Institute of Technology Gandhinagar, Ahmedabad 382355, India
e-mail: babji.srinivasan@iitgn.ac.in

© Springer Nature Singapore Pte Ltd. 2020
M. Kumar et al. (eds.), *Resilience, Response, and Risk in Water Systems*,
Springer Transactions in Civil and Environmental Engineering,
https://doi.org/10.1007/978-981-15-4668-6_2

The sensing technologies facilitate the monitoring of parameters in real time. They are expected to be flexible and accurate to maintain reliability. Networking technology aids in the transmission of the collected information to the central computing facility. These facilities carry out the work of data management, prediction and modelling. They also help to determine the critical variables, system performance and generate necessary insights to take corrective measures. The control technologies are expected to deliver the same to the physical systems in a synchronized manner. The water CPS is a comprehensive framework towards strengthening security and prevention of water losses in WDN.

2.1.1 Disruption in Water Distribution Networks

The natural calamities like earthquakes, hurricanes, floods, ageing infrastructures or man-made disasters including terrorist attacks, cyber-attacks and hazardous material released by industries lead to various failures in WDN (Workgroup 2009). Some of these disruptions like floods and hurricanes can have long-term effects on the quality of water supplied through the network (Chisolm and Matthews 2012). Apart from such physical failures, cyber-attacks and water contamination are other challenging threats to WDN (Taormina et al. 2017). Maroochy water breach (Slay and Miller 2007) exhibited the need for cyber-security of WDN and its components. The complex nature of WDN makes it prone to a high risk of failures. Many times the system interruptions prove dangerous for the system and the population as they are detected only after their failure. Potential impacts of such disruptions on WDNs comprise pipe breaks, loss of facilities/supplies, damage to infrastructure, degradation of quality, and financial and environmental losses. Such failures result in a degradation of the system's performance and delivered services. Parameters such as flow rate, pressures, pH, TDS, hardness and conductivity can be used to identify these impacts and gauge their magnitudes.

   Resilience refers to the property of the system to function with minimal loss in its targeted performance (Hosseini et al. 2016). It is the ability of the system to bounce back after the occurrence of an unfortunate disruptive event. Pipe failure is one of the most common failure scenarios noticed during disruptions in WDN. Resilience towards the case of single pipe failure in a network has been studied extensively in the literature. Ability of the network in maintaining connectivity and surplus head during such an event is studied by Di Nardo et al. (2018). Berardi et al. (2014) studied the effect of multiple failures occurring simultaneously on the supplied demand. They used multi-objective search to determine severe performance losses in the network. Studies have stated the importance of network's geometry and topology in the resilience measure of the WDN (Di Nardo et al. 2018; Meng et al. 2018). Uncertain demand has been reported as another such event responsible for the decline in network's performance. A multi-objective problem was solved for minimization of investment cost of the network while being able to retain nodal pressure under uncertain demand scenarios (Babayan et al. 2007). Diao et al. (2016) developed a tool for analysing WDNs under various disruptions. Studies performed on some networks revealed that while the networks were resilient to certain disruptions, they failed to sustain when exposed to some other events. Networks were compared mainly based on failure magnitude and duration under the scenarios of excess demand, pipe failure and substance intrusion scenarios. Ayala-Cabrera et al. (2017) proposed metrics for evaluation of resilience of water distribution networks giving special consideration to the adaptive, absorptive and restorative capacities.

The first step to achieve the resilient cyber-physical design of water distribution networks involves the ability to detect the abnormal situations. The abnormalities might be some physical, operational or contamination faults. The following study focuses on physical failures caused due to disruptions on WDN. During such events, operational parameters such as pressures, flow in the network or nodal head are instrumental in determining the performance of the network. The further sections would discuss the techniques to measure network performance in cyber-physical framework and its advantages in dealing with disruptive events.

2.2   Nodal-Level Performance of Water Distribution Networks

Conventionally, the water distribution networks are operated in demand-driven (DD) mode. In this mode, it is assumed that the pressures at the nodes are adequate to fulfil the nodal demands. Thus, the nodal pressures are the function of the nodal outflows. In case of failures, the nodal outflows are maintained even with some sustained pressure losses. However, the network gets dragged into pressure-deficient situations with an increase in the intensity of the failures. The state of the network when the design nodal demands are satisfied at desired pressures is termed as normal operation. The system may restore to its normal operation during single pipe failure but it might not restore with the increase in the number of failed pipes. The researchers have come up with the pressure-dependent demand (PDD) mode to analyse the network in such pressure-deficient situations. In PDD mode, the nodal pressures govern the nodal outflows, thus making it possible to study such scenarios of network failures. Comparative study of the network operations in DD and PDD mode has shown the advantages of analysing pressure-deficient networks in PDD mode over the DD mode (Braun et al. 2017; Nyende-Byakika et al. 2012; Gupta and Bhave 1996).

Todini (2000), Prasad and Park (2004) and Jayaraman and Srinivasan (Jayaram and Srinivasan 2008) proposed various resilience indexes to measure system performance. The system's energy, pipeline layout (branches and loops) and supply sources are the factors considered for their calculations. These indexes quantify system's ability to maintain the nodal demands during failure events in WDN. These studies can indicate the failures in the network as a whole, but they cannot identify locally affected areas or node. Thus, these indexes are not applicable in terms of selectivity. Baños et al. (2011) studied the networks ability to maintain the pressure at nodes under uncertain demands. It motivates for the exclusive consideration of network topology for performance measure. Lack of information about the affected nodes also compromises with the ability to decide upon control measures. Thus, the proposed approach tries to overcome these drawbacks by quantifying nodal-level performance measure for WDN.

Head and flow relationships have been studied in the literature by researchers, and several relations have been proposed. Bhave (1981) proposed the head–flow relation assuming that the node outflows always match the expected demand. However, partial flow condition is not considered in this study. Under normal operation, either expected flow or a no flow condition exists. Theoretically, the head value can range from [0, ∞). In contrast, assuming the flow to be unbounded, Reddy and Elango (1989) proposed head-dependent analysis. Considering the nodal heads and flow simultaneously could depict a more realistic situation. Both the head and flow have been considered simultaneously in the literature (Germanopoulos 1985; Bao and Mays 1990). Gupta and Bhave (1996) performed a comparative study of the methods available in the literature for the performance of networks under pressure deficiency. In this study, head and flow are utilized to calculate two performance metrics on real-time basis. These metrices provide information about the performance at the node and the state of network as a whole. Using this approach, various physical and operational failures in WDNs at nodal level can be identified. In this work, we have considered three failure scenarios: (i) pipe break, (ii) pipe clog and (iii) excess demand.

2.2.1 Head Ratio

From studies in the literature, it can be seen that the failures in the water distribution networks is accompanied by a drop in nodal head. The amount of drop in the nodal head can be used to identify different kinds of failures.

The following ratio could give insight into the system's current operation.

$$\phi_{H_i} = \frac{\overline{H_i^t}}{H_i} \qquad (2.1)$$

where $\overline{H_i^t}$ = mean of available head measured over time t at node i H_i= supply pressure required for normal operation at node i

The numerator $\left(\overline{H_i^t}\right)$ represents the mean of available head at node i over past t time units. It is taken as mean of the past few measurements so that the errors can be reduced in case wrong measurement is obtained at any time due to errors from the instrument or some other sources. The denominator is nodal head required to be maintained during normal operations. The above ratio is calculated at each demand node of the network. The performance of the node can be interpreted from the value that the above ratio assumes.

$$\phi_H \begin{cases} < 1 \text{ insufficient head at node } i \\ = 1 \text{ normal operation of node } i \\ > 1 \text{ excess energy available at node } i \end{cases}$$

The expression gives an insight into the current performance of node i. It compares the real-time measurements with the head values required for normal operation of the network. The desired nodal head for normal operation changes with the change in nodal demand. A simple linear metric has been selected for the purpose of real-time implementation of the cyber-physical water network in the future work.

Simulations demonstrate the effectiveness of the above ratio in failure identification. Sensitivity analysis under different noise levels and with changes in pipe resistance is also carried out to test to robustness of the above measure. Simulation results on different networks are presented in the next section.

2.2.2 Flow Ratio

As reported in the literature (Bao and Mays 1990), nodal heads alone cannot determine the network performance. Thus, we here considered a similar ratio for nodal flow. The following ratio also gives information about the current status of nodal outflows.

$$\phi_{F_i} = \frac{\overline{Q_i^t}}{Q_i} \tag{2.2}$$

where $\overline{Q_i^t}$ = mean of nodal outflow measured over time t at node i Q_i = normal expected demand at node i

The numerator $\left(\overline{Q_i^t} \right)$ represents the mean of nodal outflows at node i over past t time units. We take the mean of the past few measurements so that the measurement error can be reduced. The denominator is the nodal expected demand under normal operation. The above ratio is calculated for each node in the network. The nodal performance can be inferred from the value of the ratio above.

$$\phi_{F_i} \begin{cases} < 1 \text{ insufficient water supply at node } i \\ = 1 \text{ expected demand satisfied at node } i \\ > 1 \text{ excess water being supplied at node } i \end{cases}$$

The expression (2.2) can be used to infer the percentage of the expected quantity of water that is supplied. It is similar to the water service availability found in the literature. The flow measurements at any time are compared with the flow values required to satisfy the demand in the normal operation of the network. The simple linear form helps in real-time implementation in the cyber-physical framework.

Simulations have been performed on networks with active and passive elements to validate the proposed strategy. Simulation results and discussion have been presented in Sect. 2.3.1.

2.3 Case Studies

To demonstrate the effectiveness of the performance metrics introduced in Sect. 2.2, several simulations were performed on various networks. Networks present in the literature were chosen to evaluate their performance under different failures. The networks and some of their properties have been listed in Table 2.1. The architecture of the networks listed in Table 2.1 can be seen in Figs. 2.1 and 2.2.

Table 2.1 Summary of the networks used for simulations

Network	Hanoi	Net3
Type	Passive network Intermittent supply system	Active network Continuous supply system
Junctions	31	92
Pipes	34	117
Tanks	0	3
Reservoirs	1	2
Pumps	0	2

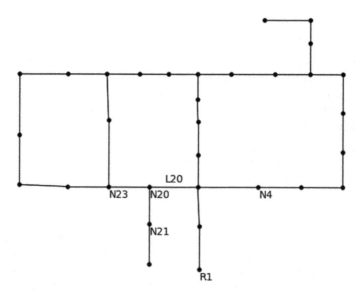

Fig. 2.1 Structure of *Hanoi* network

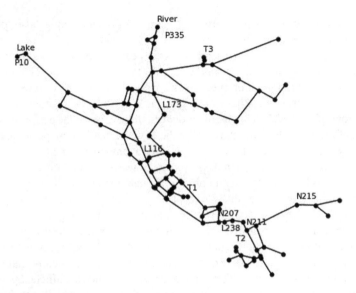

Fig. 2.2 Structure of *Net3* network

Hanoi network, popular in literature, was chosen for studying its performance during disruptions. It is gravity-fed system with no active elements such as pumps or valves. The structure of *Hanoi* network can be seen in Fig. 2.1. *Net3* is an active network that consists of tanks and pumps that help minimize the effect of disruption which will be discussed later. The structure of *Net3* can be seen in Fig. 2.2. In each of the network, the link to which the disruption is given and the nodes that are evaluated for their performance are labelled.

The performance metrics proposed in this work were evaluated by performing simulations after introducing disruptions of varied nature. Simulations were performed for three disruptive scenarios: (i) pipe break, (ii) pipe clogged and (iii) excess demand. The simulations have been performed assuming that only a single event occurs at any given time. Water Network Tool for Resilience (WNTR), a python package developed for simulating disruptions to water networks, was used for simulation purpose (Klise et al. 2018). The results are presented in this section along with necessary discussion.

2.3.1 Effect of Disruptions on Performance Measures

Pipe break is one of the disruptive scenarios for which the simulations are done to study qualitative behaviour of network. An event of pipe break would mean large quantity of water exits the pipe network through the opening. This not only suggests loss of water but insufficient supply at the demand nodes. Further, a situation of pressure deficiency occurs as a result of energy loss. The supply takes place at reduced pressure which is undesirable.

The parameters required for executing pipe break comprise of the broken link, area of the leak, discharge coefficient, and leak start and end times. The remaining parameters used for simulating pipe breaks are as given below in rows 2 to 4 of Table 2.2. Time series plots have been generated for the nodes shown in the last but one row of Table 2.2. This helps in understanding the nodal performance as a function of time. The last row shows the time at which the heat maps are generated to better evaluate the status of the entire network at any given time.

Hanoi network, an intermittent supply system, was simulated for 1 h. Link 20 of the network was broken after 30 min of normal operation and was left broken till the end of the simulation. This results in loss of a large quantity of water and also pressure deficiency in the network. The nodal demands may not be completely satisfied in case the pressure drops below nominal pressure. The drop in head ratio and flow ratio of some of the nodes below unity can be seen in Fig. 2.3a, b, respectively. The value of both ratios being unity indicates that the network operated normally prior to 30 min. It implies that all the nodal demands were satisfied at sufficient pressures. After 30 min, the ratio started dropping as a result of introduced pipe break. The head ratio reached its minimum value and stabilized at around 0.7 as can be observed in

Table 2.2 Parameters used for simulating pipe breaks	Network	Hanoi	Net3
	Link broken	20	173
	Break start time (h)	0.5	6
	Discharge coefficient	0.2	0.2
	Nodal performance	4, 20,21,23	117, 119, 151, 159
	Heat map time (h)	0.25, 0.75	12

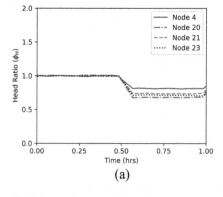
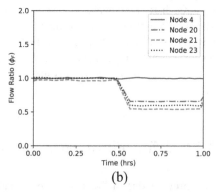

(a)   (b)

Fig. 2.3 **a** Head ratio as a measure of system's performance due to pipe breaks for entire period of simulation for some nodes in *Hanoi* network under normal noise levels and **b** flow ratio as a measure of system's performance due to pipe breaks for entire period of simulation for some nodes in *Hanoi* network under normal noise levels

Fig. 2.3a. This resulted in a lack of supply as indicated by flow ratio at pressure deficient nodes which can be observed from Fig. 2.3b.

Heat maps showing the status of the network before disruption ($t = 15$ min) and after disruption ($t = 45$ min) are shown in Figs. 2.4 and 2.5, respectively. Figure 2.4a shows the head ratio while Fig. 2.4b shows the flow ratio of the network before disruption ($t = 15$ min). All the nodes in the network are close to a value of 1 indicating normal operation. The pressure losses due to pipe break in the network at 30 min cause the network nodes to underperform as indicated by the head ratio at 45 min in Fig. 2.5a. It can be seen that a large part of the network gets affected due to just one pipe break. The corresponding nodal outflows also reduce as indicated by the flow ratio in Fig. 2.5b. The part of the network whose performance gets affected depends

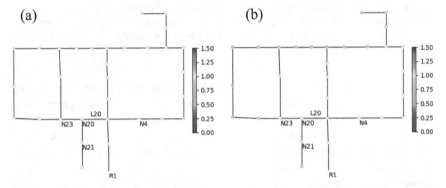

Fig. 2.4 **a** Head ratio as a measure of system's performance before disruption ($t = 0.25$ h) at every node in the *Hanoi* network under normal noise levels and **b** flow ratio as a measure of system's performance before disruption ($t = 0.25$ h) at every node in the *Hanoi* network under normal noise levels

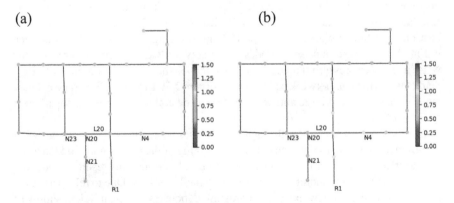

Fig. 2.5 **a** Head ratio as a measure of system's performance due to pipe breaks ($t = 0.75$ h) at every node in the *Hanoi* network under normal noise levels and **b** flow ratio as a measure of system's performance due to pipe breaks ($t = 0.75$ h) at every node in the *Hanoi* network under normal noise levels

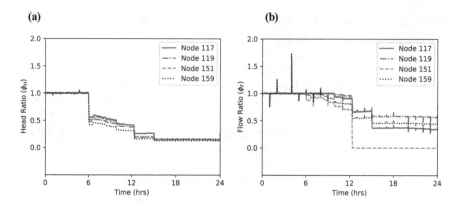

Fig. 2.6 **a** Head ratio as a measure of system's performance due to pipe breaks for entire period of simulation for some nodes in *Net3* network under normal noise levels and **b** flow ratio as a measure of system's performance due to pipe breaks for entire period of simulation for some nodes in *Net3* network under normal noise levels

on a number of factors like the topology of the network that includes node degree, connectivity, clustering coefficient and network physical attributes like elevations of the nodes, pump locations and available head in the tanks among others.

Another network *Net3* with active elements was chosen for the study. As shown in Table 2.1, it contains two pumps, two reservoirs and three tanks. In the literature, *Net3* is also used as a benchmark network to test tools and techniques. To test the approach of detecting disruptions, link 173 was broken after 6 h of normal operation till the end of the simulation. Discharge coefficient of 0.2 was used for simulation. It can be seen from Fig. 2.6a that there is a reduction in supply pressure starting 6 h. The conditions worsen as the time proceeds and drops to a new level around 15 h. However, the flow does not decrease immediately since the pressure is still higher than the nominal head required for full flow to occur. Towards the later part of the simulation, the flow starts decreasing as indicated by the flow ratio in Fig. 2.6b. The heat maps of the network at time $t = 12$ h are showing the head ratio can be seen in Fig. 2.7a. A large number of nodes in the network were observed to receive the water at a lower pressure than desired. This implies that link 173 which is of larger diameter is critical to network performance. Figure 2.7b shows the flow ratio at every node in the network. A large number of nodes can be noticed receiving an insufficient quantity of water.

Over a period of time, flow through pipes causes deposition of particles, growing of biofilms and corrosion of pipe material. Clogging of the pipe is not a sudden event but the area available for flow reduces over a period of time. This results in clogging of the pipes thereby reducing effective area available for flow. Due to this, part of the network downstream may experience pressure deficiency and insufficient supply of water. However, drop in the performance of the network does not solely depend on this. Presence of source or active elements may help minimize the effect of such an event.

(a)   (b)

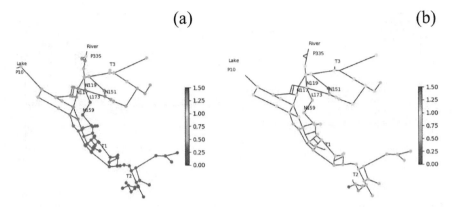

Fig. 2.7 **a** Head ratio as a measure of system's performance due to pipe breaks ($t = 12$ h) at every node in the *Net3* network under normal noise levels and **b** Flow ratio as a measure of system's performance due to pipe breaks ($t = 12$ h) at every node in the *Net3* network under normal noise levels

To simulate such a scenario, link 20 in *Hanoi* network was chosen. Clogging was introduced to the same link to which pipe break was given to compare the effect of the two on the networks' performance. Clogging of pipe can be characterized by placing a pressure reducer valve (PRV) along that link. The diameter of the valve can be kept smaller than that of the pipe. Pipe was clogged for the entire period of simulation. Table 2.3 shows the parameter used for simulating pipe clog scenarios using WNTR.

For simulation purpose, PRV diameter of 90% of actual pipe diameter was placed in link 20. Again, the time series plots showing the variation in the two ratios at some of the nodes are shown in Fig. 2.8. The head ratio is seen in Fig. 2.8a while the flow ratio in Fig. 2.8b is plotted for the entire duration of simulation. It should be noted that, while the nodes downstream of the pipe clogged are at a lower head than that required for normal operation, the nodes upstream of the pipe receive water at higher heads than required. The relative positions of the nodes and links can be observed in Fig. 2.2. The heat map in Fig. 2.9a shows the head ratio at each of the nodes in the network at time, $t = 15$ min. The corresponding flow ratio can be seen in Fig. 2.9b. It can be observed that almost half of the network underperforms indicated by the

Table 2.3 Parameters used for simulating pipe clog

Network	Hanoi	Net3
Link clogged	20	238
Fraction of pipe diameter Clogged	0.1	0.2
Nodal performance	4, 20, 21, 23	207, 211, 215
Heat map time (h)	0.25	5, 12

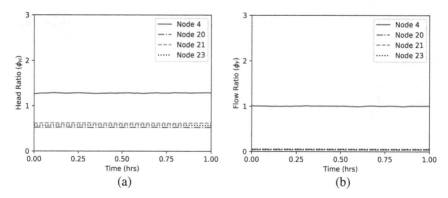

Fig. 2.8 **a** Head ratio as a measure of system's performance due to pipe clog for entire period of simulation for some nodes in *Hanoi* network under normal noise levels and **b** flow ratio as a measure of system's performance due to pipe clog for entire period of simulation for some nodes in *Hanoi* network under normal noise levels

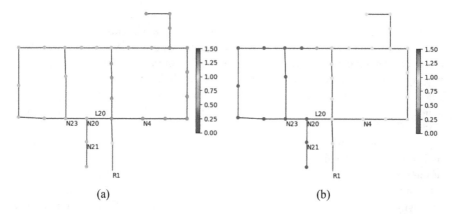

Fig. 2.9 **a** Head ratio as a measure of system's performance due to pipe clog ($t = 0.25$ h) at every node in the *Hanoi* network under normal noise levels and **b** flow ratio as a measure of system's performance due to pipe clog ($t = 0.25$ h) at every node in the *Hanoi* network under normal noise levels

head and flow ratio as a result of a single pipe failure in the network. On the contrary, many other nodes in the other half receive complete supply but at an elevated head.

Net3 was simulated for a similar disruptive event. Link 238 that acts as a bridge connecting two parts of a large network was chosen for simulation purpose. It should be noted that *Net3* contains tanks, reservoirs and pumps. Figure 2.10a shows the head ratio in the network. Notice that even though the pipe clogging was simulated for the entire 24 h duration, the network continued supplying required quantity for 6 h post disruption without significant loss in the head at which it was supplied as seen in Fig. 2.10b. Tank 2 (T2) in the downstream of the link 238 continued supplying for an extended period. Heat map plotted after 5 h of operation showing the head ratio

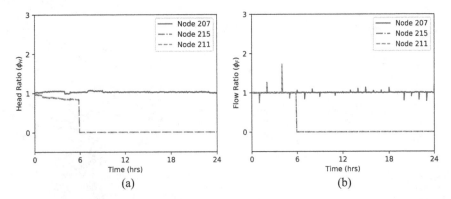

Fig. 2.10 a Head ratio as a measure of system's performance due to pipe clog for entire period of simulation for some nodes in *Net3* network under normal noise levels and **b** flow ratio as a measure of system's performance due to pipe clog for entire period of simulation for some nodes in *Net3* network under normal noise levels

in the network can be seen in Fig. 2.11a. The corresponding flow ratio at every node in the network is plotted in Fig. 2.11b. The path from reservoir to the tank passes through the link 238. Due to this, the tank does not get refilled and hence unable to satisfy nodal demands after 6 h. Hence, the performance further drops in terms of head and demand as shown in Fig. 2.12a, b.

Firefighting operations demand an excess quantity of water in a short time span. Large volume of water is extracted from the networks causing the network to loose its performance. Drawing excessive water than the capability of the network causes suction in the network. The pressure at some of the nodes depending on the networks topology and presence of various elements decreases affecting their supply. However, there may be nodes which receive more than normal demand. Hence, in excess

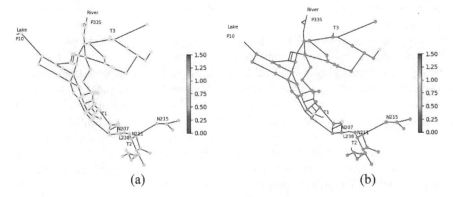

Fig. 2.11 a Head ratio as a measure of system's performance due to pipe clog ($t = 5$ h) at every node in the *Net3* network under normal noise levels and **b** flow ratio as a measure of system's performance due to pipe clog after disruption ($t = 5$ h) at every node in the *Net3* network under normal noise levels

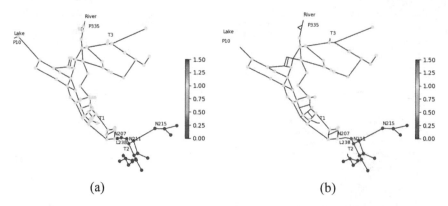

(a)   (b)

Fig. 2.12 **a** Head ratio as a measure of system's performance due to pipe clog ($t = 12$ h) at every node in the *Net3* network under normal noise levels and **b** flow ratio as a measure of system's performance due to pipe clog ($t = 12$ h) at every node in the *Net3* network under normal noise levels

demand scenario, while the supply pressure at some of the nodes goes down, the flow ratio may assume a value higher or lower than 1.

The design peak factor of the gravity-driven networks is essentially less as compared to that of the network supplied through pumping. Gravity-driven networks are mostly deployed in the areas of low population density. Also, it is assumed that there is availability of secondary water source to take care of occasional high water demands. Sometimes lake or tube wells are those alternate sources that are expected to meet the demands in such cases. In the areas of high-density population, the water demand per hour in itself is very high due to which pumping stations are necessary. Thus, mostly in urban or semi-urban areas, the water distribution networks are pressure-driven. Such networks are expected to overcome the high uncertainty in the water demand at adequate pressure. Hence, it is essential to study the behaviour of pressure-driven systems in excess demand scenario as a part of system resilience assessment.

Net3 was used for simulating excess demand scenario. As already mentioned, *Net3* contains reservoirs and pumps. Node 101 in the network was chosen for simulating excess demand being an articulation point. An articulation point, from network science perspective, is one that if removed disconnects the network into two parts. A demand pattern was created for excess demand for 4 h from 6 to 10 h. For rest of the time before 6 h and after 10 h, the default pattern was left unchanged.

From Fig. 2.13a, it can be seen that the head ratio at some of the nodes drops due to the large volume of water being drawn from another node. Thus, excess demand at one node affects the pressure at which other nodes are serviced. However, this does not help differentiate excess demand situation with another situation such as pipe break. The flow ratio shown in Fig. 2.13b reveals the situation. The heat map for excess demand is shown in Fig. 2.14a, b for the head and flow ratio after 8 h.

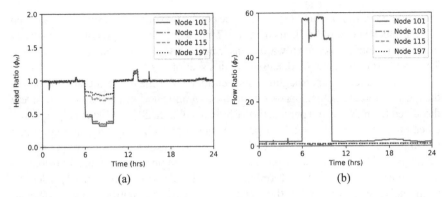

Fig. 2.13 a Head ratio as a measure of system's performance due to excess demand for entire period of simulation for some nodes in *Net3* network under normal noise levels and **b** flow ratio as a measure of system's performance due to excess demand for entire period of simulation for some nodes in *Net3* network under normal noise levels

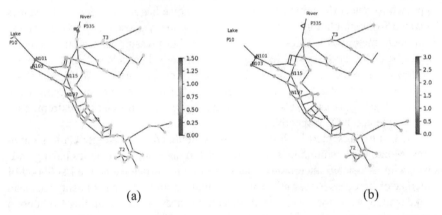

Fig. 2.14 a Head ratio as a measure of system's performance due to excess demand ($t = 8$ h) at every node in the *Net3* network under normal noise levels and **b** flow ratio as a measure of system's performance due to excess demand ($t = 8$ h) at every node in the *Net3* network under normal noise levels

Even though node 101 is directly connected to the source, a significant part of the network experience pressure deficiency after introducing excess demand.

2.3.2 Sensitivity Analysis

As discussed in the previous sections, the nodal flow analysis is a more compre-hensive approach to deal with disruptions in WDNs. However, in order to use the proposed approach, the performance metrics need to be accurate and robust at the same time. Sensor noise and pipe resistance are the two most important factors which can introduce uncertainties in its measure. Thus, a detailed analysis of the sensitivity of the performance metrics towards sensor noise and pipe resistance is carried out. The results for the same are discussed in this section.

Pipe resistance or pipe roughness as mentioned in some of the literature is one of the most important factors that determine the hydraulic performance. It is extensively discussed from the design perspective (Bhave and Gupta 2006). However, pipe resistance properties are also vital for performance monitoring, system inspection and real-time control of WDNs (Bhave 1988). Further, sensor measurements are always accompanied by high-frequency noise. It is a random phenomenon that corrupts the measurement with no useful information. High signal-to-noise ratio (SNR) is desired as the noise levels are low in the signal. Monte carlo simulations were performed to validate the approach of identifying abnormal network operations in presence of such uncertain quantities. While the network is being operated in a real scenario, the sensors would pick up noise and corrupt the measurements it would provide. Thus, the true value always remains unknown. To mimic such a situation and validate the approach to detect disruptions to the network, noise has been added to the values obtained from simulations. The noise levels have been chosen as provided by the instrument manufacturers for pressure and flow. For pressure measurements, noise level of < 1% of the span, was found in manufacturer's datasheet while that of ± 5% for flow measurements. Simulations in this work were performed with normal noise levels of 1% and 2.6% as found in the literature for pressure and flow, respectively (Dias et al. 2013). Noise levels of ± 1.5% and ± 5% were chosen as an extreme case for pressure and flow, respectively.

The results presented in this section were obtained using Monte Carlo simulations. Monte Carlo simulations are performed to understand processes having variables with associated uncertainty. The technique helps determine the likelihood of outcome of a process for variations in their input variables. Instead of using the mean or the median value for a variable, this technique uses random sampling to choose a value for a variable. The method is an effective way for determining the outcome of processes/events involving a large number of degrees of freedom. A large number of simulations are performed by randomly sampling values for each variable from probability functions and generating outputs that helps better forecasting for informed decision making. The technique finds its applications in a variety of areas like business and finances, statistical physics, research and development, transportation, and environment.

For each of the three disruptions, viz. pipe break, pipe clog and excess demand, 1000 simulations were performed to obtain extreme values of the two proposed metrics. Noise introduced varied in the range of ± 1% in pressure measurement and ± 5% in flow measurements. The pipe resistance for every pipe in the network was allowed to change up to 40% of the initial value for each case.

To analyse the effectiveness of the two metrics in the presence of high sensor noise, simulations were performed under extreme noise levels as mentioned above. Even at higher levels of noise, the approach could successfully detect the abnormal operation of the network. Noise magnitude of ± 1.5% for pressure and ± 5% for flow was added to check for the validity of the approach. Time series plots for the same nodes of *Hanoi* network under pipe break scenario as in Fig. 2.3 are shown below in Fig. 2.15a, b. It can be observed that the drop in network performance in terms of the head at which the water is supplied and the actual flow of water at the nodes can still be detected. The heat maps have been avoided owing to their similarity with the results already discussed.

Next, simulating *Net3* under extreme levels of measurement noise for both pressure and flow, similar observations can be made. The time series plots for the simulations performed under uncertain demand can be observed in Fig. 2.16a, b for head and flow ratio, respectively. The heat maps have been avoided owing to their similarity with the plots shown in Fig. 2.14.

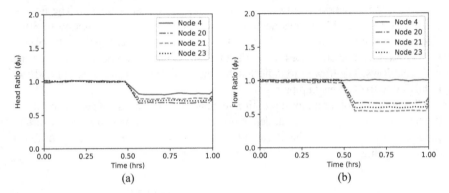

Fig. 2.15 **a** Head ratio as a measure of system's performance due to pipe breaks for entire period of simulation for some nodes in *Hanoi* network under extreme noise levels and **b** flow ratio as a measure of system's performance due to pipe breaks for entire period of simulation for some nodes in *Hanoi* network under extreme noise levels

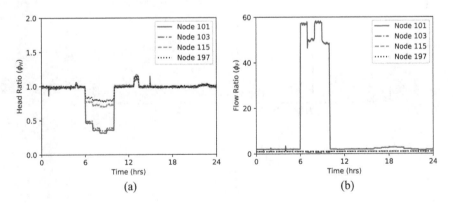

Fig. 2.16 **a** Head ratio as a measure of system's performance due to pipe breaks for entire period of simulation for some nodes in *Net3* network under extreme noise levels and **b** flow ratio as a measure of system's performance due to pipe breaks for entire period of simulation for some nodes in *Net3* network under extreme noise levels

The pipe resistance is the material property that changes over the period of time. Pipe corrosion and salts in water affect the internal linings of the pipes. The effect of the pipe resistance on the calculation of ϕ_{H_i} and ϕ_{F_i} at the nodes for two networks is shown below. Here we discuss the case of *Hanoi*, which is a passive network and *Net3* which is an active network. The margin of noise consideration for flow ($\pm$ 2.6%) and pressure ($\pm$ 1%) measurements is standard as per the literature. In the *Hanoi* network (Fig. 2.1), the link 20 is broken after 30 min of normal operation and 20% of the actual flow is flowing out of the pipe walls as leak. The supply from reservoir (R1) is able to satisfy the demand at the nodes 4 but not at nodes 20 and 23 as evident from the heat map shown in Fig. 2.17.

In the next simulation, the link 116 in the *Net3* is broken after 6 *h* of normal operation. The leak discharge coefficient is 0.4. As the carrying capacity of the pipe decreases, the networks ability to satisfy the demands also decreases. The flow at the nodes 191 and 111 shows decline when pipe resistance is increased to 30%. The metric can measure the resilience of the system at higher pipe resistance and Fig. 2.18a, b can be compared to see the effect of changes in pipe resistance on the nodal outflows. The flow ratio indicates that the available flow at nodes 111 and 191 reduces due to changes in pipe resistance.

Clogging is another major disruption responsible for the loss of serviceability of the network. Scaling, corrosion, deposition and sedimentation cause the closure in pipes to variable extents. The effect of the pipe resistance on the detection of faults due to clogging is discussed for different water distribution networks.

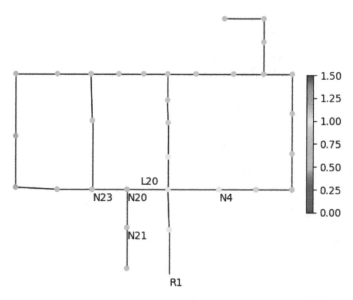

Fig. 2.17 Flow ratio as a measure of system's performance due to pipe break ($t = 0.75$ h) at every node in *Hanoi* network with 30% increase in pipe resistance

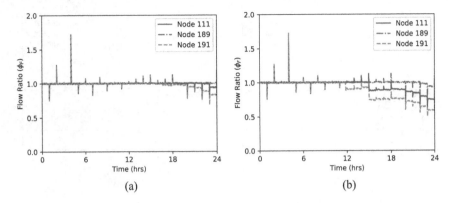

Fig. 2.18 **a** Flow ratio as a measure of system's performance due to pipe break for the entire period of simulation for some nodes in the *Net3* network with design pipe resistance and **b** flow ratio as a measure of system's performance due to pipe break for the entire period of simulation for some nodes in the *Net3* network with 30% increase in pipe resistance

In the *Hanoi* network, blockage was introduced with 10% clogging in the link 20. In case 1, the pipe resistance was kept the same as that of the design value. In case 2, the carrying capacity of all the pipes in the network is reduced by 30% of the design value (the decrease in the C value of all the pipes in the network). The clog in link 20 was introduced for the complete duration of simulation. The heat map of the network after 30 min of operation in the clogged state can be seen in Fig. 2.19a. As link 20 is clogged, the nodes in the left loop of the networks show a drop in their head ratio. This is expected from the hydraulic characteristics of the water networks. The flow

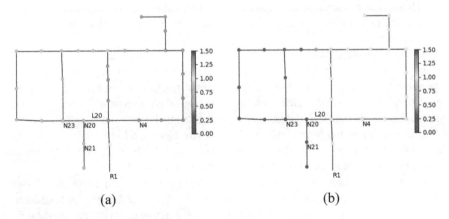

Fig. 2.19 **a** Flow ratio as a measure of system's performance due to pipe break ($t = 0.5$ h) at every node in *Hanoi* network with design pipe resistance and **b** flow ratio as a measure of system's performance due to pipe break ($t = 0.5$ h) at every node in *Hanoi* network with 30% increase in pipe resistance

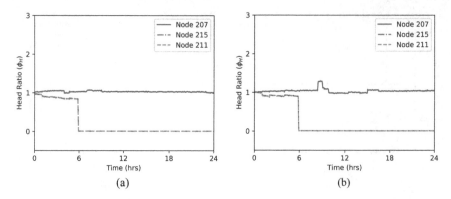

Fig. 2.20 **a** Head ratio as a measure of system's performance due to excess demand for the entire period of simulation for some nodes in the *Net3* network with design pipe resistance and **b** head ratio as a measure of the system's performance due to excess demand for the entire period of simulation for some nodes in the *Net3* network with 30% increase in pipe resistance

ratio at each node in Fig. 2.19b indicates that the entire left part of the network is facing scarcity of water supply.

The components of the active network add some kind of dynamic nature to the network. *Net3* (Fig. 2.2) is an active network with two pumps connected to reservoirs. Such kind of networks is common in urban areas where the supply is continuous. Clogging was introduced in the link 238 which acts as a bridge in the network. Due to the presence of the pump, flow ratios at the nodes in the upstream of link 238 are maintained. With the decrease in the pipe carrying capacity, more energy is required to maintain the flow. The increased head ratios in the nodes near the disrupted nodes with higher pipe resistance as compared to design values can be seen in Fig. 2.20a, b. Also, the head ratio at the node 207 away from the disrupted link is maintained in both the scenarios.

Excess demand cannot be conclusively termed as a fault but a state of network operation where the nodal demands are way beyond the design parameters. During disasters to maintain the functionality of any infrastructure, water demand increases considerably. Fire is one such scenario that causes an enormous increase in the demand in short period of time. It is expected from the water network to deliver their service in such a highly vulnerable scenario. Here, we have tried to study the response of *Net3* network in the case of excess demand. The node 101 in *Net3* was subjected to excess demand of peak factor 110 from 6 to 10 h.

Figure 2.21a, b indicates the comparative study of the effect on the flow ratio due to increased pipe resistance. In Fig. 2.21a, the flow ratio with the design pipe resistance during excess demand is nearly 59 times the normal demand. As the pipe resistance increases, the maximum value of the flow ratio drops down to 44 times during excess demand scenario. However, the flow ratio is able to determine the serviceability of the system in the degraded state with due representation of the energy loss due to pipe resistance.

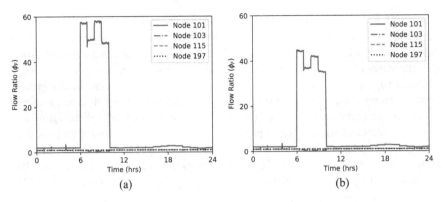

Fig. 2.21 **a** Flow ratio as a measure of system's performance due to excess demand for the entire period of simulation for some nodes in the *Net3* network with design pipe resistance and **b** flow ratio as a measure of the system's performance due to excess demand for the entire period of simulation for some nodes in the *Net3* network with 30% increase in pipe resistance

2.4   Conclusion

The infrastructure of water distribution networks (WDNs) is considered to be of *vital importance* as they accommodate the water supply needs of the society. They are identified as critical infrastructure in many countries as they affect their economy, security and social well-being. WDNs are designed with an objective to satisfy nodal demands at sufficient pressures. However, they often encounter disruptions such as breakage of pipe, uncertain demand, loss of facilities, power loss and substance intrusion. Such failures prevent the system from achieving its design objectives. The WDNs may or may not be able to withstand disruptions, resulting in a drop in its performance. They need to be upgraded so that they could adapt, absorb or recover to its normal state, in other words, they become more resilient to such disruptions. Recent inclination towards cyber-physical systems motivates its utilization for the application of resilient water distribution networks.

A water cyber-physical system is capable to identify faults, perform decision-making based on some logic and restore to its normal state all by itself. The focus in this work is on developing an approach that could give insight into the networks performance at nodal level during such disruptive events. The simulations done on both the networks, Hanoi and Net3, have demonstrated the ability of the proposed metrics to identify disruptions. Furthermore, it also informs about the nature of disruption and the affected area that can be useful in planning appropriate corrective actions. This would help to minimize the degrading effects on its performance in real time. Thus, the proposed approach is a step towards developing an intelligent water supply system for the future.

The current study focuses only on the physical failures, and however, it can be further extended to applications of water contamination and development of an integrated secured and sustainable water supply system. The major assumption in this study is that the sensor measurements at every node in the network are available. However, cost is an important aspect that limits the number of sensors that can be placed. This work can be extended to solving a sensor placement problem and to predict the states of the nodes from limited available information. With the knowledge of the disruptions and nodal states, appropriate control strategies can be designed to recover the network from failed state and prioritizing them. Loss of pressure energy is one of the common impacts of disruptions on water networks. Scheduling and control of active elements such as pumps and valves can be developed for providing energy during pressure deficient situations. Further, pumps can be operated as turbine to recover excess energy available in the network for the design to be sustainable. These could help realize a complete resilient cyber-physical framework for WDN.

References

Ayala-Cabrera D, Piller O, Deuerlein J, Herrera M (2017) Towards resilient water networks by using resilience key performance indicators. In 10th world congress of ewra-on water resources and environment,"panta rhei", Athens, Greece

Babayan AV, Savic DA, Walters GA (2007) Multiobjective optimisation of water distribution system design under uncertain demand and pipe roughness. In: Topics on system analysis and integrated water resources management, pp 161–172

Baños R, Reca J, Martínez J, Gil C, Márquez AL (2011) Resilience indexes for water distribution network design: a performance analysis under demand uncertainty. Water Resour Manage 25(10):2351–2366

Bao Y, Mays LW (1990) Model for water distribution system reliability. J Hydraulic Eng 116(9):1119–1137

Berardi L, Ugarelli R, Røstum J, Giustolisi O (2014) Assessing mechanical vulnerability in water distribution networks under multiple failures. Water Resour Res 50(3):2586–2599

Bhave P, Gupta R (2006) Analysis of water distribution networks, alpha science international ltd. Oxford, UK

Bhave PR (1981) Node flow analysis distribution systems. Transp Eng J ASCE 107(4):457–467

Bhave PR (1988) Calibrating water distribution network models. J Environ Eng 114(1):120–136

Braun M, Piller O, Deuerlein J, Mortazavi I (2017) Limitations of demand-and pressure-driven modeling for large deficient networks. Drinking Water Eng Sci 10(2):93–98

Chisolm EI, Matthews JC (2012) Impact of hurricanes and flooding on buried infrastructure. Leadership Manage Eng 12(3):151–156

Diao K, Sweetapple C, Farmani R, Fu G, Ward S, Butler D (2016) Global resilience analysis of water distribution systems. Water Res 106:383–393

Dias RP, Filho JGD, De Lucca YFL (2013) Water flow meter measurement uncertainties. Water Resour Manage 171

Di Nardo A, Di Natale M, Giudicianni C, Greco R, Santonastaso GF (2018) Complex network and fractal theory for the assessment of water distribution network resilience to pipe failures. Water Sci Technol: Water Supply 18(3):767–777

Farmani R, Walters G, Savic D (2006) Evolutionary multi-objective optimization of the design and operation of water distribution network: total cost vs. reliability vs. water quality. J Hydroinformatics 8(3):165–179

Farmani R, Walters GA, Savic DA (2005) Trade-off between total cost and reliability for anytown water distribution network. J Water Resour Plann Manage 131(3):161–171

Geem ZW (2006) Optimal cost design of water distribution networks using harmony search. Eng Optimization 38(03):259–277

Germanopoulos G (1985) A technical note on the inclusion of pressure dependent demand and leakage terms in water supply network models. Civil Eng Syst 2(3):171–179

Gupta R, Bhave PR (1996) Comparison of methods for predicting deficient-network performance. J Water Resour Plann Manage 122(3):214–217

Hosseini S, Barker K, Ramirez-Marquez JE (2016) A review of definitions and measures of system resilience. Reliab Eng Syst Safety 145:47–61

Jayaram N, Srinivasan K (2008) Performance-based optimal design and rehabilitation of water distribution networks using life cycle costing. Water Resour Res 44(1)

Klise KA, Murray R, Haxton T (2018) An overview of the water network tool for resilience (wntr). (Tech. Rep.). Sandia National Lab. (SNL-NM), Albuquerque, NM (United States)

Lin J, Sedigh S, Miller A (2009) Towards integrated simulation of cyber-physical systems: a case study on intelligent water distribution. In 2009 eighth IEEE international conference on dependable, autonomic and secure computing, pp 690–695

Meng F, Fu G, Farmani R, Sweetapple C, Butler D (2018) Topological attributes of network resilience: a study in water distribution systems. Water Res 143:376–386

Nyende-Byakika S, Ngirane-Katashaya G, Ndambuki JM (2012) Comparative analysis of approaches to modelling water distribution networks. Civil Eng Environ Syst 29(1):79–89

Prasad TD, Park N-S (2004) Multiobjective genetic algorithms for design of water distribution networks. J Water Resour Plann Manage 130(1):73–82

Slay J, Miller M (2007) Lessons learned from the maroochy water breach. In: International conference on critical infrastructure protection, pp 73–82

Taormina R, Galelli S, Tippenhauer NO, Salomons E, Ostfeld A (2017) Characterizing cyber-physical attacks on water distribution systems. J Water Resour Plann Manage 143(5):04017009

Todini E (2000) Looped water distribution networks design using a resilience index based heuristic approach. Urban water 2(2):115–122

Van Ginneken M, Netterstrom U, Bennett A (2019) More, better, or different spending? Trends in public expenditure on water and sanitation in Sub-saharan Africa

Wang Z, Song H, Watkins DW, Ong KG, Xue P, Yang Q, Shi X (2015) Cyber-physical systems for water sustainability: challenges and opportunities. IEEE Commun Mag 53(5):216–222

Workgroup C (2009) All-hazard consequence management planning for the water sector

Chapter 3
An Overview of Big Data Analytics: A State-of-the-Art Platform for Water Resources Management

Nirav Raval and Manish Kumar

3.1 Introduction

Water resources have become stressed with the enhancement of population growth, looming agricultural/industrial production, unprecedented rise in living standard, and uncontrolled climate change (Kumar et al. 2019a; Roshan et al. 2020). Water scarcity refers to the condition when sufficient water resources are not available to meet a particular region's water requirement (Kumar et al. 2019b, 2020). It is estimated that worldwide approximately four thousand million people are not provided with adequate quantity of potable water for at least one month a year. With the population projected to expand to nine thousand million by 2050, the demand for potable water is set to increase dramatically (du Plessis 2019; Patel et al. 2019; Singh et al. 2020). Hence, to meet the demand of blooming population for clean water, monitoring-based management of water resources is essential.

Advancement of the engineered sensors, data monitoring, and communication devices enables continuous monitoring of particular water system. As a result of this, near real-time series data with high frequency can be recorded. Such perpetual measurement produces bulk of data, called as big data (Gandomi and Haider 2015; Mayer-Schönberger and Cukier 2013). The term "Big data" which includes the major processes like, data acquisition, storage, extraction, and cleaning as well as analysis and interpretation were first proposed by Michael Cox and David Ellsworth in 1997. The targeted use of advanced big data analytics is emerging for effective and sustainable management of water resources in the scientific community. Application of the computer models is increasing in the field of water science and engineering because of the urgent necessity for deeper perspicacity into water systems and

N. Raval · M. Kumar (✉)
Discipline of Earth Sciences, Indian Institute of Technology Gandhinagar, Gandhinagar, Gujarat 382355, India
e-mail: manish.kumar@iitgn.ac.in

© Springer Nature Singapore Pte Ltd. 2020
M. Kumar et al. (eds.), *Resilience, Response, and Risk in Water Systems*,
Springer Transactions in Civil and Environmental Engineering,
https://doi.org/10.1007/978-981-15-4668-6_3

43

demand for providing effective solutions toward stressed water resources in a sustainable manner (Bibri and Krogstie 2017; Singh et al. 2020; Mukherjee et al. 2020). However, heterogenous nature of big water data causes difficulty in its storing, handling, and processing. Thus, to manage water resources effectively and sustainably, proper application of advanced water analytics is one of the prime necessities of the current decade.

A key contribution of this chapter is to bring forth the basic overview, characteristics, applications, challenges, and open platform/proposed model supporting water resources-based big data. Also, the future directions related to the integrations of various platforms are provided.

3.2   Big Water Data and Associated Characteristics

Technological advancement facilitates constant acquisition and processing of data at an unprecedented rate which can be further managed with the readily available software and hardware. This capability can be inclusively termed as "big data" (Adamala 2017). To characterize the big data, commonly used parameters include (i) Volume: As name suggests, volume is generally the quantity of data generated, processed, and stored. In the twenty-first century, data generation is constantly increasing as a result of which big data sizes are reported in multiple of terabytes and petabytes. For the storage, handling, and processing of this bulk data, the distributed systems are used instead of traditional database technology (Schroeck et al. 2012). (ii) Velocity: Speed with which generated data can be transferred and processed is known as velocity. At present time, streaming data (collected in real time) is one of the leading edges of big data. The modern applications and computer-based programmes/softwares enable the sorting, transmitting, and processing of generated data at faster rate (David et al. 2014). (iii) Variety: The availability of different types of data represents the variety. Water-related data is highly unstructured. The modern big data technology enables simultaneous collection and usage of structured and unstructured data. In water resource management, there has been more efforts require to integrate all types of water data from across different sections/sectors into one continuous data stream (Zikopoulos and Eaton 2011). (iv) Veracity: The quality or trustworthiness of any water-related data is known as veracity and it is directly associated with the health aspect as water is considered to be one of the primary necessities for the survival of living being. In general, it is a measure of the accuracy of the data. Quality control is one the important parameters to be considered for big data. (v) Value: It refers to the actionable perception gained from generated data. Having access to big water data will not going to complete the work unless and until it's conversion into some value has not been performed. In case of water consumption survey study, the availability of data is not sufficient to reach the decision making untill its conversion into some deliverable value. With the help of the state-of-the-art models/softwares and algorithms, large amounts of data can be converted into deductive information for final decision making (David et al. 2014; Madden 2012).

3.3   Big Data Analytical Methods

Big data analysis can be useful in enlightening the decision-making process in numerous areas such as environmental, natural disaster, and resources management. Numbers of big data analytical methods are used to infer a value/decision from the acquired big data (Chen et al. 2012; Manyika 2011). Most important of them are listed with their characteristics in Table 3.1.

3.4   Big Data and Water Resources Management

Water usage is more than double the rate of the population growth in the last century, which makes water as one of the precious resources of the present decade. This also increased the importance of effective management of water resources via the big data analytics. The major 5 "V" capabilities of big data (shown in Fig. 3.1) can help in proper perception and management of these scarce water resources.

3.4.1   Types of Water Data and Data-Sharing Methodologies

A diverse set of information that addresses the environmental, physical, ecological, social, economic, cultural, and political parameters of water usage, availability, and accessibility is known as water data. Water data can be divided into five categories: (i) water quality: The physical, biological, and chemical characteristics of water are often referred as water quality, an important parameter to determine the potability of water. To identify the water quality, single measurement is not enough but measurement of the number of water characteristics is required. It is a measure of the condition of water usually in reference to the requirements of some ecological process or anthropogenic purpose. (ii) Water quantity: It is often regarded as a rate at which volume of water is moving downstream (Wanielista et al. 1997). (iii) Water use: It includes the human consumptive uses (i.e., per capita), application by various sectors (i.e., agriculture, industry), environmental practice (i.e., evapotranspiration rates), and ecosystem services. (iv) Water extremes: Hazard and natural disaster-related data that include drought/flood monitoring and weather data. (v) Water indicators: Such indicators are generally linked to few common aspects of human or environmental health. Water indicators integrate other water-related data to provide a metric for water sustainability and utilization for human well-being (Sternlieb and Laituri 2010). Water data can be generated as primary and secondary data. The collection of water quality and quantity-related raw data can be defined as the primary data. For the measurement of the primary data, different methods are used depending upon the characteristics of water and availability of resources. Data

Table 3.1 Big data analysis methods and their characteristics

Sr. No.	Methods	Characteristics
1.	A/B or bucket or split testing	For the improvement of given objective variables, determination of the required treatment is done by comparing the control group with a variety of other test groups
2.	Association rule learning or fuzzy learning	This method compromises variety of algorithms to produce and test possible rules for determining interesting relationships
3.	Classification	It consists of supervised and unsupervised learning techniques to recognize the appropriate categories in which new data points fits
4.	Cluster analysis	As training data are not used in this method, it is considered as a type of unsupervised learning
5.	Crowd sourcing	In this method, data has been collected from the large group of people. Open-call technique has been used for this purpose
6.	Data fusion and integration	It is used for the integration and analysis of the data from variable sources in order to develop insights in more accurate and effective manner
7.	Data mining	As the name suggests it is basically the data extraction technique. It includes (i) association rule learning, (ii) classification, (iii) cluster analysis and (iv) regression methods
8.	Ensemble learning	In this type of the supervised learning, multiple predictive models have been used to obtain better predictive performance
9.	Neural networks	It is a series of algorithms that undertake to recognize underlying relationships in a set of data through a process that mimics the way the human brain operates. Various water quality parameters can be simulated using this computational method
10.	Network analysis	It is basically used to characterize relationships among discrete nodes in a network or graph

(continued)

Table 3.1 (continued)

Sr. No.	Methods	Characteristics
11.	Optimization	In this method, complex systems and processes are redesigned to improve their efficiency according to the specified objectives
12.	Pattern recognition	It is a set of machine-learning methods that assign label (some sort of output value) to instance (given input value) according to a particular algorithm
13.	Predictive modeling	In this method, a mathematical model is created/selected to best predict the probability of an outcome. Water quality predictive models can incorporate both mathematical expressions and expert scientific judgment
14.	Regression	A set of statistical methods to understand how the value of the dependent variable changes when one or more explanatory (independent) variables is modified. Ex.: Water usage can be estimated indirectly by applying multiple regression analysis
15.	Spatial analysis	It is a set of analyses methods used for the identification of geometric, topographic, and geographic informations encoded in a dataset
16.	Statistics	Statistics refers to the science of the data acquisition, organization and interpretation. For example, Response Surface Methodology (RSM) explores the relationships between several explanatory (independent) variables and one or more response variables
17.	Simulation	Modeling the behavior of complex systems, often used for forecasting, predicting, and scenario planning
18.	Time series analysis	Temporal analyses of data points are significant methods to extract significant results from the acquired datasets. For example, water quality and quantity data collected from specific time intervals to represent the real situations

which is derived directly from the sensors or hydraulic measurements are known as secondary data. Primary data can be easily shared as compared to secondary data.

Water resources-based data is highly fragmented as data is generated by number of entities and warehoused in many locations. Due to the fragmented nature, water data sharing is considered as a barrier toward big data capabilities. The data fragmentation

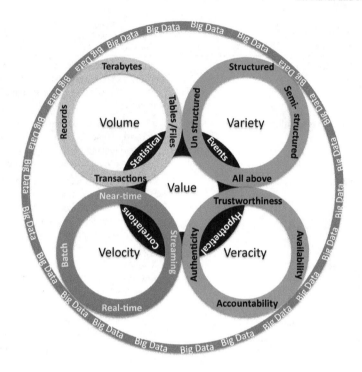

Fig. 3.1 Characteristics of Big data

problem can be overcome by using three common methodologies: (i) one-to-one: As name suggests, the data is generated by one entity and used for a single purpose. The most common example includes the academic research study or a contracted consulting project, (ii) one-to-many: In this case, data are generated by one entity and provided to many users for many purposes, and (iii) many-to-many.

3.4.2 Appositeness of Big Data to Water Resources

Science has been driven by the data but with advancement of technology, the word data has been replaced by big data. Water resources, one of the significant fields of environmental science, comprise a big data issue and flourishes increasingly. Big data helps in identifying the suitable data to resolve the problems, which are difficult to be addressed by traditional data. Some of the major applications of big data are highlighted here:

Irrigation process which requires appropriate amount of water is mainly dependent on the number of climatic factors as well as on crop and soil types. These data can be easily provided with the help of the automated sensors and continuous monitoring systems. By using the big data farming, efficiency can be improved through reducing water requirements. Variety of automated sensors, continuous monitoring systems, robotics, and computational technology provide useful information related to the water quality which enables to understand the movement of chemicals. In addition, big data can be helpful in monitoring flood, tsunami, and drought conditions as well as the melting of ice and related climatic problems can also be monitored.

In addition to the above-discussed water resources-based applications, big data techniques have been also utilized for many applications such as oceanic (e.g., oil spill pollution detection), agriculture (e.g., food monitoring and security), urban planning, management, and sustainability, climate change (global warming, acid rain), energy assessment, disease problem, ecosystem assessment, land development and use, and so on.

3.4.3 Limitations of the Big Water Data Analytics

Big data analytics help in identifying, analyzing, and interpreting the available data for the proper management of water resources. However, at present, water resources systems in many developing countries are organized with the help of hydrological data. This represents potable water accessibility and availability data from which demand for the current and future generations can be derived. Such type of the conventional datasets mostly leads to ineffective planning, design, and functioning of water management schemes. The following listed limitations need to overcome to acquire complete benefit of the big data analytics.

Because of its large volume, the quality of stored and transmitted database is one of the major concerns in big data. Errors can be introduced from the first stage of data collection to the final deposition. Most of the automated instruments are either battery operated or need some kind of power supply. Sudden failure of which is directly associated with the gap in time series data. For example, data gap usually happens during the measurement of water consumption data using the smart water meters. Water resources quality data are complex to handle, store, and process because of their heterogeneous nature. Hence, modeling is still being done using traditional simulation models supported by GIS data.

3.5 Big Water Data Platform Components and Structure

Water resources management-related conceptual framework of the big water data open platform is shown in Fig. 3.2. It basically consists of nine blocks as discussed below: (i) The first bloc, i.e., decision support tools contain decision support technique to resolve the real-world difficulties. Because of the various available techniques, the first difficulty lies in the selection of the best decision method. (ii) Knowledge-based system deals with collection and storage of water data and ultimately transfers that information to stakeholders, including professionals and experts

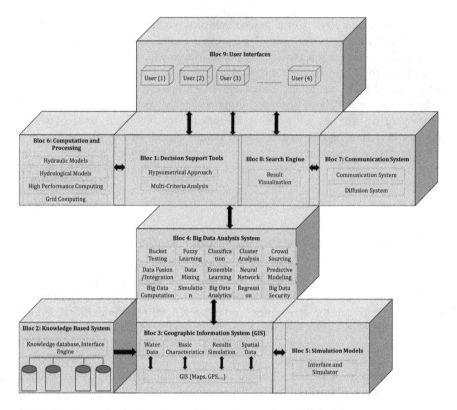

Fig. 3.2 Big data open platform for water resources management

in the field of water research. (iii) The third block geographic information system (GIS) is generally used to capture, store, analyze, and integrate complex hydrological data. ArcGis normally collects maps, applications, data, and allows users to recognize data in order to quickly deduce the best conclusion. (iv) The most important bloc of the big data platform is big data analysis system which consists number of tools to arrange, investigate, envisage, and extract useful water sources regarding information from large quantities and varieties of datasets. It requires suitable technologies (like, big data computing, analytics, mining, and security) to competently process large quantities of data.

(v) With help of the fifth bloc called simulation models, the data acquired from GIS will be linked and tried to simulate the water-associated difficulties using the simulators and interface. (vi) Sixth block of the big data platform, computation and processing, furnishes a receptacle of tools like hydraulic/hydrological models and high performance/grid computing. These tools help for the advancement of water resources prediction. (vii) After acquiring and processing the water data, the next important bloc is the communication system which makes pertinent data and information available to achieve efficiency and effectiveness. (viii) Search engine as the eighth block enables users to find the suitable information from the big water data warehouses. (ix) User interfaces as the ninth and final bloc help operators to formulate the water resources-based problems by entering related data and portraying the obtained results and graphics.

3.6 Modern Big Data Cycle in the Context of Water Resources

Some meaningful outputs from the collected data can be drawn in order to reach up to the final conclusion. In general, two main processes, i.e., data management and analytics are used for extracting meaningful results from the big water data. The term data management can be defined as the acquisition of data, its temporary storage and final preparation for suitable for analysis. Analytics refer to methods utilized to investigate and get conclusive findings from big data. Both of these processes are normally divided into five stages as shown in Fig. 3.3. Data management is the first process which needs to be performed, after acquiring the big data. From this process, the structured data can be stored and retrieved using some traditional methods such as data marts and data warehouses. Extract–load–transform (ELT) tools are used for extraction, transformation, and loading of data into the final database.

One or more analytical methods discussed in the above section have been used by water engineers and scientists for the modeling and management of water systems. (Shafiee et al. 2018) proposed the framework for the state of the flow of water data as represented in Fig. 3.4. Number of sensors have been installed in the environment for the collection of data. After proper data management, the stored data was embedded into models for analysis and interpretation. In this system (Fig. 3.4), the water data lake collects data during every stage. Analytics handles and further processes raw

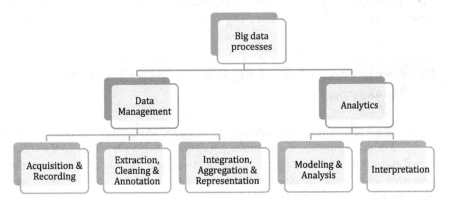

Fig. 3.3 General classification of the big data processes

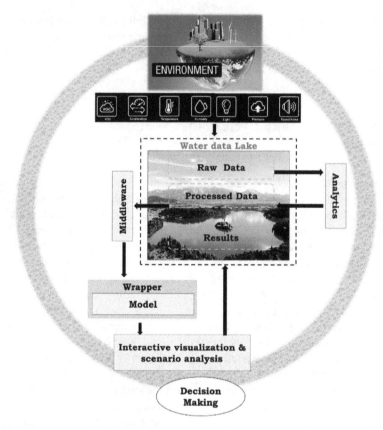

Fig. 3.4 Typical water data lake

data and finally, returns cleaned/forecasted data. Middleware pulls, aggregates, and formats data for a model. A wrapper provides communication capabilities to a model.

3.7 Future Perspectives of Big Data for Water Resources Management

At present, a number of big data platforms are available related to the water resources. Table 3.2 displays some of the common big data platforms pertaining to the water resources with their major objectives, significance, and limitations. Based on the associated limitations, the prospective applications of big data in water resources management are highlighted below.

As it has been mentioned, big data techniques have demonstrated wide applications in the decision-making process by predicting the outcomes. However, despite having access to a broad range of data sources and technical resources, the water

Table 3.2 Common big data platforms pertaining to the water resources with their major objectives, significance, and limitations

Sr. No.	Model	Objective	Importance	Limitation	References
1.	Big data platform	To solve water resources problems using big data analytics	It is important in providing effective tools to solve complex water resources systems, water modeling issues and helps in decision making	The heterogeneity of water data coming from various sources causes problem in architecture of the big data platform	Chalh et al. (2015)
2.	Cloud GIS platform	To develop the water resources and hydropower cloud GIS Platform	It has powerful ability of data mining and analysis. Big data-based cloud GIS platform of water resources and hydropower can provide the suitable decision support for the design, development, operation, and maintenance	The security of the data is the key element which restricts the development of the cloud GIS platform.	Wang and Sun (2013)
3.	Big data water resource management standards	To apply big data method in the development of water management standards	It can assemble and organize the basic data of water resources and assists the setting of water resource management standards	Supervision and data standardization are the major associated problems	Bai et al. (2017)

utility sector appears to make partial use of it for the enhancement of water quality and source distribution. With the high-density survey of big data, the risk of the error will also increase mainly due to the lack of the availability of instant processing techniques. Hence, the future perspective of big data research is not to obtain more and more data but it should mainly focus on the development of the new generation of smaller, cheaper, and accurate sensors to produce real-time data. The integration techniques can be helpful in improved decision making and management of the water resources. For example, machine learning, one of the analysis techniques, is able to extract accurate patterns and relationships from the data. At present, a number of the models, methodologies, and techniques are accessible for the planning and management of the water resources. However, none of them provides a convenient solution.

3.8 Conclusion

With advancement of the computer science and Web technology, the data generation has become increasing in day-to-day life. These large datasest ultimately pose challenges in its storage, handling, analysis, and interpretation. Water is one of the prime requirements for the survival of life and is progressively becoming a precious resource due to its inflated usage. Increased population and economic/industrial growth cause stress on available water resources. Similarly, climate change also significantly affects the water resources due to its direct effects on important hydrological processes, i.e., precipitation and evaporation. With the help of big data, each and every component of environment, such as water resources, can be managed.

The aim of the current chapter is to present an overview of big water data, associated characteristics, applications, and limitations. It also gives a summary related to the open big data platforms/proposed models supporting water resources. The authors can get the specific idea about the available models by referring to this chapter. It also highlights the future perspective required for the proper utilization of big data technique for the water resources management. Despite the increasing importance of modeling in water resources management and planning, no single methodology/tool provides an acceptable solution. Hence, more research is required in development of single but comprehensive methodology/tool. The basic available models are generally restricted to local/regional-level strategies, while the challenges are transdisciplinary and encompass knowledge from various sciences and engineering backgrounds.

References

Adamala S (2017) An overview of Big Data applications in water resources engineering. Machine Learning Res 2(1):10–18

Bai Y, Bai X, Lin L, Huang J, Fang H, Cai K (2017) Big Data technology in establishment and amendment of water management standard. Appl Ecol Environ Res 15(3):263–272

Bibri SE, Krogstie J (2017) On the social shaping dimensions of smart sustainable cities: a study in science, technology, and society. Sustain Cities Soc 29:219–246

Chalh R, Bakkoury Z, Ouazar D, Hasnaoui MD (2015) Big data open platform for water resources management. In: 2015 International Conference on Cloud Technologies and Applications (CloudTech), pp 1–8

Chen H, Chiang RH, Storey VC (2012) Business intelligence and analytics: from Big Data to big impact. MIS Q 36(4)

Cox M, Ellsworth D (1997) Application-controlled demand paging for out-of-core visualization. In: Proceedings. Visualization '97 (Cat. No. 97CB36155), pp 235–244

David H, Branko K, Amin R, Avi O, Barbara M, Katherine BM (2014) Sensing and Cyberinfrastructure for smarter water management: the promise and challenge of ubiquity. J Water Res Plan Manage 140(7):01814002

du Plessis A (2019) Current and future water scarcity and stress. In: du Plessis A (ed) Water as an inescapable risk: current global water availability, quality and risks with a specific focus on South Africa, pp 13–25. Springer International Publishing

Gandomi A, Haider M (2015) Beyond the Hype: Big data concepts, methods, and analytics. Int J Inf Manage 35(2):137–144

Kumar M (2019a) Micro-components quantification of end uses of water consumption in low income settings. Interim Project Report. 1–44

Kumar M, Chaminda T, Honda R, Furumai H (2019b) Vulnerability of urban waters to emerging contaminants in India and Sri Lanka: resilience framework and strategy. APN Science Bulletin

Kumar M, Deka JP, Kumari O (2020) Development of water resilience strategies in the context of climate change, and rapid urbanization: a discussion on vulnerability mitigation. Groundwater Sustain Dev 10:100308

Madden S (2012) From databases to Big Data. IEEE Internet Comput 16(3):4–6

Manyika J (2011) Big data: the next frontier for innovation, competition, and productivity. http://www.mckinsey.com/insights/mgi/research/technology_and_innovation/big_data_the_next_frontier_for_innovation

Mayer-Schönberger V, Cukier K (2013) Big Data: a revolution that will transform how we live, work, and think. Houghton Mifflin Harcourt

Mukherjee S, Patel AKR, Kumar M (2020) Water scarcity and land degradation nexus in the era of anthropocene: some reformations to encounter the environmental challenges for advanced water management systems meeting the sustainable development. In: Kumar M, Snow D, Honda R (eds) Emerging issues in the water environment during Anthropocene: A South East Asian Perspective. Springer Nature. ISBN 978-93-81891-41-4

Patel AK, Das N, Kumar M (2019) Multilayer arsenic mobilization and multimetal co-enrichment in the alluvium (Brahmaputra) plains of India: a tale of redox domination along the depth. Chemosphere 224:140–150

Roshan A, Kumar M (2020) Water end-use estimation can support the urban water crisis management: a critical review. J Environ Manage Ms. Ref. No.: JEMA-D-20-00036R1. https://doi.org/10.1016/j.jenvman.2020.110663

Schroeck M, Shockley R, Smart J, Romero-Morales D, Tufano P (2012) Analytics: the real-world use of big data: How innovative enterprises extract value from uncertain data, Executive Report. In: IBM Institute for Business Value and Said Business School at the University of Oxford

Shafiee ME, Barker Z, Rasekh A (2018) Enhancing water system models by integrating big data. Sustainable cities and society 37:485–491

Singh A, Patel AK, Kumar M (2020) Mitigating the risk of Arsenic and Fluoride contamination of groundwater through a multi-model framework of statistical assessment and natural remediation techniques. In: Kumar M, Snow D, Honda R (eds) Emerging issues in the water environment during Anthropocene: a South East Asian perspective. Springer Nature. ISBN 978-93-81891-41-4

Sternlieb FR, Laituri M (2010) Water, sanitation, and hygiene (WASH) indicators: measuring hydrophilanthropic quality. J Contemp Water Res Educ 145(1):51–60

Wang X, Sun Z (2013) The design of water resources and hydropower cloud GIS platform based on big data 313–322

Wanielista M, Kersten R, Eaglin R (1997) Hydrology: water quantity and quality control. In: Hydrology: water quantity and quality control

Zikopoulos P, Eaton C (2011) Understanding Big Data: analytics for enterprise class Hadoop and streaming data, 1st edn. McGraw-Hill Osborne Media

Chapter 4
Role of Physical Parameters in Developing a Geogenic Contaminant Risk Approach

Ashwin Singh, Arbind Kumar Patel, and Manish Kumar

4.1 Introduction

The resilience of water quality is a direct function of the extent and rate of urbanization (Duh et al. 2008; Ren et al. 2003; Hall et al. 1999). With a two-fold increase in population since 1978 and a projected increase of 100% in the urban population by 2030, the stress on fresh water system will be the highest ever witnessed in the centuries so far (Karnauskas et al. 2018; Chen et al. 2013). While half the population in the world still lacks access to basic sanitation, it splits open multiple routes of contaminations, taking the life of every one in three children across the globe from contaminated drinking water.

Disrupting the nutrient and energy cycles of the ecosystem through sporadic growth of urban fringes causes degradation of the natural water, leading to reduction in agents of quality rejuvenation (Groffman et al. 2004; Pickett et al. 2001; Collins et al. 2000). The excessive use of steroids, antibiotics and hormones for treating animals and promoting their growth has caused its accumulation in freshwater systems (Pei et al. 2006; Fu et al. 2017; Bayen et al. 2014; Kumar et al. 2019b; Kumar et al. 2020). Around 50 million pounds of antibiotics are produced in USA alone, out of which 50% is used for agriculture under different applications (Levy 1998). According to Elmund et al., 75% of the consumed antibiotics is excreted in the form of active metabolites, which has high chances of participating as surface runoff from catchments and finally, aiding in increasing the resistance of bacteria. According to USGS 1999, a total of 95 stream networks in the USA are contaminated with

A. Singh
Discipline of Civil Engineering, Indian Institute of Technology, Gandhinagar, Gandhinagar 382355, India

A. K. Patel · M. Kumar (✉)
Discipline of Earth Sciences, Indian Institute of Technology, Gandhinagar, Gandhinagar 382355, India
e-mail: manish.kumar@iitgn.ac.in

© Springer Nature Singapore Pte Ltd. 2020
M. Kumar et al. (eds.), *Resilience, Response, and Risk in Water Systems*,
Springer Transactions in Civil and Environmental Engineering,
https://doi.org/10.1007/978-981-15-4668-6_4

antibiotics. There have been evidence from across the globe that the immunity and resistance of the disease-causing bacteria have been rising (Chee-Sanford et al. 2001; Guardabassi et al. 1998).

Natural geogenic contamination through high phosphorous and arsenic exposed soils has life threatening consequences and therefore has put more severe risk of developing cancer through contaminated drinking water at an earlier age (Spallholz et al. 2004; Su et al. 2011; Hinwood et al. 1999). Heavy metal contamination of groundwater and its possible coupling with physical parameters like precipitation and temperature adds extreme vulnerability to the exposed population (Tyler 1975, 1990; Cheng 2003; Farombi et al. 2007; Wei and Yang 2010). Groundwater fluctuations too aid in making conditions oxic and anoxic. This further accelerates the processes of dissolution which impacts the water quality (Patel et al. 2019a, b; Kumar et al. 2019a, b; Singh et al. 2020b). The problem of developing a water quality index is therefore very complex and difficult. Integrating physical parameters like precipitation and temperature (which has a very active role in dilution and dissolution of ions and changing the aquifer dynamics through forced evaporation under confined conditions) could lead to meaningful assessment of water quality problem (Brunner et al. 2004; Mumby et al. 2004; Ritchie et al. 1990). To include emerging pollutant parameters such as Pharmaceuticals and Personal Care Products (PPCBs) and microplastics require a rigorous chemical assessment of water on a seasonal basis (Wilson et al. 2003; Ebele et al. 2017). Assessing the nature of water quality based on major ion chemistry provides only a fraction of the information about the water chemistry of the place. Therefore, there is a need to assess critically the current practice of developing regional water quality index.

Till now, an index based on oxygen equivalent approach was supposed to provide more insights to the overall problem of pollution risk assessment. But now the role of satellite imageries has increased in the field of water quality assessment. Hyperspectral imageries have also been used to indirectly infer to the health of the coastal ecosystems. Aquatic humus has been identified as one of the parameters which optically is very active and can therefore be used as an indicator of water quality using remote sensing techniques, though with additional computational difficulties. Apart from aquatic humus, chlorophyll also can be traced with the help of active and passive radars and therefore it could be used as a parameter in predicting the water quality. One limitation of excessive dependency on satellite imageries is its availability at required spatial resolution. For example, parameters like soil and sub-soil moisture, which actively carries the imprints of contamination are available usually at a spatial scale of few kilometres (downscaling it further to smaller resolution often leads to loss of information). This forces the practice of in situ data collection and makes the study more region centric than having any global connotation.

While extreme climatic events like floods and droughts do impact the water quantity for consumption, there is corresponding impact on the water quality too (Roshan and Kumar 2020; Singh et al. 2019, 2020a). There have been numerous studies suggesting that climatic extremes in highly urbanized settings have serious implications on the lives of the habitants but a parallel projection for water quality has been a missing link so far. The IPCC projections under different Representative

Concentration Pathways (RCPs) provide a method to understand the variations of atmospheric parameters under different incident energy received by the earth based on the CO_2 concentration in the environment till 2100. To integrate it into a single index will require assessing the behaviour of individual parameter in impacting the water quality. Groundwater fluctuation anomalies obtained through Gravity Recovery and Climate Experiment (GRACE) satellite have known impact in changing the carbonate/bi-carbonates strength in groundwater. Therefore, assimilating the seasonal storage anomalies is a key for developing the future water quality index. One potent threat to the surface freshwater resource has been the sediment flux intrusion from different geomorphic zones. In this regard, the sediment connectivity holds a key in understanding the possible increase or decrease in the Total Dissolved Solids (TDS). For developing a water quality index in Himalayan region having catchment in the range of 10^3–10^4 km^2, will require a rigorous connectivity analysis considering factors like connectivity response unit, stream power and morphometry of the river.

   The present chapter explores the current criteria and status of developing regional water quality index through a possible insight for incorporating the various physio-chemical parameters which traditionally have been left out owing to un-matching skills, computational complexities and assessment methods required from different fields.

4.2   Parameters Impacting the Water Quality—*Mode of Acquisition and Assessment*

The factors impacting the water quality can be sub-divided into three categories, i.e. physical, chemical and biological. All these three factors differ with respect to their mode of acquisition, analysis and interpretation. Apart from this, there co-exists a very delicate relationship between these three groups of parameters. For example, chemical parameters like arsenic which has been a known carcinogen in the lower Gangetic as well as Brahmaputra alluvial plains have been found to relate highly with the physical geological conditions. The presence of vertisols with expansive clay content of calcic and haplic origins has high correlations with arsenic concentration. Calcisols and Luvisols have also been found to have some effect on fluctuating arsenic concentration and therefore they are used as active predictors of arsenic contamination. Since arsenic is usually not included as a parameter in estimating Heavy Metal Pollution index (HPI), therefore a water source may appear to be devoid of metal contamination but may contain arsenic in cancer causing concentration. The Global Gridded Soil Information uses ensemble models for estimating soil information at a spatial resolution between 1000 and 250 metres. Information regarding the soil type, geological past and physical properties like soil gradation can be obtained with certain level of associated probability (Mukherjee et al. 2020). The Cation Exchange Capacity (CEC) of the soil does have a coupling with the soil pH in terms of KCl and H_2O. Therefore, ignoring factors like CEC and including only pH in developing water quality index have certain drawbacks with regards to the whole understanding of water chemistry. One of the most important physical parameters driving the entire water dynamics is the temperature. Temperature does change the direction of aquifer reactions and can have an active role in driving the rock–water interactions at sub-soil level.

The Soil Organic Carbon Stock (SOC) is another important factor that decides the level of pollution in the groundwater and surface water systems. In a usual groundwater setting, the Gross Primary Productivity and Net Photosynthesis (Table 4.1) are extremely low. The SOC leaching to the sub-surface coming from catchment runoffs becomes the only source of energy. This is more evident in glacial regions where snowmelt is accompanied by a corresponding jump in the SOC content of the groundwater. Excluding SOC from considering as a parameter in water quality index leads to exclusion of our understanding of bio-geochemistry of water. The Food and Agriculture Organization (FAO) therefore has maintained an inventory of total soil carbon stock in units of tonnes/hectare. Assessment and preservation of SOC are essential for another reason also. A climatic extreme event may lead to change in the landscape which will cause the carbon stock of the ground to get soaked away by the atmosphere. This will force the CO_2 concentration in the atmosphere to increase. This will further start a chain reaction where due to a blanket of water vapour, the heat received per unit of the surface will increase and will further cause the SOC to escape in the form of gas. The International Soil Reference and Information Centre (ISRIC) maintains an inventory of SOC at surface and sub-surface level. The ISRIC also maintains a global inventory of soil salinity. The soil salinity has been derived based on parallel processing of thermal imageries and soil maps. The soil salinity has a very strong relationship with the soil moisture (Table 4.1). Atmospheric sulphate is another important factor which can be derived using MERRA reanalysis products (Fig. 4.1). Wet deposition of acidic anions has the capability to alter the weatherability of the minerals, therefore higher concentration over Ganga and Brahmaputra flood plains carries the possibility of future changes in the alkali desorption processes.

Another equally important parameter causing the water chemistry to change is groundwater fluctuations. Excessive utilization of groundwater forces the CO_2 to release to the atmosphere. The release of CO_2 is accompanied with further changes in carbonate chemistry of the groundwater. In tropical countries, significant changes in the groundwater characteristics are observed with seasonal shift. This also has a significant impact in changing the saturation state of minerals like Dolomite $(CaMg(CO_3)_2)$ and Calcite $(CaCO_3)$ which has the potential to undergo carbonate weathering. Usually, in tropical climate, the concentration of carbonates/bicarbonates increases after the monsoon season but the situation could sometimes be reversed owing to strong silicate weathering of the aquifers following the undersaturation of Halite (NaCl). As far as anions like $SO_4{}^{2-}$ and Cl are concerned, they both have geogenic (minerals) and anthropogenic (agro-chemicals) occurrence and therefore are always difficult to pinpoint the exact source.

Human Health Risk Index (HRI) is yet another powerful tool that can help in developing hazard quotient based on differential risk on age and gender. In Fig. 4.2, the probability of Cancer Incidence (CI) per million population has been estimated

Table 4.1 Some of the parameters impacting water quality and their method of aquisition

Parameter	Parameter	Affect	Acquisition
Heavy Metal Pollution Index	Chemical	Direct	In situ + Laboratory analysis
Water Quality Index	Chemical	Direct	In situ + Laboratory analysis
Lead	Chemical	Direct	In situ + Laboratory analysis
Iron	Chemical	Direct	In situ + Laboratory analysis
Zinc	Chemical	Direct	In situ + Laboratory analysis
Gross Primary Productivity	Physical	Direct	MODIS—MOD17A2H
Net photosynthesis	Physical	Direct	MODIS—MOD17A2H
Enhanced vegetation index	Physical	Indirect	LANDSAT 8
Normalized difference vegetation index	Physical	Indirect	LANDSAT 8
Leaf area index	Physical	Indirect	AVHRR
Fraction of photosynthetically active radiation	Physical	Direct	AVHRR
Landcover	Physical	Indirect	Multi-sensors
Land surface temperature	Physical	Direct	MODIS—MYD11A1
Albedo	Physical	Direct	MODIS—MCD43A3
Vegetation transpiration	Physical	Indirect	PML_V2 Product
Interception from vegetationcanopy	Physical	Indirect	PML_V2 Product
Water body evaporation	Physical	Indirect	PML_V2 Product
Soil organic carbon stock	Chemical	Direct	ISRIC
Bulk density (fine soil)	Physical	Indirect	ISRIC
Clay content (<2 μm), mass fraction	Physical	Direct	ISRIC
Coarse fragments in volumetric %	Physical	Indirect	ISRIC
Silt Content (2–50 μm) in mass fraction	Physical	Direct	ISRIC
Sand Content (50–2000 μm) in mass fraction	Physical	Indirect	ISRIC
Cation Exchange Capacity of Soil	Chemical	Direct	ISRIC
Soil pH in terms of H_2O	Chemical	Direct	ISRIC
Soil pH in terms of KCl	Chemical	Direct	ISRIC
Digital Elevation Model	Physical	Indirect	ASTER
Calcic Vertisols	Physical	Direct	Soil Grid
Gleyic Luvisols	Physical	Direct	Soil Grid
Haplic *Acrisols*	Physical	Direct	Soil Grid
Haplic Calcisols	Physical	Direct	Soil Grid
Haplic Cambisols	Physical	Direct	Soil Grid

(continued)

Table 4.1 (continued)

Parameter	Parameter	Affect	Acquisition
Haplic Fluvisols	Physical	Direct	Soil Grid
Haplic Gleysols	Physical	Direct	Soil Grid
Haplic Luvisols	Physical	Direct	Soil Grid
Haplic Vertisols	Physical	Direct	Soil Grid
Leptic Phaeozems	Physical	Direct	Soil Grid
Absolute Depth of Bedrock	Physical	Indirect	Soil Grid
GRACE Groundwater Storage	Physical	Direct	Multi-model—JPL
GRACE Terrestrial Storage	Physical	Direct	GRACE
Canopy Water Storage	Physical	Indirect	CLM, VIC, NOAH, MOSAIC
Soil Moisture (Modelled)	Physical	Direct	CLM, VIC, NOAH, MOSAIC

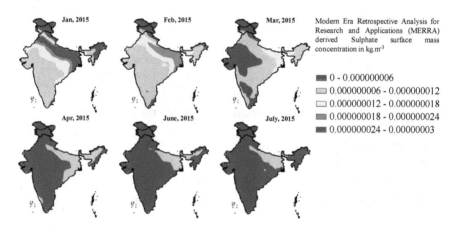

Fig. 4.1 Country-level mapping of sulphate surface concentration on a monthly basis based on reanalysis product derived through modern era retrospective analysis for research and applications (MERRA)

based on in situ collection of arsenic water samples in Darbhanga, district of Bihar. Due to less skin surface area, children are at very high risk of developing cancer. Further with lesser water intake and higher aggregation the Chronic Daily Intake is also very higher. Identifying vulnerable regions is extremely important for a heavily urbanized setting where arsenic contamination has been reaching to about 50 ppb. For example, in case of a city like Guwahati where the problem of arsenic contamination has been persistent over the years, the city future urbanization plans are not in accordance to fight the current problem of arsenic contamination (Fig. 4.3). As can be seen from the figure, the current trend of urbanization is driven towards the direction of heavily contaminated arsenic regions. This further point out our current inability in regards to developing a policy that could limit anthropogenic activities in such region. In cities like Guwahati, excessive abstraction of groundwater despite

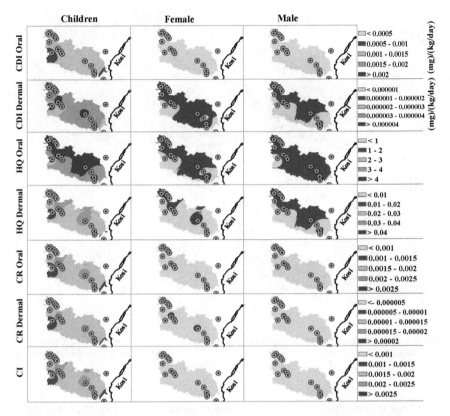

Fig. 4.2 Evaluation of (non)-carcinogenic risk through oral and dermal exposure of arsenic in the groundwater based on cancer index in the Darbhanga district of Bihar, India

the presence of Brahmaputra has been one of the leading reasons to make aquifer situation anoxic and causing increased formation of oxyhydroxides.

Another problem that arise with consistent exposure of arsenic is with regards to development of "Arsenic Resistant Genes". Arsenic has been one element which can mimic the properties of phosphorous and can accumulate in the cell cytosols during the cell conduction of glycerol. The only response in terms of efflux mechanism remains extrusion through activation of ATP binding cassettes. Further, it has been well documented that arsenic has detrimental rection with thiol functional groups to inhibit the process of energy release by ATPs.

The increased consumption of Pharmaceuticals and Personal Care Products (PPCBs) and the prescription antibiotics has increased the concentration of active metabolites in the water bodies. Various researchers already suggested devising a new holistic risk assessment mechanism for assessing water quality but till date, a multi-model approach integrating all the factors ranging from physical, chemical and biological is missing. As opposed to agro-chemicals, the discharge of PPCBs can take place through multiple domestic as well as commercial routes. The existence of

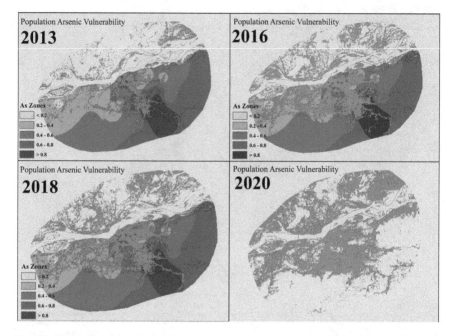

Fig. 4.3 Population risk vulnerability index based on arsenic contamination of groundwater of Guwahati city. The red pixels show the built-up areas. The different shades of blue show the arsenic vulnerability zones (Singh et al. 2019; Kumar et al. 2019b)

such large uncertainties in quantifying the contamination routes makes the study of PPCBs so difficult.

4.2.1 Method of Integrating Multiparameter Data into a Common Index

To spatially integrate the factors into a single-piece information requires the usage of interpolation techniques. The process of converting point datasets into continuous surface phenomenon demands the use of geographical information system (GIS) in conjugation. Kriging as used by many researchers in the past associate special importance to spatial autocorrelations. A simple mathematical explanation of kriging's functioning can be understood with Eq. 4.1.

$$X(e) : e \in D \tag{4.1}$$

The equation states that a variable or a function X is estimated at e number of locations. The variable e while belongs to a set of values from D. The function can either has a two- or three-dimensional space occupancy. To make the entire computation simpler, the variation of the function is assumed to be real rather than belonging to complex number group. The entire calculations can also be approximated to a stochastic process where definite time interval data is used. The main objective of kriging in mathematical terms is to estimate the value of X at e_0, while the value at $e_1, e_2, e_3......$ etc., is known. It is assumed that the value at the unknown location is a weighted sum of contributions from all the known location values.

$$X(e_0) = \sum \textbf{factor} \cdot X(e_i); \tag{4.2}$$

where i is ranging from 1 to number of known sampling sites.

Usually, the auto-correlation function is used in selecting appropriate values of **factor**.

A better technique is when regression is used along with kriging as in case of regression kriging (RK) approach. A slight difference between regression and kriging is that regression offers an edge while predicting values based on known factors having little association among themselves. Kriging on the other hand is more accurate when individual behaviour of a parameter is to be mapped spatially. In this approach, regression plays the role of estimation fitting and kriging calculates the bias or drift experienced in prediction.

$$X(e_0) = \sum \beta \cdot \text{var}(e_i) + \sum \alpha \cdot X(e_i) \tag{4.3}$$

where β is the estimated intercept through regression, α is the associated kriging weight and **var**(e_i) shows the considered variable for prediction.

The values of coefficients are determined using a suitable curve fitting technique. Usually, either ordinary least square (OLS) or generalized least square (GLS) methods are used for such purpose. For the GLS technique only the matrix of predictors and co-variance is required to make a prediction through known vectors at unknown location. The coefficient β and α can be estimated by Eqs. 4.4 and 4.5.

$$\beta = \left[\left(\textbf{var}(e_i)^\text{T} \cdot \textbf{Co} - \textbf{var}^{-1} \cdot \textbf{var}(e_i)\right)^{-1}\right] \cdot \left[\textbf{var}(e_i)^\text{T} \cdot \textbf{Co} - \textbf{var}^{-1} \cdot X(e_i)\right] \tag{4.4}$$

$$\alpha = X(e_0) \cdot \beta + \textbf{factor}_0 \cdot [X(e_i) - \textbf{var}(e_i) \cdot \beta] \tag{4.5}$$

where **var**(e_i) is the predictor matrix and **Co-var** is the co-variance matrix

Literature suggests that in the field of water quality analysis selection of an appropriate interpolation technique is the real challenge. Environmental Protection Agency (EPA) has laid guidelines to use cumulative frequency diagram (CFD) as a standard process to assess the exceedance criteria of parameters which further requires using suitable interpolation techniques. Even EPA in USEPA, 2007, has admitted that kriging could be a better technique for water quality assessment. The process becomes challenging when a hydro-dynamic process is expected to capture high variance through very small observation dataset. The Chesapeake Bay in USA is one such classic example. To assess the water quality of the bay, interpolation techniques in conjunction with hydro-dynamics numerical models are applied. Kriging being a powerful tool helps to reduce a snap-shot (grid) static data into a continuous surface model. Through a collaborative effort with USEPA, continuous temporal monitoring of the water quality of the bay is being taken for the last three decades. Sometimes, there is a requirement to reduce the dataset before doing the interpolation. In this case, factor reducing technique such as principal component analysis (PCA) can be used.

4.3   Potential of Satellite Imageries in Developing an Ensemble

While few of the chemical parameters may be required to be estimated at the site but most of the water quality parameter can indirectly be estimated using optical remote sensing techniques. For example, Coloured Dissolved Organic Matter (CDOM) can be estimated through Moderate Resolution Imaging Spectroradiometer (MODIS) sensor. When the concentration of CDOM exceeds its threshold limit, it imparts yellow to brown colour to the water. CDOM exhibits a very distinct spectral signature absorption. While wavelength near ultraviolet regimes is totally absorbed, it almost drops to zero when wavelength reaches the red band. Scientists have often used this property to numerically fit a model to understand the absorption characteristics of CDOM.

$$\alpha(\lambda) = \alpha(\lambda_o) \cdot \exp(S(\lambda - \lambda_o)) \tag{4.6}$$

where $\alpha(\lambda)$ denote the absorption in the λ band and S is the fitted slope and $\alpha(\lambda_o)$ is the reference band absorption which is usually 375 nm.

With the knowledge of $\alpha(\lambda)$ and S, the total amount of CDOM in a water system can be identified. This concept can be applied further to even understand the concept of mixing and removal of CDOM. Further, MODIS sensors can be used to estimate chlorophyll a, total phosphorus and total nitrogen. This technique was used by Harma et al., to estimate the water quality of 102 coastal ecosystems and 85 lakes in Finland. However, these techniques are prediction methods based on irradiance values; therefore, requires usage of a regression model simultaneously. Further koponen et al. 2004, developed Bio-optical Reflectance Model (BRM) based on MODIS irradiance data to study the quality of water bodies. The team further stretched their model to estimate turbidity in water system. A more detailed methodology was developed by Wang et al., for estimating the change in the water quality of reservoirs using Landsat 5 Thematic Mapper imageries. The start point of the analysis was chosen to be March, 1996, when the reservoir is chemically stable as it being the dry season. Also, the probability of atmospheric hindrance through clouds is less. The method

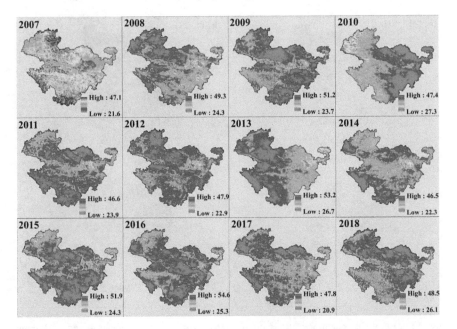

Fig. 4.4 The Land Surface Temperature (LST in Celcius) of Marathawada region in Maharashtra for the month of April has been estimated temporally and spatially using grid interpolation techniques. LST often changes the aquifer dynamics forcing evapo-transpiration of minerals during rock–water interactions

involved a combination of in situ as well satellite image-based processing. A radiometrically corrected image is correlated with known in situ collected data points for developing a regression model for water quality prediction.

In case of surface and groundwater, precipitation and evaporation are the two major phenomena along with rock–water interactions mediated via mineral interaction. Assessment of Land Surface Temperature (LST) is possible using the satellite imageries (Fig. 4.4). Based on the albedo characteristics, there exists a possibility of sodic formations which is one of the leading reasons for loss of agricultural productivity in regions like Marathwada of India. Further, there is a complimenting impact on the groundwater characteristics through a change in the mineral saturation states. Presence of soils like calcisols is considered a boon in these scenarios as they have high capability to precipitate free fluoride ions available in the groundwater. High LST-driven transpiration also helps in increased conduction of arsenic during biomass accumulation in the root zones of the crops especially rice. It has been argued that the presence of microplastics in arsenic contaminated soil may bring down its uptake by crops. The problem is often simpler for surface water as compared to the groundwater. For example, in Oak Lake, USA, due to heavy fireworks on the eve of Uncle Sam Jam event, there is high accumulation of perchlorate (Fig. 4.5). Absence of a strong circulation mechanism for lakes makes the situation more vulnerable for

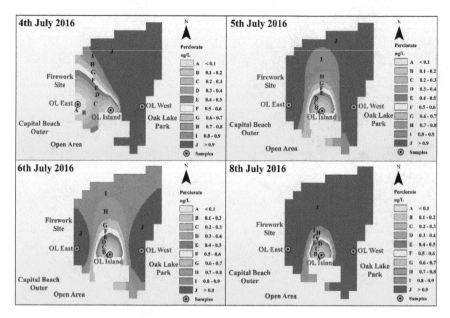

Fig. 4.5 A probable perchlorate transport plume of Oak Lake in Nebraska based on kriging interpolation technique (Modified from Kumar et al. 2019b)

future. Therefore, using a kriging-based zoning of lake based on perchlorate concentration helps in identification of contamination plume transport. The process of surface adsorption led transport of contaminant is especially very high for Ganga and Brahmaputra flood plains mainly due to high concentration of atmospheric soot particles like Black Carbon (BC) which has been proved to be great adsorbing material (Fig. 4.6). Through MERRA-based temporal and spatial mapping of India, it could be seen that situation is fast changing since 1980. If we compare it with today (2019), we can see that there exists a high vulnerability for almost all the major river basins of India with regards to black carbon led solute transport.

4.4 Understanding the Geogenic Impact and Sediment Connectivity on Water Quality

Geogenic occurrences like weathering, volcanic eruption and biological degradation are believed to be the responsible factors for the release of arsenic in the atmosphere. Anthropogenic factors including activities like uranium and gold mining have also resulted in elevating arsenic concentration above permissible level. The excessive release of arsenic in the Younger Deltaic Deposition (YDD) of West Bengal and Bangladesh has caused the situation to turn into a calamity. A probable cause for such high arsenic concentration in oxic groundwaters of Bangladesh is explained

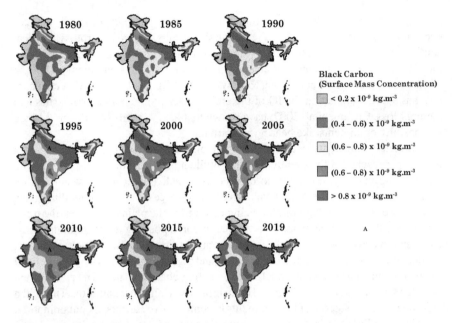

1980   1985   1990   1995   2000   2005   2010   2015   2019

Black Carbon
(Surface Mass Concentration)

$< 0.2 \times 10^{-9}$ kg.m^{-3}

$(0.4 - 0.6) \times 10^{-9}$ kg.m^{-3}

$(0.6 - 0.8) \times 10^{-9}$ kg.m^{-3}

$(0.6 - 0.8) \times 10^{-9}$ kg.m^{-3}

$> 0.8 \times 10^{-9}$ kg.m^{-3}

Fig. 4.6 Using surface mass concentration of Black Carbon of Modern Era Retrospective Analysis for Research and Application (MERRA) a possible vulnerability assessment with respect to water quality can be made

through reductive dissolution of arsenic from the surface of iron-oxyhydroxides but the hypothesis is challenged for its validity in the Gangetic aquifers of Bihar where there is considerably lower concentration of base metal sulphides. It has been speculated that the quaternary sediments of Gangetic Plain, rich in clay and organic carbon, have high initial retention capacity of arsenic. It is also not clear how the relationship of arsenic varies with depth. While Nickson et al. 1998, argue that arsenic concentration decreases with depth, studies like Acharyya (1999), state otherwize. Similarly, it has also been argued that anoxic conditions too can facilitate reductive hydrolysis at different aquifer depths (Singh et al. 2020b). The study of arsenic geochemistry is therefore region and condition specific requiring testing of all the available hypothesis for understanding the fate and occurrence of arsenic.

It is now known that 50 blocks situated in 11 districts of Bihar are affected by arsenic contamination. Various studies have confirmed the arsenic contamination of Bihar and have listed reductive hydrolysis as the probable cause. Smith et al., have confirmed that excessive abstraction of groundwater can also lead to arsenic contamination. There also exists a scientific gap in understanding the extent of arsenic contamination in the MGP. While Chakraborty et al., suggest that arsenic contamination could be a widespread problem in Bihar based on their study on groundwater of Bhojpur district, Acharya, argue that the extent is geographically very small, limited to a small linear path along the flood channel.

The problem of arsenic contamination becomes complicated with the co-occurrence of heavy metals above the permissible limits. Various neurotoxicological studies have shown that exposure to lead (Pb) along with arsenic above permissible level leads to alterations in the central monoaminergic system, memory loss in children and damage of central nervous system (CNS). Manganese (Mn) exposure also has a direct impact on the IQ of the children, leading to both verbal issues and memory loss. Chuang et al. 2007, have shown that heavy metals like Pb and Mn are responsible for ailments like hearing impairment.

Sediment connectivity determines the fate of sediment reaching the downstream from upper reaches of the stream network. While it has a dominant role in changing the river morphology, it could also be used as an active indicator of water quality degradation. Usually, the Himalayan Rivers carry excessive sediments during the monsoon season due to high rainfall being received in the plains. However, the trend continues till the post monsoon season, due to corresponding decrease in the snow cover and an active increase of glacial sediments coming due to snow melt. Sediment connectivity is a direct function of the hinterland properties (such as slope, relief and catchment characteristics), channel and vegetation characteristics. The rivers when they flood in the plains carry the metal contaminants in high concentration. This when adsorbed on to the surface of the sediments becomes active carriers of contamination. The Gangetic Plains of India have imprints of heavy industrial contamination. However, much of the arsenic contamination of Middle Gangetic Plains is assumed to be of geogenic in nature. This could be due to the sediments being transported from high arsenic-rich mineral system through active sub-surface flow. Another equally important factor is connectivity response unit (CRU). There is a high correlation of the connectivity parameters in a relatively flat terrain with the change in the LULC. Total stream power and specific stream power also impact the loading of sediment transportation.

4.5 Conclusion

The present chapter discusses various satellite data products as a possible substitute to in situ data collection of water quality parameters. This integration will help in developing a water quality assessment tool on a temporal and spatial scale. This will further aid in devising continuous pollution monitoring techniques of water resources at a global level. Through a comprehensive literature review, merits as well as demerits of various studies done in this direction has been studied. The practice of understanding the water quality problem only at a regional and time specific level will not help in formulating strategies at a pace required to mitigate the effects of environmental degradation of water resources. Therefore, there is an urgent need for critically assessing the process of developing global water quality index.

References

Acharyya SK (1999) Comment on Nickson et al. 1998, Arsenic poisoning of Bangladesh groundwater. Nature 401:545

Bayen S, Yi X, Segovia E, Zhou Z, Kelly BC (2014) Analysis of selected antibiotics in surface freshwater and seawater using direct injection in liquid chromatography electrospray ionization tandem mass spectrometry. J Chromatogr A 1338:38–43

Brunner P, Bauer P, Eugster M, Kinzelbach W (2004) Using remote sensing to regionalize local precipitation recharge rates obtained from the chloride method. J Hydrol 294(4):241–250

Chee-Sanford JC, Aminov RI, Krapac IJ, Garrigues-Jeanjean N, Mackie RI (2001) Occurrence and diversity of tetracycline resistance genes in lagoons and groundwater underlying two swine production facilities. Appl Environ Microbiol 67(4):1494–1502

Chen W, Lu S, Jiao W, Wang M, Chang AC (2013) Reclaimed water: a safe irrigation water source? Environ Dev 8:74–83

Cheng S (2003) Heavy metal pollution in China: origin, pattern and control. Environ Sci Pollut Res 10(3):192–198

Chuang HY, Kuo CH, Chiu YW, Ho CK, Chen CJ, Wu TN (2007) A case-control study on the relationship of hearing function and blood concentrations of lead, manganese, arsenic, and selenium. Sci Total Environ 387(1–3):79–85

Collins HP, Elliott ET, Paustian K, Bundy LG, Dick WA, Huggins DR, Smucker AJM, Paul EA (2000) Soil carbon pools and fluxes in long-term corn belt agroecosystems. Soil Biol Biochem 32(2):157–168

Duh JD, Shandas V, Chang H, George LA (2008) Rates of urbanisation and the resiliency of air and water quality. Sci Total Environ 400(1–3):238–256

Ebele AJ, Abdallah MAE, Harrad S (2017) Pharmaceuticals and personal care products (PPCPs) in the freshwater aquatic environment. Emerg Contaminants 3(1):1–16

Farombi EOOA, Adelowo O, Ajimoko Y (2007) Biomarkers of oxidative stress and heavy metal levels as indicators of environmental pollution in African cat fish (Clarias gariepinus) from Nigeria Ogun River. Int J Environ Res Publ Health 4(2):158–165

Fu J, Lee WN, Coleman C, Nowack K, Carter J, Huang CH (2017) Removal of disinfection byproduct (DBP) precursors in water by two-stage biofiltration treatment. Water Res 123:224–235

Groffman P, Law N, Belt K et al (2004) Ecosystems 7:393. https://doi.org/10.1007/s10021-003-0039-x

Guardabassi L, Petersen A, Olsen JE, Dalsgaard A (1998) Antibiotic resistance in Acinetobacterspp. isolated from sewers receiving waste effluent from a hospital and a pharmaceutical plant. Appl Environ Microbiol 64(9):3499–3502

Hall RI, Leavitt PR, Quinlan R, Dixit AS, Smol JP (1999) Effects of agriculture, urbanization, and climate on water quality in the northern Great Plains. Limnology Oceanography 44(3part2):739–756

Hinwood AL, Jolley DJ, Sim MR (1999) Cancer incidence and high environmental arsenic concentrations in rural populations: results of an ecological study. Int J Environ Health Res 9(2):131–141

Karnauskas KB, Schleussner CF, Donnelly JP, Anchukaitis KJ (2018) Freshwater stress on small island developing states: population projections and aridity changes at 1.5 and 2 C. Regional Environ Change 18(8):2273–2282

Kumar M (2019a) Micro-components quantification of end uses of water consumption in low income settings. Interim Project Report. 1–44

Kumar M, Chaminda T, Honda R, Furumai H (2019b) Vulnerability of urban waters to emerging contaminants in India and Sri Lanka: resilience framework and strategy. APN Science Bulletin

Kumar M, Deka JP, Kumari O (2020) Development of water resilience strategies in the context of climate change, and rapid urbanization: a discussion on vulnerability mitigation. Groundwater for Sustainable Development 10, 100308

Levy SB (1998) The challenge of antibiotic resistance. Sci Am 278(3):46–53

Mukherjee S, Patel AKR, Kumar M (2020) Water scarcity and land degradation Nexus in the era of Anthropocene: some reformations to encounter the environmental challenges for advanced water management systems meeting the sustainable development. In: Kumar M, Snow D, Honda R (eds) Emerging issues in the water environment during Anthropocene: a South East Asian perspective. Springer Nature. ISBN 978-93-81891-41-4

Mumby PJ, Skirving W, Strong AE, Hardy JT, LeDrew EF, Hochberg EJ, Stumpf RP, David LT (2004) Remote sensing of coral reefs and their physical environment. Mar Pollut Bull 48(3–4):219–228

Nickson R, McArthur J, Burgess W, Ahmed KM, Ravenscroft P, Rahmanñ M (1998) Arsenic poisoning of Bangladesh groundwater. Nature 395(6700):338

Patel AK, Das N, Kumar M (2019a) Multilayer arsenic mobilization and multimetal co-enrichment in the alluvium (Brahmaputra) plains of India: a tale of redox domination along the depth. Chemosphere 224:140–150

Patel AK, Das N, Goswami R, Kumar M (2019b) Arsenic mobility and potential co-leaching of fluoride from the sediments of three tributaries of the Upper Brahmaputra floodplain, Lakhimpur, Assam, India. J Geochem Explor 203:45–58

Pei R, Kim SC, Carlson KH, Pruden A (2006) Effect of river landscape on the sediment concentrations of antibiotics and corresponding antibiotic resistance genes (ARG). Water Res 40(12):2427–2435

Pickett ST, Cadenasso ML, Grove JM, Nilon CH, Pouyat RV, Zipperer WC, Costanza R (2001) Urban ecological systems: linking terrestrial ecological, physical, and socioeconomic components of metropolitan areas. Annu Rev Ecol Syst 32(1):127–157

Ren X, Harder H, Martinez M, Lesher RL, Oliger A, Simpas JB, Brune WH, Schwab JJ, Demerjian KL, He Y, Zhou X (2003) OH, and HO_2 chemistry in the urban atmosphere of New York City. Atmos Environ 37(26):3639–3651

Ritchie SW, Nguyen HT, Holaday AS (1990) Leaf water content and gas-exchange parameters of two wheat genotypes differing in drought resistance. Crop Sci 30(1):105–111

Roshan A, Kumar M (2020) Water end-use estimation can support the urban water crisis management: a critical review. Journal of Environmental Management. Ms. Ref. No.: JEMA-D-20-00036R1. https://doi.org/10.1016/j.jenvman.2020.110663

Singh A, Patel AK, Deka JP, Das A, Kumar A, Kumar M (2019) Prediction of arsenic vulnerable zones in the groundwater environment of a rapidly urbanizing setup. Geochemistry, Guwahati, India, p 125590

Singh A, Patel AK, Kumar M (2020a) Mitigating the risk of arsenic and fluoride contamination of groundwater through a multi-model framework of statistical assessment and natural remediation techniques. In: Emerging issues in the water environment during Anthropocene, pp 285–300. Springer, Singapore

Singh A, Patel AK, Ramanathan A, Kumar M (2020b) Climatic influences on arsenic health risk in the metamorphic precambrian deposits of Sri Lanka: a re-analysis-based critical review. Journal of Climate Change, 6(1): 15–24

Spallholz JE, Boylan LM, Rhaman MM (2004) Environmental hypothesis: is poor dietary selenium intake an underlying factor for arsenicosis and cancer in Bangladesh and West Bengal, India? Sci Total Environ 323(1–3):21–32

Su S, Zeng X, Bai L, Li L, Duan R (2011) Arsenic biotransformation by arsenic-resistant fungi Trichoderma asperellum SM-12F1, Penicillium janthinellum SM-12F4, and Fusarium oxysporum CZ-8F1. Sci Total Environ 409(23):5057–5062

Tyler G (1975) Heavy metal pollution and mineralisation of nitrogen in forest soils. Nature 255(5511):701

Tyler G (1990) Bryophytes and heavy metals: a literature review. Bot J Linn Soc 104(1–3):231–253

Wei B, Yang L (2010) A review of heavy metal contaminations in urban soils, urban road dusts and agricultural soils from China. Microchem J 94(2):99–107

Wilson PS, Roy RA, Carey WM (2003) An improved water-filled impedance tube. J Acous Soc Am 113(6):3245–3252

Chapter 5
Water Indices: Specification, Criteria, and Applications—A Case Study

Bhairo Prasad Ahirvar, Shivani Panday, and Pallavi Das

5.1 Introduction

Water is a fundamental segment and crucial substance for every single living being on Earth. Contamination of water is a major issue affecting both surface and groundwater. Water quality is getting deteriorated by indiscriminate discharge of pollutants into the water bodies from various sources. Water pollution is a key issue which causes deaths and various diseases in worldwide (Shivhare et al. 2017). Moreover, progressive pollution of the river and groundwater is critical as these are important source of drinking water (Das et al. 2016; Singh et al. 2020). Water quality monitoring is a key process to assess the pollution status of any water body. About 3.4 million people die every year due to waterborne diseases (WHO 2014). In India, about 128 million people are not getting safe drinking water. A report from World Bank, 21% of communicable diseases caused due to contaminated water (Bhattacharya et al. 2011). As a result, 1600 children below the age of 5 die every day (WHO 2014). A report from State Environment Protection Administration of China (2006) stated that out of 161 environment emergency accidents, 95 accidents (59%) were due to water pollution (Lu et al. 2008). Along these lines, it is important to research water quality by different water quality index parameters. Water Quality Index (WQI) is a well-organized method for evaluating water quality in terms of qualitative and quantitative measure (Mishra and Patel 2001; Naik and Purohit 2001; Singh et al. 2019a, b). The primary goals of the water quality index are to safeguard, maintain, and access the water quality status for drinking, farming, household and industrial purposes (Chiaudani and Premazzi 1988). Horton (1965) first suggested the idea of indices representing degree of purity in water quality. For his index, Horton has chosen ten most prevalent water quality parameters, including coliforms, dissolved

B. P. Ahirvar · S. Panday · P. Das (✉)
Department of Environmental Science, Indira Gandhi National Tribal University, Amarkantak,
Madhya Pradesh, India
e-mail: pallavienv@gmail.com

© Springer Nature Singapore Pte Ltd. 2020
M. Kumar et al. (eds.), *Resilience, Response, and Risk in Water Systems*,
Springer Transactions in Civil and Environmental Engineering,
https://doi.org/10.1007/978-981-15-4668-6_5

oxygen (D.O.), pH, specific conductance, chlorides, and alkalinity. Using a linear sum aggregation feature, the index was achieved weight from 1 to 4 in range and index score. For monitoring physical, chemical and biological parameters, water quality indices are implemented.

These parameters can obviously indicate whether or not the water is suitable for use. WQIs are an effective tool for comparing and surveying the status within the definite ecosystem or among different geographical zone (Mutairi et al. 2014; Kannan 1991; Pradhan et al. 2001). Abbasi and Abbasi (2010) described and highlighted the benefit of using WQIs in detail. Numbers of water quality indices are developed everywhere so that the general water quality indices can be examined quickly and effectively within a given moment and area (Venkateshraju et al. 2010). Dinius (1972) suggested another WQI (explication of individual importance index) as a quantitative unit that can assess the price and effect of pollution on social and human welfare. Another WQI has been formed similar to index of Horton (Brown et al. 1972). He suggested an index multiplier type where the weighting of individual parameters was allocated based on the choice of author and critical analysis-based subjective opinion. Prati et al. (1971) regarded in their research 13 parameters of equal weight, these parameter values range from 0 to 13 with values greater than eight indicating pollution of heavy metals. Inhaber (1975) formed two distinct subindices, one for industrial and domestic effluents and the other for ambient water quality. In WQI, Dee et al. (1973) used various water quality parameters (such as D.O., pH, TDS, turbidity, and temperature) to assess the environmental effect of large-scale water development initiatives. Walski and Parker (1974) developed an index specifically for recreational water. Steinhart et al. (1982) used a fresh environmental quality index to evaluate the water quality of the Ecosystem of Great Lakes using physical, chemical and biological parameters.

Several researcher have developed different water quality index (WQIs) in India, namely Bhargava Index (1985), Overall Index of Pollution (Sargaonkar et al. 2008) and The River Water Quality Index. The Scatterscore Index (2005), Chemical Water Quality Index (2006), Oregon Water Quality Index, NSF-Additive Water Quality Index, and NSF-Multiplicative Water Quality Index developed in the USA. Taiwan developed Index of River Water Index (2004). Canada developed WQI in the mid 90s by water quality guidelines. The Canadian Council of Ministers of the Environment Water Quality Index (CCMEWQI) developed by the British Columbia Ministry of Environment Land and Parks, Canada, in mid '90s', and it is adapted by various countries and all across Canada to measure water quality (Bharti and Katyal 2011).

The water quality index is an effective technique for assessing surface and ground-water health status. The application of WQIs can easily determine the pollution status of any water body in single effort. For drinking, irrigation, industrial and other purposes, each nation has its own water quality requirements. To express the information in a very simplified and logical way, WQI has the potential to reduce the large data into a single value (Semiromi et al. 2011). It requires data from various sources and provides information on a water source and provides information on a water system's general status (Karbbasi et al. 2011; Mukherjee et al. 2020). It makes aware and increases understanding capability to users of the water resources and policymakers regarding water quality issues. Some of the significant water quality indicators used in water quality assessment are evaluated in the current research and present their numerical equation, application of water quality indices (WQIs) in different fields, and their advantages and disadvantages that are used globally.

5.1.1 Categories of WQI (Tirkey et al. 2013)

Water quality indices are categorized into four groups according to their applications in different fields:

1. **Public indices**: These indices use general parameters for the evaluation of water quality. Eg., National Sanitation Foundation Water Quality Index (NSFWQI), Horton Index.
2. **Specific consumption indices**: These indices used for specific consumption of water (drinking, industrial, irrigation). Eg., Oregon Water Quality Index (OWQI), British Columbia Water Quality Index (BCWQI).
3. **Designing or planning indices**: These indexes used for the planning and development projects of water resources and management.
4. **Statistical indices**: Statistical indices use a statistical and mathematical approach to asses water quality without considering any public and expert opinion.

5.1.2 Steps in Developing Water Quality Index (Sutadian et al. 2016, 2017)

Generally, there are four steps were followed to the development of water quality index:

1. **Parameter selection**: Parameters of water quality that fulfill the minimum information requirements and are within permissible limit are chosen. Based on the type of research on the water body, parameters are also chosen.
2. **Generation of subindices values**: The subindices are generated to transform the data into a single or common scale. Minimum standards of individual water quality parameters prescribed by the country are the following to generate subindices.
3. **Weight establishment**: Each selected parameters assigned weight to their relative significance in water quality evaluation. If the WQI research is on the groundwater, parameters that have a strong impact on groundwater quality will be provided maximum weight.
4. **Aggregating of subindices to get final index**: The particular formula is implemented depending on the type of WQI to get the final index.

5.1.3   Benefits of Application of WQIs

1. It is the best tool to evaluate physico-chemical and biological parameters of any water body.
2. WQIs provide a final result in simplified form.
3. Statistical and mathematical formula approach of WQIs provides an efficient result of water quality status.
4. Qualitative as well as quantitative assessment of water body can be done by using WQIs.
5. WQIs are helpful for the government to establish water quality criteria.

5.2   Overview of WQIs and Its Applications

Water quality index is an essential tool for monitoring of water pollution. Groundwater quality index was used by Acharya et al. (2018) for the assessment of groundwater quality in South West Delhi, India. Kukrer and Mutlu (2019) used Horton's index for the assessment of surface water quality of Sarayduzu Dam Lake in Turkey. National Sanitation Foundation Water Quality Index (NSFWQI) was used by Chaurasia et al. (2018) for the analysis of groundwater quality of Korba city, Chhattisgarh, India. Zeid et al. (2018) assessed water quality of El-Our city, Egypt, by using Canadian Council of Ministers of Environment Water Quality Index (CCME WQI). Surface water quality of Lepenc river basin (Republic of Kosovo) was assessed by Bytci et al. (2018) by using modified water quality index. Irrigation Water Quality Index (IWQI) was used by Abbas et al. (2018) for the assessment of groundwater quality for irrigation purpose in Iran. Biological Water Quality Index (BWQI) was used by Araman et al. (2013) for the analysis of Melan River (Johor) water quality. Shah and Joshi (2015) used Weight Arithmetic Water Quality Index (WAWQI) for the assessment of water quality of Sabarmati River, Gujarat, India. Bascon Index was used by de Oliveira et al. (2018), for the assessment of water quality. Sutadian et al. (2017), developed West Java Water Quality Index (WJWQI) for analysis of water quality of rivers of Indonesia. Average Water Quality Index (AWQI) was used by Sahid and Iqbal (2016), for the assessment of groundwater quality of Lahore city, Punjab, Pakistan.

5.2.1   Advantages and Disadvantages of Some Selected Water Quality Indices

Water quality index (WQI)	Parameters used	Advantages	Disadvantages
National Sanitation Foundation (NSF)	pH, DO, Fecal coliform density, BOD_5, Total Phosphate, NO_3^-	1. The final result comes from a single index in less time 2. Evaluation between area and water quality changes 3. The index value is related to water use 4. It facilitates to communicate with common people	1. It provides information about common water quality, not a particular quality 2. Some data lost during data handling 3. It generates some complicated problems when dealing with uncertainty and subjectivity
Canadian Council of Ministers of the Environment (CCME)	pH, Total N, Total P, DO, TDS, EC, Heavy metals, COD	1. It represents all variables in a single number 2. Flexible in data input and parameter selection 3. It is suitable for all type of water quality assessment 4. It provides a simple form of multivariate statistical analysis 5. Professionals and common people can easily understand 6. The calculation is very easy	1. Some pieces of information lost in single variables 2. The index is very sensitive 3. The result can be easily manipulated 4. All variables have same weighting. Some parameters of water quality cannot be used

(continued)

(continued)

Water quality index (WQI)	Parameters used	Advantages	Disadvantages
Oregon water quality index (OWQI)	NH_3, NO_3^-, BOD, DO, Total P, TS, pH, temperature, fecal coliform	1. It enables the parameter with the biggest effect on the water quality index to be imparted 2. It provides meaning to distinct factors on the general quality of water at distinct moments and location 3. The formula is inclined to various circumstances and critical consequences for the nature of water	1. Does not reflect on toxics concentrations, habitat or biology modifications 2. Cannot determine for particular purposes the water quality 3. Health hazardous parameters, e.g., bacteria, metals, pesticides cannot be evaluated
Bhargava index	TDS, Total hardness, SO_4^{2-}, Ca^{2+}	1. The water quality parameters are categories into four groups (Physical, biological, inorganic and toxicants) which included all essential parameters 2. Sensitivity curve used for weighting of water quality parameters 3. Best index for river, industrial effluent, and sewage water quality assessment	1. It is mainly proposed for surface water quality 2. The parameters weightage system depends on sensitivity curve which is software-based work
Contamination index	Heavy metals, trace metals, Cl^-, EC, NO_3^-, PO_4^{2-}, Hardness	1. This index is useful for groundwater contamination assessment 2. It is a useful index for assessment of chemical contamination of groundwater and associated health issue	1. There are no fix parameters for groundwater evaluation 2. There is no place for important parameters like pesticides and fertilizers contamination in groundwater

(continued)

(continued)

Water quality index (WQI)	Parameters used	Advantages	Disadvantages
Aquatic toxicity index	pH, DO, turbidity, F^-, K^+, orthophosphate, NH_3, TDS, hazardous metals (Zn, Cu, Cr, Pb, etc.)	1. This index is used for assessment of the health of aquatic ecosystem 2. It is also helpful to determine biomagnifications from the aquatic organism	1. The index is similar to smith index 2. Basic parameters were ignored

5.2.2 Applications of Water Quality Indices in Groundwater Quality Assessment of Anuppur District of Madhya Pradesh

In the current research, WQI of groundwater of **Anuppur district of Madhya Pradesh** is assessed. The total area of Anuppur district is 3701 km^2. Total 30 ground-water samples were collected from December 2018 to March 2019 from Anuppur district (Fig 5.1 and Table 5.2). The hand pumps were pumped to remove the remaining water for at least 10 min to guarantee minimum oxygen interference. Samples of groundwater were gathered for anion and cation assessment in 250 ml polypropylene bottles. The water samples were stored in refrigerator at 4 °C temperature. pH, DO, temperature, EC, and TDS were calculated in situ using a multi-parameter probe (Multi-Parameter PCSTestr 35), while alkalinity (HCO_3^-) was measured in situ by Titrimetric method using 0.1 N H_2SO_4.

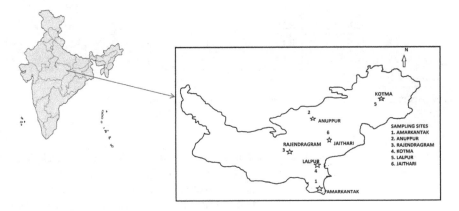

Fig. 5.1 Study area map

Table 5.1 Statistical summary of chemical constituents of groundwater samples of Anuppur

Groundwater samples

Parameters	Units	Maximum	Minimum	Average	Stdev.
pH		8.2	6.1	6.96	0.62
Salinity	mgL^{-1}	590	87	287	133.91
TDS	mgL^{-1}	542	94	263	115.47
Conductivity	$\mu S/cm$	861	97	411	241.67
NO_3^-	mgL^{-1}	8.33	0.22	3.1	2.8
Na^+	mgL^{-1}	11.4	0.03	2.55	2.08
K^+	mgL^{-1}	2.25	0.65	1.35	0.42
Ca^{2+}	mgL^{-1}	124	15	65	38.08
Mg^{2+}	mgL^{-1}	206	45	117	45.06
Hardness	mgL^{-1}	210	52	132	41.94
Cl^-	mgL^{-1}	199	25	112	46.09
Acidity	mgL^{-1}	170	38	94	34.94
Alkalinity	mgL^{-1}	210	64	124	34.31
SO_4^{2-}	mgL^{-1}	61.8	18.5	32.5	10.81

Table 5.2 Depth of groundwater in bore well of sampling sites (in ft)

Sampling sites	Depth of groundwater (ft)	Sampling sites	Depth of groundwater (ft)	Sampling sites	Depth of groundwater (ft)
AN1	155	L1	200	K1	250
AN2	158	L2	160	K2	220
AN3	153	L3	180	K3	280
AN4	162	L4	175	K4	295
AN5	170	L5	190	K5	300
AM1	252	R1	180	J1	170
AM2	260	R2	186	J2	155
AM3	250	R3	188	J3	160
AM4	280	R4	200	J4	165
AM5	320	R5	230	J5	150

AN Anuppur, *AM* Amarkantak, *L* Lalpur, *R* Rajendragram, *K* Kotma, *J* Jaithari

Correlation Analysis

Correlation provides significant relationship among variables. A high correlation coefficient implies a decent connection between factors. The zero correlation coefficient implies no connection between factors, positive estimation of r shows a positive relationship, while negative worth s demonstrates the inverse relationship. Thirteen water parameters have been used for the correlation coefficient matrix analysis.

Table 5.3 Correlation matrix of groundwater samples

Correlations

	pH	Salt	TDS	EC	HCO$_3^-$	Hardness	Ca^{2+}	Mg^{2+}	Na^{2+}	K$^+$	NO$_3^-$	Cl$^-$	SO$_4^{2-}$
pH	1												
Salt	−0.27	1											
TDS	0.32	0.5*	1										
EC	−0.06	0.75**	0.7**	1									
HCO$_3^-$	−0.19	0.58**	0.33	0.46**	1								
Hardness	0.01	−0.09	−0.08	−0.19	0.02	1							
Ca^{2+}	−0.17	0.73**	0.54**	0.88**	0.46*	−0.28	1						
Mg^{2+}	0.04	−0.21	−0.17	−0.33	−0.06	0.99**	−0.44*	1					
Na^{2+}	0.1	0.18	0.53**	0.29	−0.1	−0.18	0.21	−0.2	1				
K$^+$	−0.03	−0.37*	−0.42*	−0.51**	−0.15	0.64**	−0.52**	0.69**	−0.14	1			
NO$_3^-$	−0.11	0.08	0.01	0.3	−0.07	−0.06	0.28	−0.1	0.25	−0.05	1		
Cl$^-$	−0.03	−0.08	−0.26	−0.24	−0.21	0.14	−0.16	0.16	−0.21	0.14	−0.14	1	
SO$_4^{2-}$	0.03	−0.17	−0.39*	−0.4*	0.05	0.02	−0.38*	0.08	−0.15	0.18	−0.1	0.12	1

*Correlation is significant at the 0.05 level (2-tailed)
**Correlation is significant at the 0.01 level (2-tailed)

In Table 5.3 the stronger association corresponds to the couples TDS-salt, EC-salt, EC-TDS, Ca^{2+}-EC, Ca^{2+}-salt, Ca^{2+}-K^+, EC-K^+.

A significant positive correlation was observed for (EC-Ca^{2+}, EC-Salt, Ca^{2+}-Salt, Mg^{2+}-K^+, Hardness-K^+, Ca^{2+}-TDS).

Ca^{2+}-K^+, EC-K^+, K^+-TDS, SO_4^{2-}-TDS, SO_4^{2-}-Ca^{2+}, Mg^{2+}-Ca^{2+} show significant negative correlation.

The significant correlation between Ca^{2+}-salt and Ca^{2+}-HCO_3^- may be due to the interaction of water with dolomite and calcite minerals.

TDS of water represents any minerals, metals, and ions. Positive correlation between TDS-Ca^{2+} may be due to the presence of soluble calcium carbonate in groundwater.

There is significant high correlation between hardness and magnesium ion may be due to the presence of magnesium carbonate, but there is negative correlation between calcium and magnesium ion, and this is may be happening due to magnesium replacing the calcium in groundwater (Table 5.3).

The Brown et al. (1972) Weight Arithmetic Water Quality Index (WAWQI) is used to obtain a detailed image of the general groundwater quality. The WQI calculations were based on the Indian standard indicated for drinking water IS 10500 (2012). The WQI calculated through four steps. Eleven important water quality parameters total dissolved solids (TDS), HCO_3^-, EC, pH, hardness, NO_3^-, SO_4^{2-}, Ca^{2+}, Mg, Na, and Cl^-) were selected.

First, a weight was allocated to every 11 parameters based on its comparative significance in the general quality of drinking water (Table 5.4). NO_3^-, pH, EC, and TDS were allocated the maximum weight 4, and Ca^{2+} and Mg^{2+} were allocated the minimum weight 2 based on their comparative importance in the water quality assessment. Second, the chemical parameter relative weight (W_i) was calculated using the following equation:

Table 5.4 Water quality index (WAWQI) of groundwater samples

S. No.	Chemical parameter (mg/l)	Standard	Weight (w_i)	Relative weight (W_i)
1.	pH	8.5	4	0.111
2.	EC (μS/cm)	750	4	0.111
3.	TDS	500	4	0.111
4.	Hardness	300	3	0.083
5.	HCO_3^-	200	3	0.083
6.	Cl^-	250	3	0.083
7.	SO_4^{2-}	200	4	0.111
8.	Ca^{2+}	75	2	0.055
9.	Mg^{2+}	30	2	0.055
10.	Na^+	200	3	0.083
11.	NO_3^-	45	4	0.111
			$\sum w_i = 36$	$\sum W_i = 0.997$

$$W_i = \frac{W_i}{\sum\limits_{i-1}^{n} w_i}$$

where 'W_i' is comparative weight and 'w_i' is the weight of each parameter and 'n' means the number of parameters.

Third, a quality rating scale (Q_i) is calculated by dividing it is in each water sample's concentration by its own standard according to IS 10500 (2012) and multiplied by 100:

$$Q_i = (C_i/S_i) \times 100$$

where 'Q_i' represents the quality rating scale (QRS), 'C_i' means the concentration of each parameter in each water sample, and 'S_i' denotes the drinking water standard for each chemical parameter according to the IS 10500 (2012).

Fourth, the SI (sub-index) is first determined for each chemical parameter, which is then used to determine the WQI as per the following equation:

$$SI_i = W_i \times Q_i$$

where SI_i denotes the sub-index of the ith parameter and 'Q_i' represents the rating based on the concentration of ith parameter.

All subindices values of water sample added together to calculate final WQI as follows:

$$WQI = \sum SI_i$$

After calculating WQI value, then the values are usually classified into five categories: excellent, good, poor, very poor, and unfit water for drinking purposes as shown in Tables 5.5 and 5.6.

Irrigation water quality parameters
Permeability index (PI), magnesium hazard (MgR), sodium absorption ratio (SAR), sodium soluble percentage (SSP) and residual sodium carbonate (RSC) were calculated for the assessment of water quality (Singh et al. 2019a, b) from using following formulas:

$$PI = \frac{Na + \sqrt{HCO_3}}{Na + Ca + Mg} \times 100$$

$$KR = \frac{Na}{Ca + Mg}$$

$$SAR = Na/\sqrt{(Ca + Mg)/2}$$

Table 5.5 Water quality index of 30 groundwater samples of Anuppur district of Madhya Pradesh

Sample	pH	TDS	EC	NO_3^-	Hardness	Cl^-	Ca^{2+}	Mg^{2+}	Na^+	SO_4^{2-}	HCO_3^-	WQI
A1	7.6	542	767	4.31	100	53	110	73.1	11.3	24.5	112	67
A2	6.1	320	565	5.55	124	92	97.8	100	3.98	23	84	62
A3	8	236	630	6.14	136	82	92.9	113	1.66	21.2	72	65
A4	6.2	180	720	8.33	90	96	108	63.7	1.29	18.6	104	55
A5	6.8	450	490	2	106	142	81.7	86.1	3.38	19.8	122	62
AM1	6.4	128	97.1	1.02	74	25	24.0	68.1	1.52	50.2	124	38
AM2	7.2	168	136	0.72	94	67	30.4	81.5	0.97	49.2	108	45
AM3	7.9	175	182	0.99	110	135	23.2	104	2.33	39.8	98	52
AM4	7.7	260	270	1.01	76	121	32.8	68.0	2.06	37.2	128	49
AM5	7.2	190	180	7.05	52	167	28.0	45.1	3.05	31	64	40
L1	6.5	192	296	5.76	210	60	28.8	202	2.3	37	150	74
L2	7.03	238	290	1.62	194	96	25.6	187	0.99	29	124	71
L3	7.15	260	180	3.63	170	117	23.2	164	2.87	32.2	132	66
L4	8.16	310	190	3.26	198	85	29.6	190	2.63	35.9	108	73
L5	6.2	220	250	1.5	210	128	15.2	206	3.9	23.3	68	70
RA1	6.8	269	549	3.6	176	128	88.1	154	1.25	61.8	150	77
RA2	6.3	250	550	2	144	92	71.3	126	0.96	27	136	65
RA3	6.5	310	610	2.44	108	170	76.1	89.4	3.99	35	164	65
RA4	6.6	210	420	3.64	158	192	86.5	136	0.03	41	172	72
RA5	7.2	290	518	3.7	116	146	74.5	97.8	4.14	52	106	63
K1	6.7	404	861	7.09	100	25	124	69.8	3.13	27.4	210	68

(continued)

Table 5.5 (continued)

Sample	pH	TDS	EC	NO$_3$$^-$	Hardness	Cl$^-$	Ca^{2+}	Mg^{2+}	Na$^+$	SO$_4$$^{2-}$	HCO$_3$$^-$	WQI
K2	6.2	302	750	0.65	92	75	118	63.1	0.77	21.2	164	59
K3	6.6	355	568	1.18	116	82	117	87.5	3.63	20.2	176	64
K4	7.8	418	840	0.53	144	156	97.7	120	1.07	23.3	158	78
K5	8.2	490	488	1.51	158	78	74.5	139	0.73	18.5	124	73
JA1	6.7	94	196	5.52	174	139	61.7	159	2.13	32.7	136	66
JA2	6.8	108	138	2.78	130	170	63.3	114	1.59	35	86	54
JA3	6.9	136	129	1.48	144	153	52.1	131	0.53	33.3	104	57
JA4	6.9	123	180	0.72	110	85	57.7	95.9	1.32	28.5	116	49
JA5	6.5	146	157	0.22	166	199	42.4	155	0.68	29.6	98	63

Concentrations in mg/l
EC in µS/c

Table 5.6 Water Quality classification based on WQI

S. No.	WQI range	Type of Water	% of the sample
1	<50	Excellent	13.33
2	50.1–100	Good	86.66
3	100.1–200	Poor	0
4	200.1–300	Very poor	0
5	>300.1	Unfit for drinking	0

$$SSP = \frac{Na + K}{Na + K + Ca + Mg} \times 100$$

$$RSC = \left[HCO_3^- + CO_3^{2-}\right] - [Ca + Mg]$$

5.2.3 Entropy Weighted Irrigation Water Quality Index (EIWQI)

There are four steps to develop EIWQI. The first step involves selection of water quality parameters. The essential water quality parameters are selected on the basis of significance on the water body. In the second step, the sub-index was formulated in a scale of 0–100. Using 0–100 scale and BIS classification, best fitted curves (fig) were drawn, and their equations were used to transfer all selected parameters into a common scale.

Following equations were used for subindices (j):

$$SI(EC) = 0.012EC$$
$$SI(SAR) = 0.012SAR$$
$$SI(RSC) = 0.012RSC$$
$$SI(SSP) = 0.012SSP$$

Third step was involved the assignment of weight to all the selected parameters. Entropy weight is a scientific approach and also incorporates the variability of water quality parameters.

Following steps has been used for the calculation of entropy weight:
Step 1 Normalization of data

$$V_{ij} = \frac{a_{ij}}{a_{ij} + \cdots + a_{mj}}; \quad +\forall j \in \{1, \ldots, c\}$$

Table 5.7 Variation of irrigation water quality suitability parameters

	Anuppur
PI	6.36
Kelly's ratio	0.014
Mg ratio	119
SAR	0.25
Na %	2.11
RSC	0

where 'a_{ij}' denotes the concentration of jth parameter at ith sampling period, 'c' represents the total number of parameters, and 'm' the total number of sampling periods.

Step 2 Calculation of information entropy (E)

$$E_i = -1/\ln m \sum_{i=1}^{m} v_{ij} \ln v_{ij}; \quad \forall j \in \{1, \ldots, c\}$$

Step 3 Determination of weight (w) of each selected water quality parameter;

$$W_j = d_j/d_1 + \cdots + d_c$$

where $d_j = 1 - E_j$.

After calculation of entropy weight, the calculated values were multiplied with sub-indices. The final value of EIWQI obtained by addition of all subindices (Table 5.7)

$$\text{EIWQI} = \sum_{j=1}^{n} w_j \times \text{SIJ}$$

5.2.4 Classification of Water Quality in the Overall Pollution Index (Sargaonkar et al. 2008)

Overall index of pollution: The OIP index was adopted by Sargaonkar et al. (2008). It is useful WQI for the estimation of water pollution status of any water body in Indian conditions.

The formula for calculating the index is as follows

$$OIP = 1/n \sum_{i=1}^{n} p_i$$

where p_i represents pollution index of ith parameters and n denotes numbers of parameters.

$$P_i = V_n/V_s$$

where V_n denotes observed value of parameter and V_s denotes standard value of parameter.

On the basis of water quality status of water body, Sargaonkar et al. (2008) classified the water quality in to five categories as shown in Table 5.12.

According to OIP categories, the value or score of OIP is less than 1.9, and then, the water quality is excellent and comes under class C_1, if score is less than 3.9, then the water quality is acceptable and falls under category C_2. The OIP score is less than 7.9, 15.9 and greater than 16 show to some extent of polluted and falls under C_3, C_4, and heavily polluted (class C_5), respectively.

5.2.5   Result and Discussions

Overall status of groundwater quality is assessed by using three important water quality indexes, namely weight arithmetic water quality index (WAWQI), entropy irrigation water quality index (EIWQI), and an overall index of pollution (OIP). The result of the chemical analysis of groundwater parameters shows extensive variation (Table 5.1). The pH of the water samples ranges from 6.1 to 8.16 with an average of 6.96 and stdev. 0.62. The value of electric conductivity ranges from 97 to 861 μS/cm. Rao et al. (2002) classified EC as three types based on salt enrichment in the sample. First type when the value of EC is $\leq$1500 μS/cm, type-II when the value of EC is between 1500 and 3000 μS/cm, and type-III when the EC value exceeds $\geq$3000 μS/cm. According to the above classification of EC, the entire groundwater sample comes under the type-I (low salt enrichment). Total dissolved solids (TDS) represent the soluble salt concentration groundwater. The value of TDS ranges from 94 to 542 mg/l. The calcium concentration varied from 15 to 124 mg/l, which is slightly higher than the BIS drinking water standard. The higher concentration of calcium denotes the dissolution of carbonates and its minerals, e.g., calcite, dolomite, gypsum, and anhydrite. The magnesium ion ranges from 45 to 206 mg/l which may be due to the high dissolution of dolomite rocks and minerals. The concentration of sodium ranges from 0.03 to 11.36 mg/l. The potassium concentration ranges between 0.65 and 2.25 mg/l. The concentration of chloride ranges between 24 and 198 mg/l. The sulfate concentration in samples varied from 18 to 61 mg/l. The

Table 5.8 Classification of EIWQI

S. No.	Class	Range of EIWQI
1.	Very good	0–25
2.	Good	25–50
3.	Average	50–75
4.	Poor	Above 75

Table 5.9 Final value EIWQI

Site	EIWQI
Anuppur	20

Table 5.10 Classification of irrigation water quality parameters

S. No.	Class	Range of EC	Range of SAR	Range of RSC
1.	Low	0–1500	0–10	0–1.5
2.	Medium	1500–3000	10–18	1.5–3
3.	High	3000–6000	18–26	3–6
4.	Very high	Above 6000	Above 26	Above 6

nitrate concentration in groundwater of study area ranges from 0.22 to 8.33 mg/l, and finally, the bicarbonate concentration ranges from 64 to 210 mg/l. The present study on groundwater quality of Anuppur district, Madhya Pradesh, by using weight arithmetic water quality index (WAWQI) shows that 13.33% of the sample is in excellent quality and remaining 86.66% samples in good quality (Table 5.6). Another water quality index, entropy irrigation water quality index (EIWQI) is also applied on the groundwater samples and which shows that the overall value of EIWQI is 20 which is in the excellent category according to EIWQI classification of water quality (Tables 5.8 and 5.9). The third WQI is an overall index of pollution (OIP) is used which shows that groundwater quality is in excellent and class 1(C1) category according to the classification of water quality of OIP (Tables 5.11 and 5.12). The overall quality of groundwater of Anuppur is in good condition for drinking as well as irrigation purposes (Table 5.10).

5.2.6   WQI Studies from Worldwide

See Tables 5.13 and 5.14.

Table 5.11 Calculation of overall index of pollution at Anuppur

Station	pH	TDS	EC	NO_3^-	Hardness	Cl^-	Ca	Mg	Na	SO_4^{2-}	HCO_3^-	$\sum P_i$	OIP
AN1	7.6	542	767	4.31	100	53	111	73	11.4	24.5	112.0	6.95	0.63
AN2	6.1	320	565	5.55	124	92	98	100	3.98	23.0	84.0	5.86	0.53
AN3	8	236	630	6.14	136	82	93.0	113	1.66	21.2	72.0	5.88	0.53
AN4	6.2	180	720	8.33	90	96	108.2	64	1.29	18.6	104.0	5.82	0.53
AN5	6.8	450	490	2	106	142	81.8	86	3.38	19.8	122.0	6.04	0.55
AM1	6.4	128	97.1	1.02	74	25	24.0	68	1.52	50.2	124.0	4.61	0.42
AM2	7.2	168	136.5	0.719	94	67	30.5	82	0.97	49.2	108.0	5.06	0.46
AM3	7.9	175	182	0.99	110	135	23.2	104	2.33	39.8	98.0	5.07	0.46
AM4	7.7	260	270	1.01	76	121	32.9	68	2.06	37.2	128.0	5.34	0.49
AM5	7.2	190	180.7	7.05	52	167	28.1	45	3.05	31.0	64.0	4.62	0.42
L1	6.5	192	296	5.76	210	60	28.9	203	2.30	37.0	150.0	5.41	0.49
L2	7.03	238	290	1.62	194	96	25.7	188	0.99	29.0	124.0	5.09	0.46
L3	7.15	260	180	3.63	170	117	23.2	164	2.87	32.2	132.0	5.19	0.47
L4	8.16	310	190	3.26	198	85	29.7	191	2.63	35.9	108.0	5.53	0.50
L5	6.2	220	250	1.5	210	128	15.2	206	3.90	23.3	68.0	4.43	0.40
R1	6.8	269	549	3.6	176	128	88.2	155	1.25	61.8	150.0	7.79	0.71
R2	6.3	250	550	2	144	92	71.3	127	0.96	27.0	136.0	5.77	0.52
R3	6.5	310	610	2.44	108	170	76.2	89	3.99	35.0	164.0	6.73	0.61
R4	6.6	210	420	3.64	158	192	86.6	137	0.03	41.0	172.0	6.96	0.63
R5	7.2	290	518	3.7	116	146	74.5	98	4.14	52.0	106.0	6.97	0.63
K1	6.7	404	861	7.09	100	25	124.2	70	3.13	27.4	210.0	7.35	0.67

(continued)

Table 5.11 (continued)

Station	pH	TDS	EC	NO_3^-	Hardness	Cl^-	Ca	Mg	Na	SO_4^{2-}	HCO_3^-	$\sum P_i$	OIP
K2	6.2	302	750	0.65	92	75	118.6	63	0.77	21.2	164.0	6.40	0.58
K3	6.6	355	568	1.18	116	82	117.0	88	3.63	20.2	176.0	6.46	0.59
K4	7.8	418	840	0.53	144	156	97.8	120	1.07	23.3	158.0	7.26	0.66
K5	8.2	490	488	1.51	158	78	74.5	140	0.73	18.5	124.0	6.09	0.55
J1	6.7	94	196.5	5.52	174	139	61.7	159	2.13	32.7	136.0	5.50	0.50
J2	6.8	108	138	2.78	130	170	63.3	115	1.59	35.0	86.0	5.25	0.48
J3	6.9	136	129.5	1.48	144	153	52.1	131	0.53	33.3	104.0	5.12	0.47
J4	6.9	123	180.2	0.72	110	85	57.7	96	1.32	28.5	116.0	4.72	0.43
J5	6.5	146	157.3	0.22	166	199	42.5	156	0.68	29.6	98.0	5.05	0.46

Table 5.12 Classification of
water quality

Water quality status	Class	Class index (OIP score)
Excellent	C1	1
Acceptable	C2	2
Slightly polluted	C3	4
Polluted	C4	8
Highly polluted	C5	16

5.2.7 *Conclusion*

Water Quality Indices (WQIs) are a useful tool in the evaluation of water quality status. In the present study, 35 studies from worldwide on WQIs have been
reviewed. From the review, it is observed that the weight arithmetic water quality
index (WAWQI) is the most frequent WQI which is applied in both surface and
groundwater quality evaluations due to easy in calculations and parameter selection.
Ground Water Quality Index (GWQI) and The National Sanitation Foundation *Water
Quality Index* (NSFWQI) have applied mostly for the assessment of the groundwater
quality. Specific water quality index is an important WQI for surface water quality
assessment which includes Physico-chemical and biological parameters and heavy
metals. The selected 35 studies from worldwide reveal that there is an irregularity in
the parameters selected in the WQIs, and the same WQI includes different parameters for their study. The WQIs study should have fix parameters that can specify
the application of WQI at the ground level. The overall quality of groundwater of
Anuppur district is in good quality for drinking as well as irrigation purposes. Three
WQIs applied on the groundwater quality of Anuppur reveal that the groundwater
quality is in good condition. There is a need to include some specific water quality
parameters like pesticides, heavy metals, and toxic chemicals which are important
parameters in examining water quality for drinking and household purposes. There
should be a specific water quality index for the drinking water quality analysis which
has worldwide acceptance.

Table 5.13 List of selected water quality indices applied in surface water by various researchers

Name of WQI	Parameters	Type of work	References
WAWQI and NSFWQI	pH, Temperature, TDS, Turbidity, NO_3^--N, PO_4^{3-}, BOD, DO	Physicochemical and biological characteristics of the Narmada river water, Madhya Pradesh, India	Gupta et al. (2017)
Hortons index	DO, Salinity, pH, Temperature, EC, TSS, COD, BOD, Cl^-, PO_4^{3-}, SO_4^{2-}, S^-, Na^+, K^+, Total Hardness, HCO_3^-, Mg^{2+}, Ca^{2+}, NO_2^--N, NO_3^--N, NH_4-N, Fe, Pb, Cu, Cd, Hg, Zn, Ni	Water quality index and multivariate statistical analyses of surface water Sarayduzu Dam Lake, Turkey	Kukrer and Mutlu (2019)
NSFWQI	Temperature, Turbidity, PO_4^{3-}, NO_3^-, DO, BOD, EC, TS, pH	Water quality index and multivariate analysis of Beheshtabad River, Iran	Fathi et al. (2018)
BallsonWQI	pH, Temperature, EC, DO, TDS, TSS, Ca^{2+}, Mg^{2+}, Hardness, Cl^-, SO_4^{2-}, $PO_4^{3-}-P$, Total Phosphorus, NH_4-N, NO_3^-, NO_2^-, BOD, COD	Water quality assessment of Bagmati River and Urban water using water quality indices and dissolved oxygen (DO) as an index	Kannel et al. (2007)
Biological water quality index	BOD, COD, NH_3-N, pH, DO, TSS	Comparative study of water quality index and biological water quality index of Melana River, Johor, using physico-chemical and biological analysis	Araman et al. (2013)
Bascaron index	DO, EC, Turbidity, Temperature, pH	Application and comparison of a different method of water quality indices	de Oliveira et al. (2018)
WAWQI	pH, DO, BOD, EC, NO_3^--N, Total coliform, Vr	Water quality index of Sabarmati River water, Gujarat, India	Shah and Joshi (2015)

(continued)

Table 5.13 (continued)

Name of WQI	Parameters	Type of work	References
Canadian *water quality index* (CCME)	pH, Fecal coliform, BOD, Total coliform	Water quality index of Hoogly River water, West Bengal, using cluster analysis and artificial neural network modeling	Sinha and Das (2014)
West Java water quality index	Total Fe, B, F^-, K^+, Ca^{2+}, Hardness, Cl^-, Mg^{2+}, Mn, Total Mn, Natrium, Na^+, SO_4^{2-}, Silica reactive, RSC, SAR, CN^-, Cl_2, H_2S, Se, $KMnO_4$	Water quality index of rivers in West Java Province, Indonesia	Sutadian et al. (2017)
WAWQI	EC, pH, Temperature, Turbidity, Total Hardness, Alkalinity, DO, BOD, COD, NH_4-N, NO_3^--N, NO_2^--N, PO_4^{3-}-P	Water quality index of Karacomak Dam water in Kastamonu City, Turkey	Imneisi and Aydin (2016)
WAWQI	pH, EC, TDS, Total alkalinity, Total Hardness, TSS, Ca^{2+}, Mg^{2+}, Cl^-, NO_3^-, SO_4^{2-}, BOD	Evaluation of drinking water quality and water quality index an urban water body in Shimoga Town, Karnataka	Yogendra and Puttaiah (2008)
Index of river water quality	DO, BOD_5, suspended solid, NH_3-N	Assessment of water quality of Keya River Taiwan using water quality index (WQI) and principal component analysis (PCA)	Liou et al. (2004)
Chemical water quality index	Total N, Particulate phosphorus, Total P, Dissolved P, DO, Pb	Development of chemical index as a measure of in-stream water quality in response to land-use and land cover changes	Tsegaye et al. (2006)

(continued)

Table 5.13 (continued)

Name of WQI	Parameters	Type of work	References
Overall index of pollution	DO, pH, TDS, Turbidity, Total Hardness, Cl^-, SO_4^{2-}, NO_3^-, F^-, TC, BOD	Development of an overall index of pollution for surface water based on a general classification scheme in Indian context	Sargaonkar et al. (2008)
Specific Water Quality Index	Temp. Turbidity, EC, pH, TS, SS, TDS, BOD, DO, Total coliform bacteria, fecal coliform bacteria, Hardness, Alkalinity, SAR, Cl^-, Na^+, K^+, Mg^{2+}, Fe, Cd, Total Cr, Total Hg, Mn, Ni, Zn, Cu, Pb, NH_3-N, NO_3^--N, NO_2^--N, PO_4^{2-}-P	Application of specific water quality index using Delphi technique in Thai River, Thailand	Prakirake et al. (2009)
CCME	Turbidity, temperature, pH, EC, TDS, TSS NO_3^-, As, Be, Cd, Cu, CN, Fe, Li, Mn, Mo, Ni, Ag, Zn	Evaluation of water quality of Mackenzie River, Canada, through water quality index	Lumb et al. (2006)
Bhargava WQI	Temperature, pH, EC, Salinity, Turbidity, DO, COD, BOD_5, NO_3^-, NH_3, PO_4^{3-}, TS, hardness, total dissolved iron, Total coliform, Fecal coliform, SAR	Water quality index of Huong, Thach Han, and Kien Giang rivers of Central Vietnam	Hop et al. (2008)

Table 5.14 List of selected water quality indices applied studies in groundwater by various researchers

Name of WQI	Parameters	Type of work	References
Ground water quality index (GWQI)	pH, TDS, Salinity, Total hardness (TH), EC, HCO_3^-, Ca^{2+}, Mg^{2+}, Na^+, Cl^-, F^-, SO_4^{2-}, NO_3^-	Groundwater quality evaluation for irrigation and drinking purposes using water quality indices West Delhi, India	Acharya et al. (2018)
GWQI and NSF	pH, TDS, TH, HCO_3^-, Cl^-, SO_4^{2-}, NO_3^-, F^-, Ca^{2+}, Mg^{2+}, Fe, Mn, Zn	Drinking water quality evaluation using hydrogeochemical parameters and water quality index in the northwest of Bardhaman district of West Bengal, India	Batabyal and Chakraborty (2015)
WAWQI	pH, TH, Mg^{2+}, F^-, K^+, NH_3, HCO_3^-, TDS, Zn, Cu, Cr, Pb, Cd, Ni, SO_4^{2-}, Ca^{2+}, Mn, EC, Cl^-, Na^+	Water quality index (WQI) of groundwater in Varanasi District, U.P., India	Chaurasia et al. (2018)
CCME	pH, TDS, EC, TH, Ca^{2+}, Mg^{2+}, Na^+, K^+, HCO_3^-, SO_4^{2-}, Cl^-, NO_3^-, SAR, Soluble Na percentage, Mg hazard	Groundwater quality modeling using water quality index and GIS, Upper Egypt	Rabeiy (2017)
Irrigation water quality index	Ca^{2+}, Mg^{2+}, Na^+, Cl^-, HCO_3^-, TDS, EC, SAR	Assessment of groundwater quality for irrigation purpose using irrigation water quality index and GIS in villages of Chabahar city, Sistan and Baluchistan, Iran	Abbas et al. (2018)
National Sanitation Foundation (NSF)	Na^+, Ca^{2+}, Mg^{2+}, K^+, HCO_3^-, SO_4^{2-}, Cl^-, F^-, NO_3^-, pH, EC, TSS, TDS, TH, turbidity, taste, odor, color, and total coliform	Evaluation of drinking water of groundwater using water quality index in Greater Noida, Uttar Pradesh, India	Singh and Hussain (2016)

(continued)

Table 5.14 (continued)

Name of WQI	Parameters	Type of work	References
NSF	pH, Turbidity, TDS, TH, SO_4^{2-}, Mg^{2+}, NO_3^-, Cl^-, Ca^{2+}, F^-	Groundwater quality evaluation using water quality index in Chennai, Tamil Nadu, India	Balan et al. (2012)
WAWQI	Temperature, pH, TS, TSS, TDS, EC, HCO_3^-, Cl^-, PO_4^{3-}	Water quality index of groundwater Allahabad city	Chaurasia et al. (2016)
WAWQI	pH, EC, TH, TDS, HCO_3^-, Cl^-, SO_4^{2-}, Ca^{2+}, Mg^{2+}, K^+	Evaluation of drinking suitability of groundwater quality using water quality index of Mehsana district Gujarat, India	Patel and Vadodaria (2015)
GWQI	K^+, Na^+, Mg^{2+}, Ca^{2+}, SO_4^{2-}, Cl^-, TDS, pH	Groundwater quality evaluation using water quality index in Qazvin province, Iran	Saeedi et al. (2010)
GWQI	Turbidity, pH, TDS, TH, Cl^-, Fe	Water quality index of groundwater in Kolkata city, West Bengal, India	Das et al. (2017)
WAWQI	pH, EC, TDS, Ca^{2+}, Mg^{2+}, Na^+, K^+, Cl^-, Bicarbonate, SO_4^{2-}, NO_3^-, F^-, PO_4^{3-}	Suitability of groundwater for drinking and irrigation purpose using water quality index and GIS a technique in Modjo River Basin, central Ethiopia	Kawo and Karuppannan (2018)
WAWQI	pH, TDS, Na^+, K^+, Ca^{2+}, Mg^{2+}, Cl^-, SO_4^{2-}, NO_3^-, HCO_3	Evaluation of groundwater quality using water quality index (WQI) in Karachi, Pakistan	Adnan and Qureshi (2018)
Mauloom WQI	Ca^{2+}, Mg^{2+}, Cl^-, SO_4^{2-}, TH, F, NO_3^-, TDS, HCO_3^-	Assessment of groundwater quality using water quality indexing Greater Noida (U.P), India	Saleem et al. (2016)

(continued)

Table 5.14 (continued)

Name of WQI	Parameters	Type of work	References
NSF	DO, Fecal coliform, pH, BOD, Temperature, NO_3^-, PO_4^{3-}, Turbidity, TSS	Groundwater quality assessment using water quality index in Maikunkele area of Nigeria	Yisa et al. (2012)
Average water quality index	Turbidity, Ca^{2+}, pH, HCO_3^-, Cl^-, TH, TDS	Evaluation of groundwater quality using averaged water in Lahore City, Pakistan	Sahid and Iqbal (2016)
The Scatterscore index	As, Al^{2+}, Ba, Ca^{2+}, Cd, Cr, Cu, Fe, Hg, Mn^{2+}, Na^+, Ni, NH_3, N, Pb, Se, Zn, pH, Acidity, Cl^-, SO_4^{2-}, HCO_3^-, Turbidity	Water quality evaluation using scatter score: reconnaissance method in acid mine drainage area	Kim and Cardone (2005)
Ground river water quality index	pH, EC, TDS, TH, Ca^{2+}, Mg^{2+}, Na^+, K^+, HCO_3, SO_4^{2-}, Cl^-, NO_3^-, F^-, Dissolved PO_4^{3-}, SiO_2	Water quality index of surface and groundwater around the coal mining area, Korba, Central India	Singh et al. (2017)

References

Abbas A, Radfard M, Amir HM, Ramin N, Yousefi M, Soleimani H, Mahmood A (2018) Ground-water quality assessment for irrigation purposes based on irrigation water quality index and it's zoning with GIS in the villages of Chabahar, Sistan and Baluchistan, Iran. Data Brief 19:623–631

Abbasi T, Abbasi SA (2010) Water quality Indices. Elsevier, Amsterdam

Acharya S, Sharma SK, Vinita K (2018) Assessment of groundwater quality by water quality indices for irrigation and drinking in South West Delhi, India. Data Brief 2019–2028

Adnan K, Qureshi FR (2018) Groundwater quality assessment through water quality index (WQI) in new Karachi town, Karachi, Pakistan. Asian J Water Environ Pollut 15(1):41–46

Araman ZN, Said IM, Salmiati Azman S, Hussin MHM (2013) Comparison between water quality index (WQI) biological water quality index (BWQI) for water quality assessment: a case study of Melan River, Johar. Malays J Anal Sci 17(2):224–229

Balan IN, Shivakumar M, Madan Kumar PD (2012) An assessment of groundwater quality using water quality index in Chennai, Tamilnadu, India. Chron Young Sci 3(2):146–150

Batabyal AK, Chakraborty S (2015) Hydrogeochemistry and water quality index in the assessment of groundwater quality for drinking uses. Water Environ Res 87(7):607–617

Bharti N, Katyal D (2011) Water quality indices used for surface water vulnerability assessment. Int J Environ Sci 2(1). ISSN 0976-4402

Bhattacharya M, Joon V, Jaiswal V (2011) Water handling and sanitation practices in rural community of Madhya Pradesh: a knowledge, attitude and practice study. Indian J Prev Soc Med 42:93–97

Brown RM, McClelland NJ, Deininger RA, O'Connor MF (1972) A water quality index—crossing the Psychological barrier. In: Jenkis SH (ed) Proceedings of international conference on water pollution research, Jerusalem, vol 6, pp 787–797

Bytci PS, Cadraku HS, Zhushi Etemi FN, Ismaili MA, Fetoshi OB, Shala Abazi AM (2018) RASAYAN J Chem 11(2):653–660

Chaurasia GL, Singh SB, Singh S, Gupta MK (2016) Water quality index and correlation study for the assessment of groundwater quality of Allahabad city. Green Chem Technol Lett 1(1):71–76

Chaurasia AK, Pandey HK, Tiwari SK, Prakash R, Pandey P, Arjun R (2018) Groundwater quality assessment using water quality index (WQI) in parts of Varansi district, Uttar Pradesh, India. J Geol Soc India 92:76–82

Chiaudani G, Premazzi G (1988) Water quality criteria in environmental management. Report EUR 11638 EN, Commission of the European Communities, Luxembourg

Das P, Sarma KP, Jha PK, Roger Herbert R Jr, Kumar M (2016) Understanding the cyclicity of chemical weathering and associated CO_2 consumption in the Brahmaputra River Basin (India): the role of major rivers in climate change mitigation perspective. Aquat Geochem 1–27. https://doi.org/10.1007/s10498-016-9290-6

Das KR, Bhoominathan SD, Kanagraj S, Govindraju M (2017) Development a water quality index (WQI) for the Loktak Lake in India. Appl Water Sci

de Oliveira ARM, Borges AC, Antonio TM, Moyses N (2018) Viability of the use of minimum water quality indices: a comparison of methods. J Braz Assoc Agric Eng. ISSN: 1809-430

Dee N, Baker J, Drobny N, Duke K, Whitman I, Fahringer D (1973) An environmental evaluation system for water resource planning. Water Resour Res 9(3):523–535

Dinius S (1972) Social accounting system for evaluating water resources. Water Resour Res 8:1159–1177

Fathi E, Zamani-Ahmadmahmoodi R, Zare-Bidaki R (2018) Water quality evaluation using water quality index and multivariate methods, Beheshtabad River, Iran. Appl Water Sci 8:210

Gupta N, Pandey P, Hussain J (2017) Effect of physicochemical and biological parameters on the quality of river water of Narmada, Madhya Pradesh, India. Water Sci Sci Direct 11–23

Hop NV, To TC, Tung TQ (2008) Classification and zoning of water quality for three main rivers in Binh TRI Thien region (Central Veitnam) based on water quality index. Asian J Sci Technol Dev 25(2):434–444

Horton RK (1965) An index-number system for rating water quality. J Water Pollut Control Fed 37(3):300–306

Imneisi IB, Aydin M (2016) Water quality index (WQI) for main source of drinking water (Karacomak Dam) in Kastamonu City, Turkey. J Environ Anal Toxicol 6(5). ISSN: 2161-0525

Inhaber H (1975) An approach to a water quality index for Canada. Water Res 9:821–833

Kannan K (1991) Fundamentals of environmental pollution. S. Chand and Company Ltd., New Delhi

Kannel PR, Lee S, Lee YS, Kannel SR, Khan SP (2007) Application of water quality indices and dissolved oxygen as indicators for river water classification and urban impact assessment. Environ Monit Assess 132:93–110

Karbbasi AR, Hosseini MM, Bhagvand A, Nazariha M (2011) Development of water quality index (WQI) for Gorganrood River. Int J Environ Res 5(4):1041–1046

Kawo NS, Karuppannan S (2018) Groundwater quality assessment using water quality index and GIS technique in Modjo River basin, Central Ethopia. J Afr Earth Sci 300–311

Kim AG, Cardone CR (2005) Scatterscore: a reconnaissance method to evaluate changes in water quality. Environ Monit Assess 111:277–295

Kukrer S, Mutlu E (2019) Assessment of surface water quality using water quality index and multivariate statistical analyses in Saraydüzü Dam Lake, Turkey. Environ Monit Assess 191:71

Liou SM, Lo SL, Wang SH (2004) A generalized water quality index for Taiwan. Environ Monit Assess 96:35–52

Lu W, Xie S, Zhou W, Zhang S, Liu A (2008) Water pollution and health impact in China: a mini review. Open Environ Sci 2:1–5

Lumb A, Halliwell D, Sharma T (2006) Application of CCME water quality index to monitor water quality: a case of the Mackenzie River Baasin Canada. Environ Monit Assess 113:411–429

Mishra PC, Patel RK (2001) Study of the pollution load in the drinking water of Rairangpur, a small tribal dominated town of North Orissa. Indian J Environ Eco-Plan 5(2):293–298

Mukherjee S, Patel AKR, Kumar M (2020) Water scarcity and land degradation nexus in the era of anthropocene: some reformations to encounter the environmental challenges for advanced water management systems meeting the sustainable development. In: Kumar M, Snow D, Honda R (eds) Emerging issues in the water environment during anthropocene: a South East Asian perspective. Springer, Berlin. ISBN 978-93-81891-41-4

Mutairi NA, Abahussain A, Battay AE (2014) Application of water quality index to assess the environmental quality of Kuwait Bay. AABES conference paper

Naik S, Purohit KM (2001) Studies on water quality of river Brahmani in Sundargarh district, Orissa. Indian J Environ Eco-Plan 5(2):397–402

Patel YS, Vadodaria GP (2015) Groundwater quality assessment using water quality index. In: 20th International conference on hydraulics, water resources and river engineering

Pradhan SK, Patnaik D, Rout SP (2001) Ground water quality index for ground water around a phosphatic fertilizer plant. Indian J Environ Prot 21:355–358

Prakirake C, Chaiprasert P, Tripetchkul S (2009) Development of specific water quality index for water supply in Thailand. Songklanakarin J Sci Technol 31(1):91–104

Prati L, Pavanello R, Pearsin P (1971) Assessment of surface water quality by a single index of pollution. Water Res 5:741–751

Rabeiy RE (2017) Assessment and modeling of groundwater quality using WQI and GIS in upper Egypt area. Environ Sci Pollut Res. https://doi.org/10.1007/s11356-017-8617-1

Rao NS, Rao JP, Devadas DJ, Rao KS (2002) Hydro-geochemistry and groundwater quality in a developing urban environment of a semi-arid region, Guntur, Andhra Pradesh. Geol Soc India 59:159–166

Saeedi M, Abessi O, Sharifi F, Meraji H (2010) Development of groundwater quality index. Environ Monit Assess 163:327–335

Sahid SU, Iqbal J (2016) Groundwater quality assessment using averaged water quality index: a case study of Lahore city, Punjab, Pakistan. IOP Conf Ser: Erath Environ Sci 44:042031

Saleem M, Hussain A, Mahmood G (2016) Analysis of groundwater quality using water quality index. A case study of Greater Noida Uttar Pradesh, India. Civ Environ Eng 3:1237927

Sargaonkar AP, Gupta A, Devotta S (2008) Dynamic weighting system for water quality index. Water Sci Technol 1261–1271

Semiromi FB, Hassani AH, Torabian A, Karbassi AR, Hosseinzadeh L (2011) Water quality index development using fuzzy logic: a case study of the Karoon River of Iran. Afr J Biotechnol 10(50):10125–10133

Shah K, Joshi GS (2015) Evaluation of water quality index for River Sabarmati, Gujarat, India. Appl Water Sci 1349–1358

Shivhare N, Khan S, Patel N, Joshi A, Dutt B (2017) Effect of nallahs on groundwater in Indore city. Int J Eng Sci Res Technol. ISSN: 2277-9655

Singh S, Hussain A (2016) Water quality index development for groundwater quality assessment of Greater Noida sub-basin, Uttar Pradesh, India. Civ Environ Eng 3:1177155

Singh R, Syed TH, Kumar S, Kumar M, Venkatesh AS (2017) Hydrogeochemical assessment of surface and groundwater of Korba field, Central India: environmental implications. Arab J Geosci 10:318

Singh KR, Goswami AP, Kalamdhad AS, Kumar B (2019) Development of irrigation water quality index incorporating information entropy. Environ Dev Sustain pp 1–14

Singh A, Patel AK, Deka JP, Das A, Kumar A, Kumar M (2019) Prediction of arsenic vulnerable zones in groundwater environment of rapidly urbanizing setup, Guwahati, India. Geochem 125590. https://doi.org/10.1016/j.chemer.2019.125590

Singh A, Patel AK, Kumar M (2020) Mitigating the risk of Arsenic and Fluoride contamination of groundwater through a Multi-Model framework of statistical assessment and natural remediation techniques. In: Kumar M, Snow D, Honda R (eds) Emerging issues in the water environment during anthropocene: a South East Asian Perspective. Springer, Berlin. ISBN 978-93-81891-41-4

Sinha K, Das P (2014) Assessment of water quality index using cluster analysis and artificial neural network modeling: a case study of the Hooghly River basin, West Bengal, India. Desalin Water Treat 1–9

State Environmental Protection Administration of China (2006) Report on the State of the Environment in China

Steinhart CE, Schierow LJ, Sonzongi CW (1982) An environmental quality for the Great Lakes. Water Resour Bull 18(6)

Sutadian AD, Muttil N, Yilmaz AG, Perera BJC (2016) Development of river water quality indices—a review. Environ Monit Assess 188–58

Sutadian AD, Muttil N, Yilmaz AG, Perera BJC (2017) Development of a water quality index for rivers in West Java Province, Indonesia. Ecol Indic 85:966–982

Tirkey P, Tanushree B, Suklayan B (2013) Water quality indices—important tools for water quality assessment: a review. Int J Adv Chem 1(1):15–28

Tsegaye T, Shepard D, Islam KR, Johnshon A, Tadesse W, Atalay A, Marzen L (2006) Development of chemical index as a measure of in-streamwater quality in response to land-use and land cover changes. Water Air Soil Pollut 174:161–179

Venkateshraju K, Ravikuamr P, Soma Shankar RK, Prakash KL (2010) Physicochemical and bacteriological investigations of the river Cauvery of Kollegal stretch in Karnataka, Kathamandu University. J Sci Eng Technol 6(1):50–59

Walski TM, Parker FL (1974) Consumers water quality index. J Environ Eng Div 100(3):593–611

WHO (2014) Drastic consequences of diarrheal diseases

Yisa J, Jimoh TO, Oyibo OM (2012) Underground water assessment using water quality index. Leonardo J Sci 21:33–42 (ISSN: 1583-0233)

Yogendra K, Puttaiah ET (2008) Determination of water quality index and suitability of an urban waterbody in Shimoga Town, Karnataka. In: Proceedings of Taal2007: the 12th World Lake conference, pp 342–346

Zeid SAM, Seleem EM, Salman SA, Abdel-Hafiz MA (2018) Water quality index of groundwater and assessment for different usage in El-Obour City, Egypt. J Mater Environ Sci 9(7):1957–1968. ISSN: 2028-2508

Part II
Water Resilence: Vulnerability and Response

Chapter 6
Review on Mixture Toxicity of Pharmaceuticals in Environmental Waters and Wastewater Effluents

Y. B. P. Kahatagahawatte and Hiroe Hara-Yamamura

6.1 Introduction

In the last few decades, an increasingly large number of groups of pharmaceuticals have been detected in various environmental water media, municipal wastewaters, and treated wastewaters, at the very low concentrations mostly raging from ng/L to μg/L: anti-inflammatory drugs, lipid regulators, β blockers, synthetic hormones, antibiotics, antiepileptic drugs, analgesics, antidepressants, stimulants, antihistamines, antiseptics, antineoplastic agents, anxiolytic agents, anticancer drugs (Azuma et al. 2019; Cizmas et al. 2015; Deo 2014; Gogoi et al. 2018; Heberer 2002; Kimura et al. 2005; Luo et al. 2014; Murray et al. 2010; Reungoat et al. 2010; Zhang et al. 2018). Since any pharmaceutical is originally designed for a certain therapeutic use, their impacts on ecosystem and adverse health effects have been highly concerned, particularly in the context of direct and indirect potable water reuse (Angelakis et al. 2018; Jones et al. 2004).

Previous studies suggested limited adverse effects of ambient pharmaceuticals at their "single" exposure (Debroux et al. 2012; Kumar et al. 2019a, b). In the actual water environment and wastewater effluents, however, pharmaceutical compound is unlikely to exist alone. They should, in most cases, coexist with other micropollutants, transformation products, and any other by-products of wastewater treatment, or natural organic matters derived from drinking water, etc. (Fig. 6.1), and the presence of those wastewater matrices may affect the biological impacts of a compound. Indeed, only 1% of the nonspecific toxicity measured by the bioluminescence inhibition assay in the reverse osmosis concentrate, and 0.0025% in the treated water were reported to be explained by the quantified 106 emerging pollutants (Escher et al. 2011).

Y. B. P. Kahatagahawatte · H. Hara-Yamamura (✉)
Faculty of Geosciences and Civil Engineering, Institute of Science and Engineering, Kanazawa University, Kanazawa, Japan
e-mail: hiroeyh@se.kanazawa-u.ac.jp

© Springer Nature Singapore Pte Ltd. 2020
M. Kumar et al. (eds.), *Resilience, Response, and Risk in Water Systems*,
Springer Transactions in Civil and Environmental Engineering,
https://doi.org/10.1007/978-981-15-4668-6_6

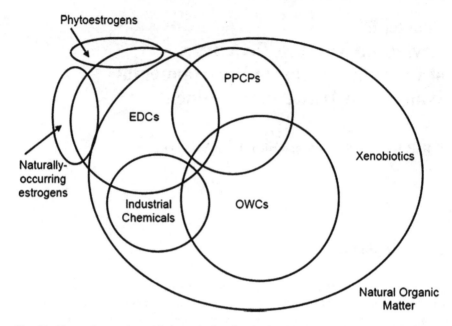

Fig. 6.1 Nomenclature of xenobiotics and other dissolved residuals found in water. PPCPs: pharmaceuticals and personal care products; EDCs: endocrine-disrupting compounds; OWCs: organic wastewater contaminants. Adopted from Debroux et al. (2012)

In this chapter previous findings on human health risk and toxicity posed by pharmaceuticals in environmental water and wastewater were reviewed. First, human health risk assessment studies of pharmaceuticals in environmental water were overviewed for potential water reuse scenarios, and then, previous researches on mixture toxicity of pharmaceuticals for aquatic organisms were summarized since relatively abundant literatures have become available in the field of ecotoxicology. Finally, recent researches on application of the TSB bioassay (TSB assay) for evaluation of chemical mixture or complex environmental samples were reviewed as a potential tool to investigate mixture toxicity of pharmaceutical in human model system.

6.2 Human Health Risk Posed by Pharmaceuticals in Environmental Water

Human health risk posed by pharmaceuticals in environment water is of particular concern in the context of water recycling, where municipal wastewater which did or did not undergo advanced water treatment processes (e.g., rapid sand filtration, ozonation, membrane separation) are intentionally reused for agricultural, landscape, recreational, domestic, and potable use, or unintentionally entered into the drinking water source (*de fact* reuse). A review published in 2012 well covered the past important findings on human health risk posed by non-regulated xenobiotics including pharmaceuticals and personal care products (PPCPs) (Debroux et al. 2012). The authors reviewed papers from the late 2019s and the early 2020s to conclude that no adverse health effects were expected from the presence of these compounds in drinking water at typical concentrations, which may occur as a result of wastewater impacts to drinking water sources and potable reuse. However, the following two points were suggested as limitations.

- Risk assessments were conducted for the selected compounds, not all compounds that may be present, including all pharmaceuticals and other xenobiotics in use, their metabolites, and their wastewater treatment degradation products
- Detailed toxicological data are only available for some of the many potential constituents of concern, which may result in bias risk assessments toward compounds with high availability of data

In this section, recent updates on human health risk assessments of pharmaceuticals in water environment, which were published in the last five years, were reviewed. Seemingly, the two gaps of knowledge suggested by Debroux et al. (2012)—incomplete detection of some pharmaceuticals, metabolites, and treatment degradation products, and biased accumulation of toxicological data—have not been closed yet.

6.2.1 Agricultural Irrigation

Among the various water reuse applications, agricultural irrigation has the longest history from the Bronze Age and is even expanding in its variety with improving effluent quality (Angelakis et al. 2018). Accordingly, relatively abundant literatures have become available for human health risk posed by this application, via dietary intake of PPCPs in vegetables, crops, and tree fruit irrigated by wastewater effluents or grown in the biosolids-amended soil. Although their experimental conditions such as growing medium, exposure concentration, exposure time, plant species were different, thus the estimated human exposure values varied widely, they resulted in a common conclusion that the human exposure to majority of PPCPs in the edible plant tissues was likely to be small (Wu et al. 2015).

Recent researches have brought new insights based on their field studies, which were not frequently conducted in the past. Prosser and Sibley (2015) investigated the uptake of triclosan and triclocarban by crops including radish, carrot, green bell pepper, tomato, cucumber, and lettuce in the biosolid-amended soil. The estimated consumption of the two antibacterial agents accounted for 0.13–0.39% and 0.73–1.5% of an acceptable daily intake (ADI) for an adult and toddler, respectively. On the other hand, Malchi et al. (2014) investigated the uptake of carbamazepine and caffeine by vegetables irrigated by wastewater to show that an adult would require to consume hundreds of kilograms of sweet potatoes or carrots daily to reach the threshold of toxicological concern (TTC) level. Similarly, Wu et al. (2014) estimated that total annual PPCPs exposure due to the treated wastewater-irrigated vegetables was 3.69 μg per capita, which was more than three orders of magnitude smaller than that in a single medical dose for one compound (typically in the 10–200 mg range).

Franklin et al. (2016) investigated uptake of sulfamethoxazole, trimethoprim, ofloxacin, and carbamazepine in wheat plants irrigated with treated wastewater, and estimated daily intake of PPCPs via grain was 166–332 ng, which was six orders of magnitude smaller than a typical daily dose (400–800 mg). Riemenschneider et al. (2016) studied the uptake of 28 micropollutants and carbamazepine metabolites in ten different field-grown vegetable species. The daily human intake of nine pharmaceuticals was calculated to 0.003–15 ng/kg of body weight. The estimated annual intakes for carbamazepine, gabapentin, ciprofloxazin, and diclofenac are about 0.001% of the minimum daily doses (typically 10–200 mg). Based on TTC concept, at least 9 kg of vegetable materials are allowed to eat before the TTC level would be reached in terms of carbamazepine, trans-DiOH-carbamazepine, and gabapentin. On the contrary, the TTC level of EP-carbamazepine and ciprofloxacin would be exceeded for a 70 kg person by the daily consumption of only one potato or half an eggplant.

So far, only one study examined the health risk of PPCPs in long-term irrigation. Christou et al. (2017) irrigated tomato crops using wastewater for three years to reveal that the concentration of sulfamethoxazole was the highest in the soil (0.98 μg/kg) throughout the studies period, followed by trimethoprim (0.62 μg/kg) and diclofenac (0.35 μg/kg). The calculated fruit bioconcentration factors were extremely high for diclofenac in the second (108) and third year (132) of the experimental period, compared to sulfamethoxazole (0.5–5.4) and trimethoprim (0.2–6.4). The estimated TTC and hazard quotient (HQ) values suggested a de minimis risk to human health caused by the consumption of tomato fruits irrigated with the wastewater in this study.

González García et al. (2019) predicted the uptake of the three non-steroidal anti-inflammatory drugs (i.e., ibuprofen, ketoprofen, and naproxen) by treated wastewater-irrigated lettuce based on a steady-state plant uptake model with added phloem transport. These pharmaceutical compounds are weak acids, and predicted concentrations in roots were higher than in the edible leaves, mainly due to phloem transport downwards. The highest concentration predicted in the leaves of the three varieties of lettuce were 28 ng/gdw for ibuprofen, 80 ng/gdw for ketoprofen, and 57 ng/gdw for naproxen, leading to a daily uptake of 575 ng/d, 786 ng/d, and 560 ng/d, respectively. The daily dietary intake of the selected pharmaceuticals was estimated to be far below usual therapeutic doses.

6.2.2 Landscape Irrigation

Landscape irrigation is another applications of reclaimed wastewater, which involves the irrigation of golf courses, parks, school grounds, etc., being widely practiced in the cities of the southwest US, China, and the middle eastern countries (Aleisa and Al-Zubari 2017; Semerjian et al. 2018; Wang et al. 2017; Wu et al. 2009). As for the human health risk posed by PPCPs in this practice, however, very limited number of researches have been available so far. Semerjian et al. (2018) assessed health risk of ten PPCPs toward children playing green areas, adult landscape workers, and adult users of athletic and golf courses irrigated by treated wastewater. The estimated risk quotients (RQs) were the highest to lowest in the order of acetaminophen, metoprolol, ciprofloxacin, erythromycin, ofloxacin, sulfadiazine, sulfamethoxazole, sulfapyridine, risperidone, and sulfamethazine, and all PPCPs for all receptors exhibited safe exposure (RQ < 1) through both dermal and ingestion exposure pathways.

6.2.3   Potable Water Reuse

Potable water reuse is one of the rapidly growing applications of reclaimed water. At present, more than 20 planned indirect potable reuse (IPR) projects are operated in the world, and over half of those are located in the western coast in US (USEPA 2018). Turning our eyes into the orient, the largest amount of reclaimed water is produced at the Changi NEWater plant, Singapore (i.e., 0.45 million m^3/d), and totally, five NEWater plants cover 40% of the domestic water needs, which may increase up to 55% by 2060 (PUB 2018). The extreme droughts in California and Texas in US led to the further exploitation of potable water reuse. Direct potable reuse (DPR) is temporally implemented in Big Spring and Wichita Falls, Texas, and a permanent DPR plant in El Paso, Texas, is now waiting for regulatory approval (USEPA 2018).

The risk assessment studies demonstrated that potential exposure of major active pharmaceutical ingredients from drinking water would not pose any appreciable risk to human health, based on the comparison of PPCPs concentrations in river water samples and predicted no effect concentrations (PNECs) calculated from acceptable daily intakes (ADIs). Bercu et al. (2008) evaluated the risks from residues of atomoxetine, duloxetine, and olanzapine, whose PNECs were calculated for children as 25.7, 19.1, 35.9 µg/L, respectively. The margins of safety ranged from 147 (duloxetine-children) to 642 (olanzapine-adults), while all PEC/PNEC ratios were more than two orders of magnitude below the value of 1.

Similarly, Cunningham et al. (2009) assessed human health of 44 active pharmaceutical ingredients marketed by GlaxoSmith Kine, representing approximately 22 general pharmacological classes with a broad spectrum of therapeutic activities. The calculated PEC/PNEC ratios varied from 7×10^{-2} to 6×10^{-11}, which indicated that the selected compounds do not appear to pose an accreditable risk to human health from drinking water and fish consumption. According to Schwab et al. (2005), 26 active pharmaceutical ingredients and their metabolites, representing 14 different drug classes, were assessed to exert no appreciable human health risk when the surface water is used as drinking water source. The calculated approximate margins of safety for exposures from drinking water and eating fish by children ranged from 30 to 38,000, based on a comparison of maximum reported MECs to PNECs.

6.3 Mixture Toxicity of Pharmaceuticals on Aquatic Organisms

In response to the rapid increase of pharmaceutical usage and release into the water environment without identification of the toxicity and a targeted treatment procedure, previous researches raised an alarm that the pharmaceutical impacts on the aquatic organisms would have become non-negligible level. In this section, evaluation methods used for mixture toxicity of pharmaceuticals were briefly summarized first, and then, previous researches on mixture toxicity of pharmaceuticals on aquatic organisms were further discussed by therapeutic classes. Some studies investigated the mixture toxicity of pharmaceuticals from the same therapeutic category such as non-steroidal anti-inflammatory drugs (NSAIDs), antibiotics, synthetic steroidal hormones, and lipid regulators, or those from various categories specially blended in a laboratory (Backhaus 2014). Table 6.1 shows a summary of the previous studies on mixture toxicity of pharmaceuticals toward the aquatic organisms.

The research findings obtained so far are mostly about their individual toxicity, and there are only few researches addressed the combination effect of pharmaceutical mixtures to the aquatic organisms. Among those researches, several studies have demonstrated synergism or antagonism effects caused by pharmaceutical mixtures at the environmental relavant concentrations (i.e., 0.012–333 ng/L) (Coors et al. 2018; Li and Yu-Chen Lin 2015). However, most of these experiments have been conducted in a laboratory scale, and very limited pharmaceutical compounds and simple mixtures were used for the toxicity evaluation, which did not fully reflect the complex combined effects of pharmaceuticals from same and/or different therapeutic classes in the environmental samples. In addition, environmental chemical mixtures may occur in a variety of doses with more complex response patterns, such as dose–dependent, endpoint (conventional and multiple biological level) dependent synergism/antagonism, thus indicating the necessity to go beyond the established concepts (Hendricksen et al. 2007).

6.3.1 Models of Mixture Toxicities

Mixture toxicity of pharmaceuticals has been evaluated using different reference models with variety of predictions and statistical approaches. Concentration addition (CA) and the independent action (IA) are the models which predict the effects of the mixture by linking the single compound toxicity based on the dose-response curves. CA is highly applicable for the compounds whose mode of actions are similar with each other towards the target site or narcosis-type compounds which induces baseline toxicity toward aquatic species. On the contrary, IA is based on the idea of dissimilar mode of action of compounds in a mixture which shows different effects toward target sites. These two are principal concepts used for years, and several researchers

Table 6.1 Summary of past investigations on toxicity of different pharmaceutical mixtures toward

Combination of pharmaceuticals	Endpoint	Toxicity	Reference
1. *Daphnia magna*			
Ibuprofen, Diclofenac	Immobilization	Showed synergistic effect	Cleuvers (2003)
Clofibrinic acid, Carbamazepine			
Diclofenac, ibuprofen, naproxen, acetylsalicylic acid	Immobilization	Showed synergistic effect	Cleuvers (2004)
Sertraline and fluoxetine	Immobilization	Cause additive effects, no indications of synergism or antagonism	Christensen et al. (2007)
Sertraline and citalopram			
Fluoxetine and citalopram			
Bezafibrate, clofibric acid, gemfibrozil, and fenofibric acid, a metabolite of fenofibrate	Immobilization	No any negative effect on D. magna	Rosal et al. (2010)
Carbamazepine, diclofenac, 17a-ethinylestradiol and metoprolol	Life history and morphological parameters over six generations	Reduced the age and increased the body length at first reproduction	Dietrich et al. (2010)
Diclofenac, ibuprofen, and clofibric acid	Survival and reproduction	No significant toxicity	Han et al. (2006)
Ciprofloxacin, 17α-ethinylestradiol, and 5-fluorouracil	Immobilization and reproduction	Mixture caused adverse effects where, individually cause no harm to organisms in environmentally relevant concentrations	Affek et al. (2018)
Fluconazole, fluoxetine, metoprolol, climbazole tris(2-chloropropyl) phosphate, 5-methylbenzotriazole, methylparaben propiconazole	Reproduction	Showed no adverse effects	Coors et al. (2018)
2. *Pseudokirchneriella subcapitata*			
Erythromycin, levofloxacin	Growth rate inhibition	Synergistic effect was predominant	Gonzá Lez-Pleiter et al. (2013)
Norfloxacin, erythromycin			
Erythromycin, tetracycline			
Levofloxacin, norfloxacin			
Levofloxacin, tetracycline			
Tetracycline, norfloxacin			

(continued)

Table 6.1 (continued)

Combination of pharmaceuticals	Endpoint	Toxicity	Reference
Erythromycin, levofloxacin, norfloxacin and tetracycline			
Sertraline, fluoxetine	Growth rate inhibition	Cause additive effects, no indications of synergism or antagonism	Christensen et al. (2007)
Sertraline, citalopram			
Fluoxetine, citalopram			
Perfluorooctane sulfonic acid (PFOS), triclosan	Growth rate inhibition	Antagonism effect	Boltes et al. (2011)
PFOS, 2,4,6-trichlorophenol			
PFOS, gemfibrozil			
PFOS, bezafibrate			
Triclosan,2,4,6-trichlorophenol			
Bezafibrate, gemfibrozil			
PFOS, triclosan,2,4,6-trichlorophenol		Very strong synergistic effect	
PFOS, bezafibrate, gemfibrozil		Antagonism effect	
Benzalkonium chloride [BAC] and 5-fluorouracil [5-FU	Growth rate inhibition	Synergistic effect	Elersek et al. (2018)
3. *Desmodesmus subspicatus*			
Ibuprofen and diclofenac	Growth rate inhibition	Showed synergistic effect	Cleuvers (2003)
Clofibrinic acid and carbamazepine			
Diclofenac, ibuprofen, naproxen, and acetylsalicylic acid	Growth rate inhibition	Showed synergistic effect	Cleuvers (2004)
Ciprofloxacin, 17α-ethinylestradiol and 5-fluorouracil	Growth rate inhibition	Mixture caused adverse effects where individually cause no harm to organisms in environmentally relevant concentrations	Affek et al. (2018)
4. *Cyanobacterium Anabaena*			

(continued)

Table 6.1 (continued)

Combination of pharmaceuticals	Endpoint	Toxicity	Reference
Erythromycin, levofloxacin norfloxacin and erythromycin erythromycin and tetracycline levofloxacin, norfloxacin levofloxacin, tetracycline tetracycline and norfloxacin erythromycin, levofloxacin, norfloxacin and tetracycline amoxicillin	Bioluminescence inhibition	Showed synergistic effect	Gonzá Lez-Pleiter et al. (2013)
Gemfibrozil, bezafibrate, bezafibrate, fenofibric acid, fenofibric acid, gemfibrozil, gemfibrozil, bezafibrate, fenofibric acid	Bioluminescence inhibition	Strong synergism at the lowest effect levels and a very strong antagonism at high effect levels	Rodea-Palomares et al. (2009)
Bezafibrate, clofibric acid, gemfibrozil, and fenofibric acid, a metabolite of fenofibrate	Bioluminescence inhibition	Very toxic	Rosal et al. (2010)

1. *Daphnia magna*, 2. *Pseudokirchneriella subcapitata*, 3. *Desmodesmus subspicatus*, and 4. *Cyanobacterium Anabaena*

reported that these approaches gave accurate and reliable results. (Cleuvers 2004, 2003; Coors et al. 2018; Geiger et al. 2016; Thrupp et al. 2018; Xiong et al. 2019).

González-Pleiter et al. (2013) suggested that the combination index (CI) method could predict deviations from additivity more accurately than the classical CA and IA model. This model involves dose-response curves for each compound and their combinations in multiple diluted concentrations by using the median effect equation; thus, it does not depend on the mode of action of individual compounds. CI value elicits clear idea about the behaviors of tested endpoints of aquatic organism where the CI < 1, =1, and >1 indicate synergism, additive effect, and antagonism, respectively. CI model has been used for many researchers in the ecological toxicity filed as well (Boltes et al. 2011; Rodea-Palomares et al. 2009; Rosal et al. 2010).

6.3.2 Mixture Toxicity of Non-steroidal Anti-inflammatory Drugs

The two studies which examined NSAIDs for their mixture toxicity to the aquatic organisms reported synergistic effect even at concentrations at which the single substance showed no or only very slight effects (Cleuvers 2004, 2003). Binary mixtures of ibuprofen and diclofenac and a mixture of diclofenac, ibuprofen, naproxen, and acetylsalicylic acid have been studied over the endpoint of acute immobilization of *Daphnia magna* and algal inhibition of *Desmodesmus subspicatus*. The concept of CA and IA was adopted for these analyses, where each compound was mixed at the half of the calculated effect concentrations ($EC_5/2$, $EC_{10}/2$, $EC_{20}/2$, $EC_{50}/2$, and $EC_{80}/2$ or $EC_{90}/2$) for binary mixtures, or at a quarter of the calculated effect concentrations ($EC_5/4$, $EC_{10}/4$, $EC_{20}/4$, $EC_{50}/4$, and $EC_{80}/4$) for quaternary mixtures. In the *Daphnia* test, the binary mixture effect was higher than the measured single compound toxicity in every EC value. As an example, at the $EC_{80}/2$ level, the concentrations responsible for the singly measured effects of ibuprofen (~12% immobilized daphnids) and diclofenac (~2% immobilized daphnids) caused a strong mixture effect of about 95% immobilization of the daphnids. Furthermore, the binary mixture exhibited even stronger toxicity than the predicted toxicity by concentration addition. However, the quaternary mixture of diclofenac, ibuprofen, naproxen, and acetylsalicylic acid showed a stronger toxicity only in $EC_{50}/4$ and $EC_{80}/4$. In the algal test, both binary and quaternary mixtures have shown the inhibition of algal growth rate in each EC value and have followed the concept of concentration addition. The concentrations examined these studies (i.e., 1–320 mg/L) were still much higher than the actual environmental concentrations (i.e., ng/L ~ μg/L) (Stumpf et al. 1999; Ternes 1998).

6.3.3 Mixture Toxicity of Antibiotics

Synergistic effects of antibiotics were reported for binary mixtures of erythromycin and levofloxacin (EC_{50}: >20.4 mg/L), norfloxacin and erythromycin (EC_{50}: 18.2 mg/L), erythromycin and tetracycline (EC_{50}: 0.27 mg/L), levofloxacin and norfloxacin (EC_{50}: 15 mg/L), levofloxacin and tetracycline (EC_{50}: 4.6 mg/L), tetracycline and norfloxacin (EC_{50}: 9.2 mg/L), and a quaternary mixture of erythromycin, levofloxacin, norfloxacin, and tetracycline (EC_{50}: 16.0 mg/L), in the algal test done for the endpoints of relative growth inhibition of *Pseudokirchneriella subcapitata*. However, some of those binary mixtures showed antagonistic effects on *Cyanobacterium Anabaena*, which are the pairs of erythromycin and levofloxacin (EC_{50}: 2.5 mg/L), norfloxacin and erythromycin (EC_{50}: 4.5 mg/L), levofloxacin and norfloxacin (EC_{50}: 4.3 mg/L), and other binary mixtures showed synergistic effects. A quaternary mixture showed increased synergistic effects over their binary mixtures toward the two aquatic organisms (González-Pleiter et al. 2013).

  Guo et al. (2016) evaluated mixture toxicity of three veterinary antibiotics of frequent use (trimethoprim, tylosin, and lincomycin) toward the algae and cyanobacteria in European surface waters. The EC_{50} value was reported as 0.248 μmol/L, where the combination of the three compounds could cause adverse impacts on primary production and nutrient cycle of algal communities. This research also supported the applicability of the CA model with the accurate predicted results to the examined endpoints. A mixture of sulfamethazine and sulfamethoxazole exhibited higher toxicity toward *Scenedesmus obliquus* with the EC_{50} value of 0.89 mg/L than their single exposure, where the EC_{50} values of each compound marked 1.23 mg/L and 0.12 mg/L, respectively. The observed mixture toxicity was more in line with the predictions of CA model rather than the IA model, indicating the similar mode of action of sulfonamides. The binary mixture can be categorized as very toxic to aquatic environment according to the EU directive 93/67/EEC, which classifies chemicals as per their EC_{50} values into three categories: very toxic ($EC_{50} < 1$ mg/L), toxic ($EC_{50} = 1$–10 mg/L), and harmful ($EC_{50} = 10$–100 mg/l) (Xiong et al. 2019). The antibiotic mixture showed toxicity even at the low concentrations which were likely to be present in the aquatic environment (Xiong et al. 2019).

6.3.4 Mixture Toxicity of Synthetic Steroidal Hormones

Another study investigated the egg production of the fathead minnow, *Pimephales promelas*, exposed to the mixture of five different synthetic steroidal hormones such as EE2 (estrogen), trenbolone (androgen), beclomethasone dipropionate (glucocorticoids), desogestrel, and levonorgestrel (progestogens). The results showed that the steroidal mixture of the lowest concentration (i.e., 351.44 ng/L) exhibited a 50% reduction of egg production, and that of the medium concentration (i.e., 1159.4 ng/L) exhibited completely ceased egg production in all fish pairs after seven days of exposure, while any single exposure of the selected compounds would produce no statistically significant effect (Thrupp et al. 2018). This study suggested that small effects can add up to reach a statistically and biologically significant response when there is simultaneous exposure to multiple chemicals in fish.

6.3.5 Mixture Toxicity of Pharmaceuticals from Different Therapeutic Classes

Quaternary mixture of drugs such as metformin (antidiabetic), bisoprolol (β blocker), ranitidine (histamine), and sotalol (β blocker) did not induce any adverse joint effects on development and behavior of *Danio rerio* embryos, at the four different total concentrations of 0.1, 1.0, 10, and 100 mg/L, which included the concentrations usually found in surface freshwaters (Godoy et al. 2019). Another study of Dietrich et al. (2010) have resulted that both pharmaceutical mixture and single compound did not provoke any significant effects on daphnia. They performed a multigenerational study over six generations using *D. magna* with a mixture of pharmaceuticals from different therapeutic categories such as carbamazepine (antiepileptic drug), diclofenac (anti-inflammatory drug), 17α-ethinylestradiol (synthetic hormone), and metoprolol (beta blocker) at the average concentrations found in rivers and streams of southern Germany.

6.3.6   Mixture Toxicity of Pharmaceuticals in Wastewater Effluents

Among few number of studies which examined the effluents of a municipal wastewater treatment plants, Coors et al. (2018) demonstrated that there were no adverse effects on the reproduction of *Daphnia Magna* by the exposure of a pharmaceutical mixture consisted of fluconazole, fluoxetine, metoprolol, climbazole, tris(2-chloropropyl) phosphate (TCPP), 5-methyl benzotriazole (5-MBT), methylparaben, and propiconazole. Moreover, a wastewater sample containing 30 compounds did not exhibited toxicity to *Daphnia magna*, with immobilization rate of <5%. In the bioluminescence inhibition test using *Vibrio fischeri*, a luminescence increased by nearly 50% during the first five minutes and decayed thereafter, which resulted in unusual negative toxicity values. This is probably due to the salinity correction of the sample. In addition, wastewater sample was found to be very toxic to *Anabaena* CPB4337, inhibiting its luminescence by 84.24% (Rosal et al. 2010).

A mixture of 11 pharmaceuticals (ibuprofen, naproxen, gemfibrozil, bezafibrate, carbamazepine, sulfapyridine, oxytetracycline, novobiocin, trimethoprim, sulfamethoxazole, and caffeine) incurred a slight decrease in morphology of the freshwater *cnidarian Hydra attenuata* at 0.1, 10 and, 100-times high concentrations of ambient concentrations but a significant increase at 1000-times high concentration. Toxicity was investigated using both lethal (based on morphology) and sub-lethal (based on morphology, feeding behavior, hydranth number, and attachment) endpoints over 96 h exposure of the mixture (Quinn et al. 2008). Chronic exposure of fish to pharmaceutical mixtures and pharmaceuticals in discharged wastewater affected reproduction and induced histopathological changes which may cause negative impacts on reproductive capacity and health (Galus et al. 2013).

Furthermore, a study done by using a mixture of pharmaceuticals present in a wastewater effluent (i.e., atenolol, metoprolol, propranolol, caffeine, diphenylhydantoin, fluoxetine, hydrochlorothiazide, diclofenac, ketoprofen, and naproxen) demonstrated a synergistic effect over the rainbow trout. The lowest observed effect concentrations and concentrations causing EC50 were within the range 0.05–54.61 μg/L which can be found in the effluent (Fernández et al. 2013).

6.4   Transcriptomics-Based Bioassay for Mixture Toxicity of Pharmaceuticals

Due to the limitation of conventional approach targeting on a few key chemicals, the effect-based approach has been gaining its popularity as a monitoring method of water environment (Altenburger et al. 2015; Bruneau et al. 2016; Zhang et al. 2014). Such kind of approaches include the transcriptomics-base (TSB) bioassay which is capable of giving broad understanding about complex mixtures induce toxicity to by evaluating various biological pathways at genome-scale on aquatic organisms or human cells (Wang et al. 2018). Previous studies demonstrated that transcriptomic response of the organisms could represent the toxic modes of action of the complex chemical mixtures and environmental samples. So far, DNA microarray-based transcriptomic analysis has been widely applied to municipal wastewaters and treated effluents, in the most cases, using aquatic organisms as a test organism. In this section, recent researches on application of the TSB bioassay for evaluation and characterization of chemical mixtures and environmental samples were reviewed as a potential tool to investigate mixture toxicity of pharmaceuticals.

6.4.1 Methods of Transcriptomic Analysis

DNA microarray-based transcriptomic analysis is a powerful gene expression analysis approach which has been applied to evaluation of wastewaters and treated effluents quality, using aquatic organisms (Martinović-Weigelt et al. 2014; Prokkola et al. 2016) and human cells (Fukushima et al. 2014; Hara-Yamamura et al. 2013, 2020) as a test organism. The multiple-endpoints gene alteration-based (MEGA) assay is a rapid, cost-effective, and high-throughput real-time PCR-based assay (qPCR assay) that targets cellular responses to contaminants present in wastewater effluents at the transcriptional level, therefore optimum and powerful screening test for monitoring the toxicity of wastewater effluents (Fukushima et al. 2017).

More recently, the novel RNA sequencing technique (RNA-seq), instead of the DNA microarray, has been applied in this field to understand more extensive response of organism to the effluents. The RNA-seq approach could provide information regarding the mode of action of pollutants and then be useful for the identification of which parameters must be studied at higher integration level in order to diagnose sites where the presence of complex and variable mixtures of chemicals is suspected (Bertucci et al. 2018). However, high-throughput analysis of transcriptomics in chemical assessment is also limited due to the high cost and lack of "standardized" toxicogenomic methods (Wang et al. 2018).

Reduced zebrafish transcriptome ampliseq embryo approach is a novel concept developed by Wang et al. (2018) to represent the whole transcriptome and to profile bioactivity of chemical and environmental mixtures in zebrafish embryo. It provides an efficient and cost-effective tool to prioritize toxicants based on responsiveness of biological pathways (Wang et al. 2018). Digital gene expression profiling (DGE) based on Illumina Genome Analyzer sequencing platform was also suggested as an effective tool to identify global gene alteration affected by environment contaminants in hepatic cells, with twice lower cost and less time required (Zhang et al. 2014). Validation of the transcriptomic profiles from this method can be conducted by qPCR assay as used in microarray analysis.

6.4.2 Application to Human Model Systems

Effluent wastewaters consist of dissolved organic matters for which humic substances usually account a large percentage in its concentration (Artifon et al. 2019). The humic substances have long been recognized as a mitigator of toxicants. However, Hara-Yamamura et al. (2020) provided a new insight regarding toxic unknowns in effluent wastewater based on the transcriptomic analysis of human cells using affymetrix human focus microarray. Human hepatoblastoma HepG2 cell was exposed to wastewater effluents from activated sludge process (AS) and advanced membrane bioreactors (MBR) at the same dissolved organic carbon level (i.e., 30 mg/L). The qPCR assay of selected marker genes suggested that responsible constituents for potentially adverse, abnormal transcriptomic response in HepG2 could have hydrophobic nature and act with metal-DOM complexes in 1 k Da or smaller size fraction. The humic substances with acidic nature were further suggested as an inducer of effluent toxicity, probably by mediating the accumulation of micropollutants on cell membrane. Fukushima et al. (2014) also applied the DNA microarray analysis to HepG2 cells exposed to chlorinated wastewater samples after AS and MBR processes to conclude that biological impacts of chlorinated effluents were dependent on the effluent organic matter characteristics such as disinfection by-products (DBPs) and O/C ratio.

Aquatic environment has been threatened by semi-volatile organic compounds (SVOCs) and volatile organic compounds (VOCs) from petrochemicals which are not completely removed by wastewater treatment processes (Zhang et al. 2014). Transcriptomic analysis of mice liver revealed that the exposure to effluent mixture of polycyclic aromatic hydrocarbons (PAHs), phthalic acid esters (PAEs), and organochlorine compounds (OCCs) related to disruption of fatty acid metabolism, glycerolipid metabolism, arachidonic acid metabolism, linoleic acid metabolism, PPAR signaling pathway, and adipocytokine signaling pathway related to lipid metabolism, with altered genes and serum metabolites in liver and hepatotoxicity on mice (Zhang et al. 2014). Similarly, Prokkola et al. (2016) demonstrated that exposure of effluent sample with di-n-butyl phthalate (DBP), which is the second most common phthalate, to three-spined sticklebacks male fish at the median concentration in wastewater effluents in Germany and Denmark (i.e., 0.7–2.4 µg/L) induced adverse genetic response on retinoid metabolism, creatine kinase activity, and cell adhesion.

6.4.3 Application to Aquatic Organisms

Berninger et al. (2014) reported the adverse impacts on hypothalamic–pituitary–gonadal (HPG) axis in female fathead minnows by the exposure to the wastewater effluent, based on alteration of pathways associated with oocyte meiosis, TGF-beta signaling, gonadotropin-releasing hormone (GnRH), and epidermal growth factor

receptor family (ErbB), and gene sets associated with cyclin B-1 and metallopro-
teinase. The authors studied the digestive gland of *C. fluminea* by the RNA sequencing
analysis and identified a set of 3181 transcripts in response to wastewater effluent at
the whole-transcriptome level. Enrichment of gene sets playing key roles in the teleost
brain–pituitary–gonadal–hepatic (BPGH) axis function indicated that wastewater
treatment plants serve as an important source of endocrine-active chemicals that
alter the cholesterol and steroid metabolism (Martinović-Weigelt et al. 2014).

Jackman et al. (2018) studied TH-disruption by endocrine-disrupting chemicals
in wastewater effluents from two different treatment systems: anaerobic membrane
bioreactor (AnMBR) and membrane-enhanced biological phosphorous removal
(MEBPR) with pharmaceutical cocktail over TH-response gene transcripts of olfactory system and olfactory bulb of *Rana (Lithobates) catesbeiana* tadpoles using
RNA-seq analyzing method. AnMBR effluent had no effect on biomarkers; *dio2*,
Heket, *st3*, and *trpv1* transcript levels; however, *thibz-all* expression was increased
at the high effluent concentrations (AnMBR: 1.9-fold, MEBPR: 1.4-fold). Majority of transcripts showed no change in abundance relative to the MEBPR effluents.
Thus, it appears that the cause of the endocrine disruption was already in the inflow-
ing stock wastewater, is independent of measurable PPCP chemistry, and was not
removed upon treatment (Jackman et al. 2018). In addition, oxidative stress-related
genes (i.e., glutathione reductase and glutathione peroxidase) in the liver, and immune
system-related genes (i.e., complement component 1, and macrophage-inducible C-
type lectin) in gonad of female large-mouth bass were also altered, after exposure to a
synthetic estrogen 17Alpha-ethinylestradiol (EE2) (Colli-Dula et al. 2014). Overall,
these studies promote our understanding on molecular responses to antiandrogens
and estrogens in wastewater effluents over fish testis, providing a basis for further
studies on their roles in endocrine and reproductive disturbances.

Transcriptomic analysis of *Lemna minor* revealed that psychoactive drug mixture (i.e., valproic acid, citalopram, carbamazepine, cyamemazine, hydroxyzine,
oxazepam, norfluoxetine, lorazepam, fluoxetine, and sertraline) induced transient
alteration of marker genes of *cyp4*, *sod*, and *mdr1* under the exposure concentrations
normally found in effluent wastewater (ng/L ~ μg/L), while any significant alteration
of *cat* and *pi-gst* marker genes were not observed (Bourioug et al. 2018). Perfluorinated compounds (PFCs) are commonly used in consumer products such as stain,
water, and grease repellents in carpets and clothing or in cooking utensils as nonstick
coatings. Thus, PFCs can be found in municipal wastewater effluents as complex
mixtures and wastewater treatment plants have been recognized as important direct
pollution sources to the aquatic environments. PFCs are persistent, and some of
the substances bioaccumulate and biomagnify in the environment and bind to the
serum albumin, L-FABP, and membrane structures in the liver. Houde et al. (2014,
2013) conducted experiments with twelve perfluorinated acidic compounds including perfluorooctanoic acid (PFOA) and perfluorooctanesulfonic acid (PFOS) in the
downstream of municipal wastewater treatment plant by using the gene expression
analysis over liver, gill, muscle, and blood of adult northern pike fish. Transcriptomic responses were tissue-specific and indicated significant up-regulation of genes
encoding metallothionein in blood, and metallothionein, glutathion-S-transferase,

superoxide dismutase, and cytochromes P450 1A1 in gill tissue of the fish collected in the municipal wastewater treatment plant (Houde et al. 2013).

Eleven perfluoroalkyl substances (PFASs) in wastewater effluents on the fish also suppressed immune response genes, which suggested that the exposure to wastewater effluents impaired the fishy immune system, increasing their susceptibility to pathogens and/or parasites (Houde et al. 2014). Transcriptomic results of yellow perch living in St. Lawrence River, where the flame retardants, metals, and perfluoroalkyl substances have been found in sediments, water, and fish, also down regulated the genes related to lipid, glucose, and retinoid as well as a decrease in retinoid storage (Bruneau et al. 2016). Furthermore, genes related to lipid metabolism, retinol metabolism, detoxification processes, cellular proliferation, and membrane/cellular transport were also impacted by the PFCs in wastewater effluent (Houde et al. 2014).

Ings et al. (2011) demonstrated that tertiary-treated municipal wastewater elicited multiple stress-related pathways such as an organismal and cellular stress response in trout, which may lead to an enhanced energy demand in the exposed fish. The study was undertaken on rainbow trout, *Oncorhynchus mykiss* at an upstream control or 100, 50, and 10% municipal wastewater treatment sites. The DNA microarray analysis revealed that expressions of stress-related genes, hormone receptors, glucose transporter 2, protein expression of glucocorticoid receptor, heat shock proteins 70 and 90, and cytochrome P4501A1 and genes related to immune function were altered (Ings et al. 2011). Similarly, exposure to the textile mill effluent induced up-regulation of oxidative stress-related genes such as *AHP1, ATX1*, GRX1, *TRX1*, and *TRX2* in *Saccharomyces cerevisiae* (Kim et al. 2006).

Interestingly, Hasenbein et al. (2014) suggested that transcriptomic analysis can be utilized not only to investigate the effects of complex contaminant mixtures, but also to identify the contaminant sources. The authors investigated transcriptional responses of larval delta smelt exposed to water samples collected at Department of Water Resources Field Station at Hood in the downstream of the Sacramento Regional Wastewater Treatment Plant (SRWTP), by the DNA microarray. Then, transcriptomic profiles were compared between 9% effluent samples from SRWTP, water from the Sacramento River at Garcia Bend (SRGB)—upstream of the effluent discharge, and SRGB water spiked with 2 mg/L total ammonium (9% effluent equivalent). Results indicated that transcriptomic profiles from Hood are similar to 9% SRWTP effluent and ammonium spiked SRGB water, but significantly different from SRGB.

6.5   Conclusions

In this review, previous researches on mixture toxicity of pharmaceuticals in environmental water and wastewater were discussed in the scopes of human health risk assessment, mixture toxicity of pharmaceuticals to aquatic organisms, and application of transcriptomic-based bioassay (TSB assay) to evaluation of mixture toxicity. Conclusions of this review are summarized below.

1. Our present knowledge on the human health risk of residual pharmaceuticals in environmental water and wastewater effluents (as a potential source of drinking water) were exclusively based on the single exposure of the compound, and the fundamental information on their impacts as a mixture is still very limited.
2. Mixture toxicity of pharmaceuticals to aquatic organisms has been studied over several decades. The toxicity test using the same aquatic organism for the same or similar pharmaceutical mixtures sometimes ended up with contradictory conclusions. The antagonistic or synergistic effects of pharmaceuticals at the environmentally relevant concentrations are often undetectable with conventional endpoints.
3. Transcriptomics-based bioassays (TBS assay) targeted on the gene expression changes, an early biological response, are useful in charactering potential adverse impacts of chemical mixtures and environmental at the sub-lethal concentrations. The TBS assay has a potential to represent the mixture toxicity of pharmaceuticals at the environmentally relevant concentrations.
4. Residues in the treated wastewaters contain not only pharmaceuticals but also various kinds of other micropollutants and dissolved organic matters with natural or anthropogenic origins. Future studies on mixture toxicity of pharmaceuticals are required to consider mare mixture of pharmaceuticals to the matrix effect of such constituents.

References

Affek K, Załęska-Radziwiłł M, Doskocz N, Dębek K (2018) Mixture toxicity of pharmaceuticals present in wastewater to aquatic organisms. Desalin Water Treat 117:15–20. https://doi.org/10.5004/dwt.2018.21964

Aleisa E, Al-Zubari W (2017) Wastewater reuse in the countries of the Gulf Cooperation Council (GCC): the lost opportunity. Environ Monit Assess 189. https://doi.org/10.1007/s10661-017-6269-8

Altenburger R, Ait-Aissa S, Antczak P, Backhaus T, Barceló D, Seiler TB, Brion F, Busch W, Chipman K, de Alda ML, de Aragão Umbuzeiro G, Escher BI, Falciani F, Faust M, Focks A, Hilscherova K, Hollender J, Hollert H, Jäger F, Jahnke A, Kortenkamp A, Krauss M, Lemkine GF, Munthe J, Neumann S, Schymanski EL, Scrimshaw M, Segner H, Slobodnik J, Smedes F, Kughathas S, Teodorovic I, Tindall AJ, Tollefsen KE, Walz KH, Williams TD, Van den Brink PJ, van Gils J, Vrana B, Zhang X, Brack W (2015) Future water quality monitoring—adapting tools to deal with mixtures of pollutants in water resource management. Sci Total Environ 512–513:540–551. https://doi.org/10.1016/j.scitotenv.2014.12.057

Angelakis AN, Asano T, Bahri A, Jimenez BE, Tchobanoglous G (2018) Water reuse: from ancient to modern times and the future. Front Environ Sci 6. https://doi.org/10.3389/fenvs.2018.00026

Artifon V, Zanardi-Lamardo E, Fillmann G (2019) Aquatic organic matter: classification and interaction with organic microcontaminants. Sci Total Environ 649:1620–1635. https://doi.org/10.1016/j.scitotenv.2018.08.385

Azuma T, Otomo K, Kunitou M, Shimizu M, Hosomaru K, Mikata S, Ishida M, Hisamatsu K, Yunoki A, Mino Y, Hayashi T (2019) Environmental fate of pharmaceutical compounds and antimicrobial-resistant bacteria in hospital effluents, and contributions to pollutant loads in the surface waters in Japan. Sci Total Environ 657:476–484. https://doi.org/10.1016/j.scitotenv.2018.11.433

Backhaus T (2014) Medicines, shaken and stirred: a critical review on the ecotoxicology of pharmaceutical mixtures. https://doi.org/10.1098/rstb.2013.0585

Bercu JP, Parke NJ, Fiori JM, Meyerhoff RD (2008) Human health risk assessments for three neuropharmaceutical compounds in surface waters. Regul Toxicol Pharmacol. https://doi.org/10.1016/j.yrtph.2008.01.014

Berninger JP, Martinović-Weigelt D, Garcia-Reyero N, Escalon L, Perkins EJ, Ankley GT, Villeneuve DL (2014) Using transcriptomic tools to evaluate biological effects across effluent gradients at a diverse set of study sites in Minnesota, USA. Environ Sci Technol 48:2404–2412. https://doi.org/10.1021/es4040254

Bertucci A, Pierron F, Gourves PY, Klopp C, Lagarde G, Pereto C, Dufour V, Gonzalez P, Coynel A, Budzinski H, Baudrimont M (2018) Whole-transcriptome response to wastewater treatment plant and stormwater effluents in the Asian clam, Corbicula fluminea. Ecotoxicol Environ Saf 165:96–106. https://doi.org/10.1016/j.ecoenv.2018.08.090

Boltes K, Rosal R, García-Calvo E (2011) Toxicity of mixtures of perfluorooctane sulphonic acid with chlorinated chemicals and lipid regulators. Chemosphere. https://doi.org/10.1016/j.chemosphere.2011.08.041

Bourioug M, Mazzitelli JY, Marty P, Budzinski H, Aleya L, Bonnafé E, Geret F (2018) Assessment of Lemna minor (duckweed) and Corbicula fluminea (freshwater clam) as potential indicators of contaminated aquatic ecosystems: responses to presence of psychoactive drug mixtures. Environ Sci Pollut Res 25:11192–11204. https://doi.org/10.1007/s11356-017-8447-1

Bruneau A, Landry C, Giraudo M, Douville M, Brodeur P, Boily M, Gagnon P, Houde M (2016) Integrated spatial health assessment of yellow perch (Perca flavescens) populations from the St. Lawrence River (QC, Canada), part B: cellular and transcriptomic effects. Environ Sci Pollut Res 23:18211–18221. https://doi.org/10.1007/s11356-016-7001-x

Christensen AM, Faaborg-Andersen S, Ingerslev F, Baun A (2007) Mixture and single-substance toxicity of selective serotonin reuptake inhibitors toward algae and crustaceans. Environ Toxicol Chem 26:85–91. https://doi.org/10.1897/06-219R.1

Christou A, Karaolia P, Hapeshi E, Michael C, Fatta-Kassinos D (2017) Long-term wastewater irrigation of vegetables in real agricultural systems: concentration of pharmaceuticals in soil, uptake and bioaccumulation in tomato fruits and human health risk assessment. Water Res 109:24–34. https://doi.org/10.1016/j.watres.2016.11.033

Cizmas L, Sharma VK, Gray CM, McDonald TJ (2015) Pharmaceuticals and personal care products in waters: occurrence, toxicity, and risk. Environ Chem Lett 13:381–394. https://doi.org/10.1007/s10311-015-0524-4

Cleuvers M (2003) Aquatic ecotoxicity of pharmaceuticals including the assessment of combination effects. Toxicol Lett 142:185–194. https://doi.org/10.1016/S0378-4274(03)00068-7

Cleuvers M (2004) Mixture toxicity of the anti-inflammatory drugs diclofenac, ibuprofen, naproxen, and acetylsalicylic acid. Ecotoxicol Environ Saf 59:309–315. https://doi.org/10.1016/S0147-6513(03)00141-6

Colli-Dula RC, Martyniuk CJ, Kroll KJ, Prucha MS, Kozuch M, Barber DS, Denslow ND (2014) Dietary exposure of 17-alpha ethinylestradiol modulates physiological endpoints and gene signaling pathways in female largemouth bass (Micropterus salmoides). Aquat Toxicol 156:148–160. https://doi.org/10.1016/j.aquatox.2014.08.008

Coors A, Vollmar P, Sacher F, Polleichtner C, Hassold E, Gildemeister D, Kühnen U (2018) Prospective environmental risk assessment of mixtures in wastewater treatment plant effluents—theoretical considerations and experimental verification. Water Res. https://doi.org/10.1016/j.watres.2018.04.031

Cunningham VL, Binks SP, Olson MJ (2009) Human health risk assessment from the presence of human pharmaceuticals in the aquatic environment. Regul Toxicol Pharmacol. https://doi.org/10.1016/j.yrtph.2008.10.006

Debroux JF, Soller JA, Plumlee MH, Kennedy LJ (2012) Human health risk assessment of non-regulated xenobiotics in recycled water: a review. Hum Ecol Risk Assess 18:517–546. https://doi.org/10.1080/10807039.2012.672883

Deo RP (2014) Pharmaceuticals in the surface water of the USA: a review. Curr Environ Health Rep 1:113–122. https://doi.org/10.1007/s40572-014-0015-y

Dietrich S, Ploessl F, Bracher F, Laforsch C (2010) Single and combined toxicity of pharmaceuticals at environmentally relevant concentrations in Daphnia magna—a multigenerational study. Chemosphere 79:60–66. https://doi.org/10.1016/j.chemosphere.2009.12.069

Elersek T, Ženko M, Filipič M (2018) Ecotoxicity of disinfectant benzalkonium chloride and its mixture with antineoplastic drug 5-fluorouracil towards alga *Pseudokirchneriella subcapitata*. PeerJ. https://doi.org/10.7717/peerj.4986

Escher BI, Lawrence M, MacOva M, Mueller JF, Poussade Y, Robillot C, Roux A, Gernjak W (2011) Evaluation of contaminant removal of reverse osmosis and advanced oxidation in full-scale operation by combining passive sampling with chemical analysis and bioanalytical tools. Environ Sci Technol 45:5387–5394. https://doi.org/10.1021/es201153k

Fernández C, Carbonell G, Babín M (2013) Effects of individual and a mixture of pharmaceuticals and personal-care products on cytotoxicity, EROD activity and ROS production in a rainbow trout gonadal cell line (RTG-2). J Appl Toxicol 33:1203–1212. https://doi.org/10.1002/jat.2752

Franklin AM, Williams CF, Andrews DM, Woodward EE, Watson JE (2016) Uptake of three antibiotics and an antiepileptic drug by wheat crops spray irrigated with wastewater treatment plant effluent. J Environ Qual 45:546–554. https://doi.org/10.2134/jeq2015.05.0257

Fukushima T, Hara-Yamamura H, Urai M, Kasuga I, Kurisu F, Miyoshi T, Kimura K, Watanabe Y, Okabe S (2014) Toxicity assessment of chlorinated wastewater effluents by using transcriptome-based bioassays and Fourier transform mass spectrometry (FT-MS) analysis. Water Res 52:73–82. https://doi.org/10.1016/j.watres.2014.01.006

Fukushima T, Hara-Yamamura H, Nakashima K, Tan LC, Okabe S (2017) Multiple-endpoints gene alteration-based (MEGA) assay: a toxicogenomics approach for water quality assessment of wastewater effluents. Chemosphere 188:312–319. https://doi.org/10.1016/j.chemosphere.2017.08.107

Galus M, Jeyaranjaan J, Smith E, Li H, Metcalfe C, Wilson JY (2013) Chronic effects of exposure to a pharmaceutical mixture and municipal wastewater in zebrafish. Aquat Toxicol 132–133:212–222. https://doi.org/10.1016/j.aquatox.2012.12.016

Geiger E, Hornek-Gausterer R, Saçan MT (2016) Single and mixture toxicity of pharmaceuticals and chlorophenols to freshwater algae *Chlorella vulgaris*. Ecotoxicol Environ Saf 129:189–198. https://doi.org/10.1016/j.ecoenv.2016.03.032

Godoy AA, de Oliveira ÁC, Silva JGM, Azevedo CCDJ, Domingues I, Nogueira AJA, Kummrow F (2019) Single and mixture toxicity of four pharmaceuticals of environmental concern to aquatic organisms, including a behavioral assessment. Chemosphere 235:373–382. https://doi.org/10.1016/j.chemosphere.2019.06.200

Gogoi A, Mazumder P, Tyagi VK, Tushara Chaminda GG, An AK, Kumar M (2018) Occurrence and fate of emerging contaminants in water environment: a review. Groundw Sustain Dev. https://doi.org/10.1016/j.gsd.2017.12.009

González García M, Fernández-López C, Polesel F, Trapp S (2019) Predicting the uptake of emerging organic contaminants in vegetables irrigated with treated wastewater—implications for food safety assessment. Environ Res 172:175–181. https://doi.org/10.1016/j.envres.2019.02.011

González-Pleiter M, Gonzalo S, Rodea-Palomares I, Leganés F, Rosal R, Boltes K, Marco E, Fernández-Piñas F (2013) Toxicity of five antibiotics and their mixtures towards photosynthetic aquatic organisms: Implications for environmental risk assessment. Water Res 47:2050–2064. https://doi.org/10.1016/j.watres.2013.01.020

Guo J, Selby K, Boxall ABA (2016) Assessment of the risks of mixtures of major use veterinary antibiotics in european surface waters. Environ Sci Technol. https://doi.org/10.1021/acs.est.6b01649

Han GH, Hur HG, Kim SD (2006) Ecotoxicological risk of pharmaceuticals from wastewater treatment plants in Korea: occurrence and toxicity to *Daphnia magna*. Environ Toxicol Chem 25:265–271. https://doi.org/10.1897/05-193R.1

Hara-Yamamura H, Nakashima K, Hoque A, Miyoshi T, Kimura K, Watanabe Y, Okabe S (2013) Evaluation of whole wastewater effluent impacts on HepG2 using DNA microarray-based transcriptome analysis. Environ Sci Technol 47:5425–5432. https://doi.org/10.1021/es4002955

Hara-Yamamura H, Fukushima T, Tan LC, Okabe S (2020) Transcriptomic analysis of HepG2 cells exposed to fractionated wastewater effluents suggested humic substances as potential inducer of whole effluent toxicity. Chemosphere 240:124894. https://doi.org/10.1016/j.chemosphere.2019.124894

Hasenbein M, Werner I, Deanovic LA, Geist J, Fritsch EB, Javidmehr A, Foe C, Fangue NA, Connon RE (2014) Transcriptomic profiling permits the identification of pollutant sources and effects in ambient water samples. Sci Total Environ 468–469:688–698. https://doi.org/10.1016/j.scitotenv.2013.08.081

Heberer T (2002) Occurrence, fate, and removal of pharmaceutical residues in the aquatic environment: a review of recent research data. Thomas Toxicol Lett 131:5–17

Houde M, Douville M, Despatie SP, De Silva AO, Spencer C (2013) Induction of gene responses in St. Lawrence River northern pike (*Esox lucius*) environmentally exposed to perfluorinated compounds. Chemosphere 92:1195–1200. https://doi.org/10.1016/j.chemosphere.2013.01.099

Houde M, Giraudo M, Douville M, Bougas B, Couture P, De Silva AO, Spencer C, Lair S, Verreault J, Bernatchez L, Gagnon C (2014) A multi-level biological approach to evaluate impacts of a major municipal effluent in wild St. Lawrence River yellow perch (*Perca flavescens*). Sci Total Environ 497–498:307–318. https://doi.org/10.1016/j.scitotenv.2014.07.059

Ings JS, Servos MR, Vijayan MM (2011) Hepatic transcriptomics and protein expression in rainbow trout exposed to municipal wastewater effluent. Environ Sci Technol 45:2368–2376. https://doi.org/10.1021/es103122g

Jackman KW, Veldhoen N, Miliano RC, Robert BJ, Li L, Khojasteh A, Zheng X, Zaborniak TSM, van Aggelen G, Lesperance M, Parker WJ, Hall ER, Pyle GG, Helbing CC (2018) Transcriptomics investigation of thyroid hormone disruption in the olfactory system of the *Rana [Lithobates] catesbeiana* tadpole. Aquat Toxicol 202:46–56. https://doi.org/10.1016/j.aquatox.2018.06.015

Jones OAH, Voulvoulis N, Lester JN (2004) Potential ecological and human health risks associated with the presence of pharmaceutically active compounds in the aquatic environment. Crit Rev Toxicol 34:335–350. https://doi.org/10.1080/10408440490464697

Kim HJ, Rakwal R, Shibato J, Iwahashi H, Choi JS, Kim DH (2006) Effect of textile wastewaters on Saccharomyces cerevisiae using DNA microarray as a tool for genome-wide transcriptomics analysis. Water Res 40:1773–1782. https://doi.org/10.1016/j.watres.2006.02.037

Kimura K, Hara H, Watanabe Y (2005) Removal of pharmaceutical compounds by submerged membrane bioreactors (MBRs). Desalination 178:135–140. https://doi.org/10.1016/j.desal.2004.11.033

Kumar M, Chaminda T, Honda R, Furumai H (2019a) Vulnerability of urban waters to emerging contaminants in India and Sri Lanka: resilience framework and strategy. APN Sci Bull 9(1). https://doi.org/10.30852/sb.2019.799

Kumar M, Ram B, Honda R, Poopipattana C, Canh VD, Chaminda T, Furumai H (2019b) Concurrence of antibiotic resistant bacteria (ARB), viruses, pharmaceuticals and personal care products (PPCPs) in Ambient Waters of Guwahati, India: urban vulnerability and resilience perspective. Sci Total Environ 693:133640. https://doi.org/10.1016/j.scitotenv.2019.133640

Li S-W, Yu-Chen Lin A (2015) Increased acute toxicity to fish caused by pharmaceuticals in hospital effluents in a pharmaceutical mixture and after solar irradiation. Chemosphere. https://doi.org/10.1016/j.chemosphere.2015.06.010

Luo F, Gitiafroz R, Devine CE, Gong Y, Hug LA, Raskin L, Edwards EA (2014) Metatranscriptome of an anaerobic benzene-degrading, nitrate-reducing enrichment culture reveals involvement of carboxylation in benzene ring activation. Appl Environ Microbiol 80:4095–4107. https://doi.org/10.1128/AEM.00717-14

Malchi T, Maor Y, Tadmor G, Shenker M, Chefetz B (2014) Irrigation of root vegetables with treated wastewater: evaluating uptake of pharmaceuticals and the associated human health risks. Environ Sci Technol 48:9325–9333. https://doi.org/10.1021/es5017894

Martinović-Weigelt D, Mehinto AC, Ankley GT, Denslow ND, Barber LB, Lee KE, King RJ, Schoenfuss HL, Schroeder AL, Villeneuve DL (2014) Transcriptomic effects-based monitoring for endocrine active chemicals: assessing relative contribution of treated wastewater to downstream pollution. Environ Sci Technol 48:2385–2394. https://doi.org/10.1021/es404027n

Murray KE, Thomas SM, Bodour AA (2010) Prioritizing research for trace pollutants and emerging contaminants in the freshwater environment. Environ Pollut 158:3462–3471. https://doi.org/10.1016/j.envpol.2010.08.009

Prokkola JM, Katsiadaki I, Sebire M, Elphinstone-Davis J, Pausio S, Nikinmaa M, Leder EH (2016) Microarray analysis of di-n-butyl phthalate and 17α ethinyl-oestradiol responses in three-spined stickleback testes reveals novel candidate genes for endocrine disruption. Ecotoxicol Environ Saf 124:96–104. https://doi.org/10.1016/j.ecoenv.2015.09.039

Prosser RS, Sibley PK (2015) Human health risk assessment of pharmaceuticals and personal care products in plant tissue due to biosolids and manure amendments, and wastewater irrigation. Environ Int 75:223–233. https://doi.org/10.1016/j.envint.2014.11.020

PUB (2018) NEWater [WWW Document]. https://www.pub.gov.sg/watersupply/fournationaltaps/newater. Accessed 9.18.18

Quinn B, Gagné F, Blaise C (2008) Evaluation of the acute, chronic and teratogenic effects of a mixture of eleven pharmaceuticals on the cnidarian, Hydra attenuata. Sci Total Environ 407:1072–1079. https://doi.org/10.1016/j.scitotenv.2008.10.022

Reungoat J, Macova M, Escher BI, Carswell S, Mueller JF, Keller J (2010) Removal of micropollutants and reduction of biological activity in a full scale reclamation plant using ozonation and activated carbon filtration. Water Res 44:625–637. https://doi.org/10.1016/j.watres.2009.09.048

Riemenschneider C, Al-Raggad M, Moeder M, Seiwert B, Salameh E, Reemtsma T (2016) Pharmaceuticals, their metabolites, and other polar pollutants in field-grown vegetables irrigated with treated municipal wastewater. J Agric Food Chem 64:5784–5792. https://doi.org/10.1021/acs.jafc.6b01696

Rodea-Palomares I, Petre AL, Boltes K, Leganés F, Perdigón-Melón A, Rosal R, Ferná Ndez-Piñ As F (2009) Application of the combination index (CI)-isobologram equation to study the toxicological interactions of lipid regulators in two aquatic bioluminescent organisms. Water Res 44:427–438. https://doi.org/10.1016/j.watres.2009.07.026

Rosal R, Rodea-Palomares I, Boltes K, Fernández-Piñas F, Leganés F, Gonzalo S, Petre A (2010) Ecotoxicity assessment of lipid regulators in water and biologically treated wastewater using three aquatic organisms. Environ Sci Pollut Res 17:135–144. https://doi.org/10.1007/s11356-009-0137-1

Schwab BW, Hayes EP, Fiori JM, Mastrocco FJ, Roden NM, Cragin D, Meyerhoff RD, D'Aco VJ, Anderson PD (2005) Human pharmaceuticals in US surface waters: a human health risk assessment. Regul Toxicol Pharmacol. https://doi.org/10.1016/j.yrtph.2005.05.005

Semerjian L, Shanableh A, Semreen MH, Samarai M (2018) Human health risk assessment of pharmaceuticals in treated wastewater reused for non-potable applications in Sharjah, United Arab Emirates. Environ Int 121:325–331. https://doi.org/10.1016/j.envint.2018.08.048

Stumpf M, Ternes TA, Wilken RD, Silvana Vianna Rodrigues, Baumann W (1999) Polar drug residues in sewage and natural waters in the state of Rio de Janeiro, Brazil. Sci Total Environ 225:135–141. https://doi.org/10.1016/S0048-9697(98)00339-8

Ternes TA (1998) Occurrence of drugs in German sewage treatment plants and rivers. Water Res 32:3245–3260. https://doi.org/10.1016/S0043-1354(98)00099-2

Thrupp TJ, Runnalls TJ, Scholze M, Kugathas S, Kortenkamp A, Sumpter JP (2018) The consequences of exposure to mixtures of chemicals: Something from 'nothing' and 'a lot from a little' when fish are exposed to steroid hormones. Sci Total Environ 619–620:1482–1492. https://doi.org/10.1016/j.scitotenv.2017.11.081

USEPA (2018) Potable Reuse Compendium

Wang Z, Li J, Li Y (2017) Using reclaimed water for agricultural and landscape irrigation in China: a review. Irrig Drain 66:672–686. https://doi.org/10.1002/ird.2129

Wang P, Xia P, Yang J, Wang Z, Peng Y, Shi W, Villeneuve DL, Yu H, Zhang X (2018) A reduced transcriptome approach to assess environmental toxicants using zebrafish embryo test. Environ Sci Technol 52:821–830. https://doi.org/10.1021/acs.est.7b04073

Wu L, Chen W, French C, Chang A (2009) Safe application of reclaimed water reuse in the Southwestern United States. Safe Appl. Reclaimed Water Reuse Southwest, United States. https://doi.org/10.3733/ucanr.8357

Wu X, Conkle JL, Ernst F, Gan J (2014) Treated wastewater irrigation: uptake of pharmaceutical and personal care products by common vegetables under field conditions. Environ Sci Technol 48:11286–11293. https://doi.org/10.1021/es502868k

Wu X, Dodgen LK, Conkle JL, Gan J (2015) Plant uptake of pharmaceutical and personal care products from recycled water and biosolids: a review. Sci Total Environ 536:655–666. https://doi.org/10.1016/j.scitotenv.2015.07.129

Xiong JQ, Kim SJ, Kurade MB, Govindwar S, Abou-Shanab RAI, Kim JR, Roh HS, Khan MA, Jeon BH (2019) Combined effects of sulfamethazine and sulfamethoxazole on a freshwater microalga, *Scenedesmus obliquus*: toxicity, biodegradation, and metabolic fate. J Hazard Mater 370:138–146. https://doi.org/10.1016/j.jhazmat.2018.07.049

Zhang Y, Deng Y, Zhao Y, Ren H (2014) Using combined bio-omics methods to evaluate the complicated toxic effects of mixed chemical wastewater and its treated effluent. J Hazard Mater 272:52–58. https://doi.org/10.1016/j.jhazmat.2014.02.041

Zhang H, Ihara M, Hanamoto S, Nakada N, Jürgens MD, Johnson AC, Tanaka H (2018) Quantification of pharmaceutical related biological activity in effluents from wastewater treatment plants in UK and Japan. Environ Sci Technol 52:11848–11856. https://doi.org/10.1021/acs.est.8b03013

Chapter 7
Microplastic Vulnerability in the Sediments of the Sabarmati River of India

Arbind Kumar Patel, Chandrasekhar Bhagat, Kaling Taki, and Manish Kumar

7.1 Introduction

Demands of plastics are increasing constantly and hence its production also; since the 1950s, there is an increase of 37% over the last decades (Plastics Europe 2017). Recently, microplastics are a great concern for study because the use of plastics has increased in different sectors like in transportation, telecommunications, clothing, footwear, etc. The production of plastics has increased to 260 Mt/year in the twenty-first century (Plastic Europe 2013). Most plastics have high persistence in the environment, hence they remain in initial form in the environment and the degradation rate is too slow (Hopewell et al. 2009). Because of their chemical and physical properties, plastic particles are able to contaminate the soil as well as the water on a global scale. National Oceanic and Atmospheric Administration (US NOAA) defines the microplastic as the plastic particle whose size is less than 5 mm is considered as a microplastics (McCormick et al. 2014; Thompson et al. 2004; Arthur et al. 2009; Kumar et al. 2009). Many studies have proven that the abundance of microplastic is increasing in the sea as well as river sediments (Besley et al. 2017; Lazure and Desmare 2012; Van Cauwenberghe et al. 2015, Mukherjee et al 2020). Microplastics are differentiated into two types: primary microplastics and secondary microplastics. The plastics whose size is lesser than 5 mm initially are known as the primary microplastics, whereas the plastics which are having a size greater than 5 mm and are

A. K. Patel · M. Kumar (✉)
Discipline of Earth Sciences, Indian Institute of Technology Gandhinagar, Gandhinagar, Gujarat 382355, India
e-mail: manish.kumar@iitgn.ac.in

C. Bhagat · K. Taki
Department of Civil Engineering, Indian Institute of Technology Gandhinagar, Gandhinagar, Gujarat 382355, India

© Springer Nature Singapore Pte Ltd. 2020
M. Kumar et al. (eds.), *Resilience, Response, and Risk in Water Systems*,
Springer Transactions in Civil and Environmental Engineering,
https://doi.org/10.1007/978-981-15-4668-6_7

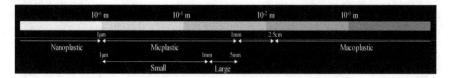

Fig. 7.1 Nomenclature of plastic debris based on size (MSFD GES Technical Subgroup on Marine Litter, 2013)

disintegrated from large debris which is known as secondary microplastics. They also differentiated into nanoplastics, microplastics, and microplastics and their sizes are shown in Fig. 7.1. Microplastic attracts more in recent decades and popularly known as an emerging pollutant. First time microplastic is recorded in the 1970s (Carpenter et al. 1972). As per the published literature microplastics has spread all over the world in different environmental matrices. The dominant type of microplastics in the river water and sediments is pallets, fibers, fragments, polystyrene, and polyethylene terephthalate (Hidalgo-Ruz et al. 2012; Imhof et al. 2012; Ng and Obbard 2006).

The immense interests about microplastics (MP) in the environment are associated with poisonous chemicals and consequent acquaintance of these toxic chemicals to the several types of micro-organisms that consume the plastics debris (Bakir et al. 2014; Bejgarn et al. 2015; Browne et al. 2011). MP debris is proficient in concentrating hydrophobic organic contaminants such as PAHs, PCBs, and DDTs (Gauquie et al. 2015; Hirai et al. 2011), rising their concentration up to the order of 106 (Mato et al. 2001). MP debris could also accrue metals from the surrounding and the accumulation has been revealed by lab experiments (Holmes et al. 2014) and environmental monitoring (Rochman et al. 2014). Metal contamination is increasing and common in the various environment matrixes and is resulting from several sources like the industrial and domestic waste discharges, mining, smelting, and e-wastes (Wang et al. 2013). Though, the data on the content of metals accumulation by MP in the actual environment system is very limited.

Accordingly, more and more research efforts should be contributed to inspect the occurrence and characteristics, especially the interface with heavy metals, of MP in the aqueous environment so as to additional measure the potential environmental risks. Furthermore, coastal sediment samples could reveal the result of long lasting interfacial contact between waters and land surface (Yu et al. 2016), and thus deliver crucial information on the transportation and fate of contaminants and alike types of MPs in the water column as in sedimentary habitations have been formerly shown (Thompson et al. 2004), Advising that density (mass/volume ratio) is not an influential factor impelling the distribution of MPs and sediment samples are good demonstration for the long lasting accruing the result of MPs. In this work, thus, MPs were recovered in the surface sediment samples from the Sabarmati River Ahmedabad, Gujarat India

Surface water is used for drinking purposes all over the world but recent studies show that contaminant concentration in the water has increased (Singh et al. 2020). The sources of microplastic in river water are improper waste disposal, insufficient waste management, and urban runoffs (Barnes et al. 2009), and microplastics in washing machine effluent from synthetic textiles. These are not removed by treatment plants due to small size and buoyancy (Browne et al. 2011). There are chances to detect the traces of microplastics in the Sabarmati river sediments because of improper waste dumping practices in the city. In this work, thus, MPs were recovered in the surface sediment samples from the Sabarmati River Ahmedabad, Gujarat India and the main objective is to investigate the abundance microplastics in the sediments along the Sabarmati River and also finding the relation with sediment texture/grain size, organic matter.

7.2   Materials and Methods

Study Area

The study was conducted, on the Sabarmati River, Gujarat, India, in the month of October 2017. The Sabarmati River flowing through the Ahmedabad city and carries huge amount of liquid as well as solid waste and hence this river is know for the one of the largest waste flowing rivers in the state of Gujarat India. The Ahmadabad city is one of the major city in the state of Gujarat and well known for business activity in the country. River Sabaramati originates in the Aravalli Range of the Udaipur District of Rajasthan and meets the Gulf of Khambhat of the Arabian Sea after traveling 371 km in a south-westerly direction across Rajasthan and Gujarat. Our study stretch is near to Ahmedabad city in the state of Gujrat, India. Ahmedabad is one of the major cities in the Gujarat state, mainly popular for trade, business, and factories of plastics, metals, pharmaceuticals, etc. This city also is known for the diamond business. Four sites are selected along the stretch of river as shown in Fig. 7.2 (A_2, A_6, A_7, and A_8). The river is flowing from Gandhinagar to Ahmedabad as shown in Fig. 7.2.

7.3   Methodology

River (Sabarmati river) sediment samples (n = 4, A2, A6, A7, and A8) collected and were analyzed for microplastic using the methodology adopted by Nel et al. (2018) and Klein et al. (2015). Grain size distribution is carried out by dry sieving, and removal of organic matter carried out by the ignition test, so that sample is free from the organic matter. To quantify the abundance of microplastics, the river sediments were sampled at four (n = 4) different locations at upstream and downstream of the Ahmadabad city as shown in Fig. 7.2. At every location, 2.5 kg sediments were collected using a stainless-steel scoop from the upper 5 cm layer. Then the sediments were dried in the open air to reduce the moisture for 24 hrs and then oven-dried at 50 °C for 48–72 h, until it will reach to constant weight, and sorting of sediment carried out according to size (75 μm–212 μm and 212 μm–4 mm). To minimize the

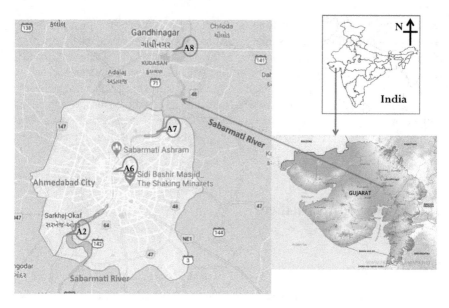

Fig. 7.2 Map showing sampling locations of Sabarmati River, Gujarat, India

risk of sediments sample contamination for example clothing fibers during on-board recovery, samples were held against the wind to avoid any airborne contamination. In the lab, sediment samples are first passed through the series of metal sieves (5 mm, 2 mm, and 212 μm) and hand sieving is done, with M-Q water. Pieces of biological organic material sized >5 mm were rinsed and discarded to minimize the error in estimation. All fractions were kept at 5 °C in the fridge to avoid the temperature influence. A 1 kg of the sieved sediment was shifted into a glass beaker by rinsing with local M-Q water and subsequently stored in 10% NaCl in glass jars for conservation. The mixture of the slurry was stirred exuberantly to dis-aggregate and suspend plastic particles for 1 h completely and permitted them to settle for 24 h (one day). The supernatant mixture was filtered through Whatman GF/A (Reddy et al. 2006) and filters were dried at room temperature and sealed in Petri dishes. Microplastics upon the mesh were carefully rinsed with distilled water into clean Petri dishes. The samples were then visually sorted at ×50 magnifications, whereby all possible microplastic particles were enumerated and The SEM (scanning electron microscope) analysis carried out for all sediments samples. Microplastic particles were identified by possessing unnatural coloration and/or unnatural (Hidalgo-Ruz et al. 2012).

Filtration, as shown in Fig. 7.4 of supernatant, is carried with help of 0.45 μm filter so that microplastic get to catch on the surface of the filter (we take it granted that microplastic is always greater than 0.45 μm).

Weight of wet filter paper (W_1) is taken and after that, filter gets oven-dry, and the weight of filter paper is a note (W_2). The quantity of microplastic is found out with the help of a simple formula given below.

Methodology for extraction of Microplastics (Klien et.al 2015)

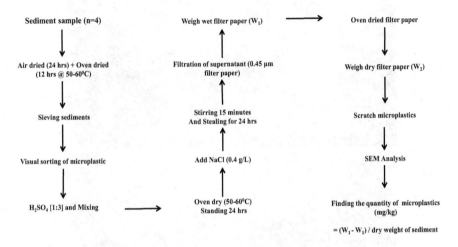

Fig. 7.3 Flowchart to quantify the microplastics in sediments samples

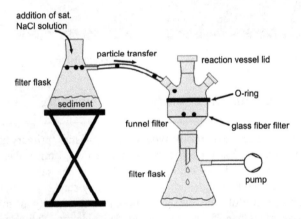

Fig. 7.4 Schematic of vacuum-enhanced density separation; floating plastic debris is transferred by the addition of saturated sodium chloride solution by means of a vacuum directly to the glass fiber filter. Diagram is taken from Klein et al. (2015)

$$\text{Quantity of microplastic(mg/kg)} = \frac{(W_1) - (W_2)\text{mg}}{\text{weight of sediment sample(kg)}}$$

SEM analysis is carried out to know the surface roughness and to observe the microstructure of microplastics. Morphology of different sediment as well as microplastic particles is as shown in Fig. 7.5.

Grain size distribution by dry sieving and organic matter content by loss of ignition test according to Konare et al. (2010).

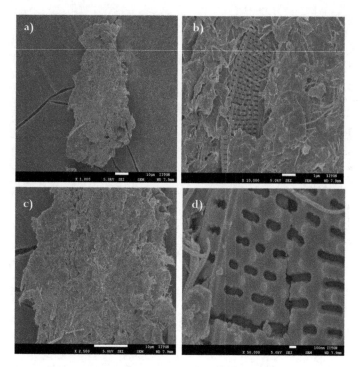

Fig. 7.5 SEM analysis to comprehend morphology of different microplastics

7.3.1   Sediment Grain Size Analysis of Sabarmati

Sieve analysis is performed to know the types of microplastics (primary microplastics less than 5 mm and secondary microplastics greater than 5 mm). Also, grain size analysis was performed to know the types of sediment. The samples of Sabarmati river sediments were containing sandy gravel, gravelly sand, and slightly gravelly sand. This analysis is useful to decide the methodology to quantify the microplastics. If sediment's size is less than 75 μm, it is classified into clay soil, which carries charge on its surface and hence this influences the transport of microplastic in the river. Cations derived from mineral weathering and pollution sources are preferentially adsorbed onto clay (negatively charged surface), which has the highest surface area-to-volume ratio of any particle size class. This suggests that since there is least clay fraction, therefore, the probability of finding the pollutants in labile forms is maximum, thus, posing a higher risk of exposure. Figure 7.6 shows that different soil composition is in the sediment sample, which helps to identify the type of soil and composition of the soil. From Fig. 7.6, it is inferred that the maximum sediment sample is classified under the sandy gravel soil means mixed soil is available throughout the stretch of Sabarmati River throughout the Ahmedabad city.

Figure 7.7a sample A2 showing the gap-graded or poorly graded soil and Fig. 7.7b classified in well-graded soil as the range of soil particle is good and hence the

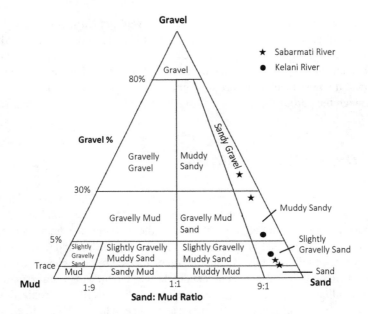

Fig. 7.6 Composition of the riverbed sediment based on their grain size

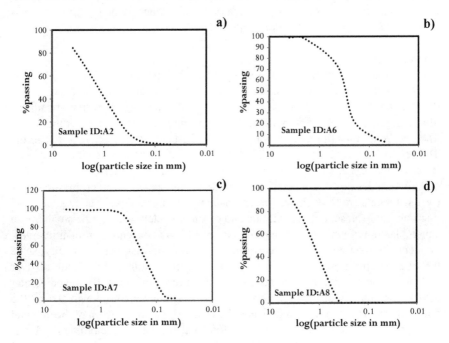

Fig. 7.7 Graduation curve for four different samples (**a–d**)

porosity of sediment sample is less, Fig. 7.7c also well-graded sediment and having less permeability and less porosity, Fig. 7.7d shows the poorly graded sediment and having high porosity and high permeability means having potential to transport the pollutants (microplastics) along with the soil particles. This analysis helps to understand the pollutants transport phenomenon in the surface water as well as the flow through the porous media.

7.3.2 Organic Matter in Rivers

Organic matter (Humus) has the ability to interact with oxides, hydroxides, mineral, and organic compounds, including toxic pollutants, to form water-soluble and water-insoluble complexes. Through the formation of these complexes, humic substances can dissolve, mobilize and transport metals and organics in soils and waters, or accumulate in certain soil horizons. Accumulation of such complexes can contribute to a reduction in toxicity. The organic matter content is found little in the Sabarmati River and it also influences the transport mechanism of pollutants like microplastics. The organic matters carry charges on the surface and hence, it has the ability to attract the pollutant which carries charges and hence, represence of organic matter enhances the contaminant transport process and increases the resistance to the treatment process as they have to make the ionic bond with the pollutants.

Microplastics
The maximum concentration of microplastic (212 μm–4 mm) is found to be 581.70 mg/kg at A2 location at the downstream of the city and 134.53 mg/kg at the location A6 (75–212 μm) as shown in Fig. 7.8. Table 7.1 show that abundance of microplastics in the different River sediments in the world and also for this study. The maximum quantity of microplastics (660 Particles. kg^{-1}) are reported in the Thames River, UK (Horton et al. 2017). Sarkar et al. 2019 conducted the study on River Ganga, eastern India and reported that abundance of microplastic is 99.27–409.86 particles. kg^{-1} and found that the dominant species of microplastic in the river sediments are Polyethylene terephthalate and polyethylene. The larger size of microplastic is present at the extreme end of study stretch whereas the smaller size smicroplastic is found just behind the last point. The concentration of microplastic along the river bed is increased from upstream to downstream of the city; since the river passing from the city, it accumulate all the residue on the bank of river and also the waste discharge in the river without proper treatment (Ram and Kumar 2020). The quantity of the major size of microplastic is found to be greater on the downstream as they may get a float and deposited at the downstream of the city and create pollution in the river. The treated effluent also enhances the problems of pollution just because there is no sufficient efficiency of the treatment plant to remove microplastics from wastewater. That is why the wastewater treatment plants are the important point source of emerging pollutants, i.e. microplastics.

Table 7.1 Microplastics abundance reported in the different river sediment of the world

Types of microplastic	Size	Quantity	Location	References
Polyethylene, polypropylene, polystyrene	<5 mm	1 gm. Kg^{-1} or 4000 particles. Kg^{-1}	Rhine River Germeny	Klein et al. 2015
Polyethylene terephthalate and polyethylene	mesoplastics (<5 mm) and microplastics (>5 mm)	11.48 to 63.79 ng/g or 99.27–409.86 particles. kg^{-1}	River Ganga, Eastern India	Sarkar et al. 2019
Sheet shaped and fibers	1–4 mm	660 Particles. kg^{-1}	Thames River, UK	Horton et al. 2017
Fiber and plastic debris	2.8 mm–11 μm	296.5 particles. L^{-1}	Kelvin River, UK	Blair et al. 2019
Fiber and plastic debris	2 mm–5 mm	550 particles. L^{-1}	Canadian lakes and rivers	Anderson et al. 2016
polyethylene (PE), Polypropylene (PP), copolymer, and paint particle	0–5mm	178–554 particles. kg^{-1}	Beijiang River China	Wang et al. 2017
Plastic debris and fiber	4 mm–75 μm	134.53 mg. kg^{-1} to 581.70 mg. kg^{-1}	Sabarmati River, Gujarat India	This study

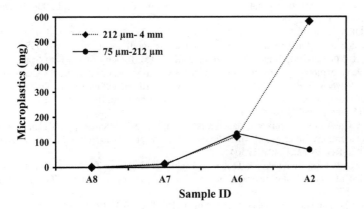

Fig. 7.8 Scattered plot showing the microplastic concentration in different sampling location

7.4 Conclusion

The gravimetric analysis to quantify microplastics is an easy and simple technique. The soil gradation results infer that sediment grain size distribution of Sabarmati are sandy gravel, gravelly sand, and slightly gravelly sand suggesting least clay fraction and therefore the probability of finding the pollutants in labile forms is maximum thus posing a higher risk of exposure. Concentration of microplastic is found to be having increasing trend from upstream to downstream of the city. The SEM analysis is confirmed that the microplastic is also present in the sediments, and hence it is difficult to be removed and monitored. The sources of microplastics are non-point source and hence difficult to manage and the important- and dominant-point sources are waste disposal site (landfill) and wastewater treatment plants. This study provided the status of microplastics contamination in the Sabarmati River and the result proves that the disposal of pollutants is going on the river directly in many ways. Microplastic concentration of both the sizes (75–212 μm and 212 μm–4 mm) is found higher in downstream sites.

References

Anderson JC, Park BJ, Palace VP (2016) Microplastics in aquatic environments: implications for Canadian ecosystems. Environ Pollut 218:269–280

Arthur C, Baker JE, Bamford HA (2009) Proceedings of the international research workshop on the occurrence, effects, and fate of microplastic marine debris, 9–11 Sept 2008. University of Washington Tacoma, Tacoma

Barnes DKA, Galgani F, Thompson RC, Barlaz M (2009) Accumulation and fragmentation of plastic debris in global environments. Philos Trans R Soc B 364(1526):1985–1998

Bejgarn S, MacLeod M, Bogdal C, Breitholtz M (2015) Toxicity of leachate from weathering plastics: an exploratory screening study with Nitocra spinipes. Chemosphere 132:114–119

Besley A, Vijver MG, Behrens P, Bosker T (2017) A standardized method for sampling and extraction methods for quantifying microplastics in beach sand. Mar Pollut Bull 114:77–83. https://doi.org/10.1016/j.marpolbul.2016.08.055

Blair RM, Waldron S, Phoenix VR, Gauchotte-Lindsay C (2019) Microscopy and elemental analysis characterization of microplastics in sediment of a freshwater urban river in Scotland, UK. Environ Sci Pollut Res 26(12):12491–12504

Browne MA, Crump P, Niven SJ, Teuten E, Tonkin A, Galloway T, Thompson R (2011) Accumulation of microplastic on shorelines worldwide: sources and sinks. Environ Sci Technol 45(21):9175–9179

Carpenter EJ, Anderson SJ, Harvey GR, Miklas HP, Peck BB (1972) Polystyrene spherules in coastal waters. Science 178:749–750

Gauquie J, Devriese L, Robbens J, De Witte B (2015) A qualitative screening and quantitative measurement of organic contaminants on different types of marine plastic debris. Chemosphere 138:348–356

Hidalgo-Ruz V, Gutow L, Thompson RC, Thiel M (2012) Microplastics in the marine environment: a review of the methods used for identification and quantification. Environ Sci Technol 46(6):3060–3075

Hirai H, Takada H, Ogata Y, Yamashita R, Mizukawa K, Saha M, Kwan C, Moore C, Gray H, Laursen D, Zettler ER (2011) Organic micropollutants in marine plastics debris from the open ocean and remote and urban beaches. Mar Pollut Bull 62(8):1683–1692

Holmes LA, Turner A, Thompson RC (2014) Interactions between trace metals and plastic production pellets under estuarine conditions. Mar Chem 167:25–32

Hopewell J, Dvorak R, Kosior E (2009) Plastics recycling: challenges and opportunities. Philos Trans R Soc Lond Ser B Biol Sci 364:2115–2126

Horton AA, Walton A, Spurgeon DJ, Lahive E, Svendsen C (2017) Microplastics in freshwater and terrestrial environments: evaluating the current understanding to identify the knowledge gaps and future research priorities. Sci Total Environ 586:127–141

Imhof HK, Schmid J, Niessner R, Ivleva NP, Laforsch C (2012) A novel, highly efficient method for the separation and quantification of plastic particles in sediments of aquatic environments. Limnol Oceanogr-Meth 10:524–537

Klein S, Worch E, Knepper TP (2015) Occurrence and spatial distribution of microplastics in river shore sediments of the Rhine-Main area in Germany—environmental science & …. ACS Publications

Konare H, Yost RS, Doumbia M, McCarty GW, Jarju A, Kablan R (2010) Loss on ignition: measuring soil organic carbon in soils of the Sahel, West Africa. Afr J Agric Res 5:3088–3095

Kumar M, Furumai H, Kurisu F, Kasuga I (2009) Understanding the partitioning processes of mobile lead in soakaway sediments using sequential extraction and isotope analysis. Water Sci Technol 60(8):2085–2091

Lazure P, Desmare S (2012) Caractéristiques et état écologiqueManche - Mer du Nord, Etat physique et chimique, caractéristiques physiques, Courantologie https://www.ifremer.fr/sextant_doc/dcsmm/documents/Evaluation_initiale/MMN/EE/MMN_EE_06_Courantologie.pdf. Accessed 01/15/2019

Mato Y, Isobe T, Takada H, Kanehiro H, Ohtake C, Kaminuma T (2001) Plastic resin pellets as a transport medium for toxic chemicals in the marine environment. Environ Sci Technol 35(2):318–324

McCormick A, Hoellein TJ, Mason SA (2014) Microplastic is an abundant and distinct microbial habitat in an urban river science & technology. ACS Publications

Mukherjee S, Patel AKR, Kumar M (2020) Water scarcity and land degradation nexus in the era of anthropocene: some reformations to encounter the environmental challenges for advanced water management systems meeting the sustainable development. In: Kumar M, Snow D, Honda R (eds) Emerging issues in the water environment during anthropocene: a South East Asian perspective. Springer, Berlin. ISBN 978-93-81891-41-4

Nel HA, Dalu T, Wasserman RJ (2018) Sinks and sources: assessing microplastic abundance in river sediment and deposit feeders in an Austral temperate urban river system. In: Science of the total environment. Elsevier, Amsterdam

Ng KL, Obbard JP (2006) Prevalence of microplastics in Singapore's coastal marine environment. Mar Pollut Bull 52:761–767

Plastic Europe (2013) Plastics—The Facts 2013: an analysis of European latest plastics production, demand and waste data. Plastic Europe, pp 1–40

Plastics Europe (2017) Plastics—the Facts 2017: an analysis of European plastics production, demand and waste data. https://www.plasticseurope.org/application/files/1715/2111/1527/Plastics_the_facts_2017_FINAL_for_website.pdf. Accessed 06/29/2018

Ram B, Kumar M (2020) Correlation appraisal of antibiotic resistance with fecal, metal and microplastic contamination in a tropical Indian river, lakes and sewage. NPJ Clean Water 3(1):1–12

Reddy MS, Basha S, Adimurthy S, Ramachandraiah G (2006) Description of the small plastics fragments in marine sediments along the Alang-Sosiya ship-breaking yard, India. Estuar Coast Shelf Sci 68(3–4):656–660

Rochman CM, Hentschel BT, Teh SJ (2014) Long-term sorption of metals is similar among plastic types: implications for plastic debris in aquatic environments. PLOS one 9(1):e85433

Sarkar DJ, Sarkar SD, Das BK, Manna RK, Behera BK, Samanta S (2019) Spatial distribution of meso and microplastics in the sediments of river Ganga at eastern India. Sci Total Environ 694:133712

Singh A, Patel AK, Ramanathan A, Kumar M (2020) Climatic influences on arsenic health risk in the metamorphic precambrian deposits of Sri Lanka: a re-analysis-based critical review. J Clim Change 6(1):15–24

Thompson RC, Olsen Y, Mitchell RP, Davis A, Rowland SJ, John AW, McGonigle D, Russell AE (2004) Lost at sea: where is all the plastic? Sci 304(5672):838–838

Van Cauwenberghe L, Devriese L, Galgani F, Robbens J, Janssen CR (2015) Microplastics in sediments: a review of techniques, occurrence and effects. Mar Environ Res 111:5–17. https://doi.org/10.1016/j.marenvres.2015.06.007

Wang SL, Xu XR, Sun YX, Liu JL, Li HB (2013) Heavy metal pollution in coastal areas of South China: a review. Mar Pollut Bullet 76(1–2):7–15

Wang J, Peng J, Tan Z, Gao Y, Zhan Z, Chen Q, Cai L (2017) Microplastics in the surface sediments from the Beijiang River littoral zone: composition, abundance, surface textures and interaction with heavy metals. Chemosphere 171:248–258

Yu X, Peng J, Wang J, Wang K, Bao S (2016) Occurrence of microplastics in the beach sand of the Chinese inner sea: the Bohai Sea. Environ Pollut 214:722–730

Chapter 8
Evaluation of Water Quality of Community Managed Water Supply Schemes (CMWSS) in Galle District

M. N. M. Shayan, G. G. Tushara Chaminda, K. C. Ellawala, and W. B. Gunawardena

8.1 Introduction

Dr. Sarath Amarasiri, according to book Caring for water (2008), states that the quantity of water which is easily accessible to the man is to be 0.75% globally. As a consequence of this, globally, 1.2 billion people live without access to safe water supply and 700 million are living without water supply, mainly in developing countries. Therefore, clean access to drinking water enhances the public health and save lives (Maggie and Menachem 2007; Kumar et al. 2017). On the other hand, it is an indisputable logic which delinates that, a considerable amount of people around the world are facing the problems of water scarcity (Water Stress). As a consequence, about 50% of population in the developing countries encounter water-borne diseases such as diarrhea, cholera, ascaris, hookworm (Murcoot 2001). And millions of people suffer from preventable illnesses and die every year (World Health Organization 2004).

In the context of Sri Lanka, it is an island gifted with natural water sources still water scarcity is not obvious. Although Sri Lanka is self-sufficient in drinking water, due to population growth the potable water demand has been raised immensely. In Sri Lanka, the sole body which is responsible for drinking water supplying and maintenance of quality is governed by National Water Supply and Drainage Board (NWSDB). According to statistics of NWSBD reports of (2011), NWSDB and local authority water supply schemes (WSS) supply water to 34% and 10% of the country's population, respectively. The water supplied by the NWSDB is up to its standard as mentioned in the guidelines of SLS 614 in 2013. But NWSDB caters this water

M. N. M. Shayan (✉) · G. G. Tushara Chaminda · K. C. Ellawala
Department of Civil and Environmental Engineering, University of Ruhuna, Galle, Sri Lanka
e-mail: shayanpera2@gmail.com

W. B. Gunawardena
Department of Civil Engineering, University of Moratuwa, Moratuwa, Sri Lanka

© Springer Nature Singapore Pte Ltd. 2020
M. Kumar et al. (eds.), *Resilience, Response, and Risk in Water Systems*,
Springer Transactions in Civil and Environmental Engineering,
https://doi.org/10.1007/978-981-15-4668-6_8

mainly for municipalities and its suburbs (NWSDB 2011). The supply of water to rural areas by the NWSDB is comparatively less. Therefore, community managed water supply schemes (CMWSS) became popular schemes in which the pertinent community where potable water is inadequate, people collaborate to gain potable water by gathering their own resources and extracting water from an in situ source which has been located to the close proximity of that community (Montgomery et al. 2009). As of definition given by Mimrose et al. (2011), Community Managed Water Supply can be explained as a public water system that serves at least 25 residents throughout the year and may consist of one or multiple wells or reservoirs (actually from in situ sources). As some literature reviews that community which have maximum of 5000 feeders are to be fallen under this scheme (NWSDB 2011).

In southern province, community water supply nutures 10% from total supply with approximately about 700 community-based organizations (CBO) (Southern Province annual report 2016). In Galle, which is a component city of southern province, at present, mainly people in 12 Divisional Secretariats (DS) out of total 19 DS have formed 172 CMWSS to cater the water needs of 10,814 families catering 7,376,600 L of water per day. Among these schemes, 56 are shallow well schemes, 111 are stream-based schemes while remaining five schemes are deep well schemes (Community Supply Department 2014; Patel et al 2019). Although these rural people receive water for their daily consumption, it does not mean that they endure hygienic potable water which meets the specifications given by Sri Lanka Standards (Mimrose et al. 2011; Kumar et al. 2019). Although these rural people receive water for their daily consumption, it does not mean that they endure hygienic potable water which meets the specifications given by Sri Lanka Standards Institute which is the legal organization which is responsible to look after the quality requirement of potable water. As of some reviews, quality could be of two versions. That is organoleptic quality (i.e., sensorial information from taste, odor, color, and turbidity) and inorganoleptic quality where it is not perceptional to human beings through his sensation. Therefore, consumers can sense the quality of the water through these organoleptic qualities. On the other hand, water can be contaminated with chemical and microbiological contaminants which may enter orally, through drinking water and tend to suffer from health-related issues. In Sri Lankan contest, chemical contamination is not often like in India, Nepal. In almost all small community water supply systems, fecal bacteria are likely to be found (Hofkes and Hiusman 1983). At the same time, in Sri Lankan context too, this condition is valid and common in CMWSS. As some local researchers have found that some areas in Sri Lanka, there is a probability of contamination of heavy metals and chemical contaminants such as nitrates mainly where the water sources are of close proximity to agricultural lands and regions. So as these CMWSS do not adopt treatment trains in advance, there is a probability of contamination of such above-mentioned contaminants. When considering about Galle district, it seems to worse as mentioned above, only five schemes adopt treatments among 172 CMWSS and they also use primary methods for the treatment. On the other hand, the total system of CMWSS in Galle district neither has regular nor once in year, surveillance program to investigate the quality of the potable water and its acceptability in advance. Because most of the schemes (about 97.5% of total CMWSS) are providing the raw water of the in situ source without any preliminary treatment and schemes where it is subjected to treatment (about 3.5% of total CMWSS) are also not addicting pertinent treatment techniques (Community Supply Department 2014). On the other hand, the total system of CMWSS in Galle district, neither has regular nor once in year surveillance program to investigate the quality of the potable water and its acceptability in advance. As a result of that, issues related to health and other social issues bound with health have been arised in these communities (Community supply Department 2016). Therefore, in this study, the main objective is to evaluate the hygienic quality of community managed water supply schemes in Galle District based on physical, chemical, and biological parameters and to develop a water quality index (WQI) based on overall quality.

8.2 Materials and Methods

8.2.1 Study Area

In Galle district, all the 172 community water supply schemes were selected for the study. Water samples from all schemes were tested. These 172 distributes around following DS divisions shown in Table 8.1. These all schemes are distributed among 12 DS divisions and main sources of these schemes are streams, wells, and deep wells. Table 8.1 delineates the distribution of samples with respect to divisional secretariat wise and source wise in Galle district.

Table 8.1 Distribution of community supply schemes in Galle district

Divisional Secretariat (DS)	Number of stream schemes	Number of shallow well schemes	Number of deep well schemes	Total amount of schemes
Akmeemana	0	4	0	04
Ambalangoda	0	4	0	04
Baddegama	7	4	2	13
Balapitiya	0	6	0	06
Benthota	0	5	0	05
Elpitiya	10	4	0	14
Imaduwa	0	7	1	08
Nagoda	13	11	0	24
Neluwa	34	0	2	36
Niyagama	9	8	0	17
Thawalama	34	0	0	34
Yakkalamulla	4	3	0	07
Total	111	56	5	172

8.2.2 Data Description

All the 172 samples were subject to testing under physical parameters such as temperature, turbidity, and conductivity. Chemical parameters such as pH, fluoride, Fe, total nitrate, total hardness, heavy metals, Mn, Cd, As, and Pb. Biological parameters such as E-Coli and T-Coli. Moreover, all the tests were followed ASTM guidelines.

8.2.3 Evaluation of Water Quality

Water quality index (WQI) is a universal language to communicate the in situ quality of a water source. As literature reviews, following are the advantages of a developed WQI for a region or a country.

- The use of an index can translate water quality monitoring data into a form that the public and policy makers can easily interpret and utilize (House 1990; and Yogendra and Puttaiah 2008).
- Indices facilitate quantification, simplification, and communication of complex data allowing for an effective way to convey environmental information (Swamee and Tygai 2007).
- WQI is valuable and unique rating to depict the overall water quality status in a single term that is helpful for the selection of appropriate treatment technique to meet the concerned issues (Shweta.et al. 2013).
- These indices assess the appropriateness of the quality of the water for a variety of uses (Cude 2001).

However, a huge number of water quality indices viz. Weight Arithmetic Water Quality Index (WAWQI), National Sanitation Foundation Water Quality Index (NSFWQI), Canadian Council of Ministers of the Environment Water Quality Index (CCMEWQI), Oregon Water Quality Index (OWQI), etc., have been formulated by several national and international organizations (Shweta et al. 2013). So there are a lot of adhered techniques of developing a WQI for a region or a country. In the world, numerous indices have been introduced. Among them, Canadian Council of Ministers of the Environment Water Quality Index (CCMEWQI) is a compatible index due to following reasons (Terrado et al. 2010).

- A unique value is used to represent large number of data.
- Have flexibility in the selecting input parameters and objectives.
- Statistical techniques are used to interpret the complex multivariate data.
- Clear and intelligible idea is given to general public
- Easy to calculate
- Tolerance to missing data
- Suitable for analysis of data coming from automated sampling.
- Combine various measurements in a variety of different measurement units in a single metric.

The following steps will depict the format of adhered mathematical steps to generate the proposed water quality index.

Step 01—Calculation of $F1$ and $F2$

$F1$ (**Scope**) represents the percentage of variables that do not meet their objectives at least once during the time period under consideration ("failed variables"), relative to the total number of variables measured:

$$F1 = \left(\frac{\text{Number of failed variables}}{\text{Total Number variables}} \right) \times 100$$

$F2$ (**Frequency**) represents the percentage of individual tests that do not meet objectives ("failed tests"):

$$F2 = \left(\frac{\text{Number of failed tests}}{\text{Total number of tests}} \right) \times 100$$

Step 2—Calculation of excursion

Excursion is the number of times by which an individual concentration is greater than (or less than, when the objective is a minimum) the objective.

When the test value must not exceed the objective:

$$\text{excursion} = \left(\frac{\text{Failed Test Value } i}{\text{Objective } j} \right) - 1$$

When the test value must not fall below the objective:

$$\text{excursion} = \left(\frac{\text{Objective } j}{\text{Failed Test Value } i} \right) - 1$$

Step 3—Calculation of normalized sum of excursions (NSE)

The normalized sum of excursions NSE is the collective amount by which individual tests are out of compliance. This is calculated by summing the excursions of individual tests from their objectives and dividing by the total number of tests (both those meeting objectives and those not meeting objectives).

$$\text{NSE} = \frac{\sum \text{excursion } i}{\text{number of tests}}$$

Step 4—Calculation of $F3$

$F3$ is calculated by an asymptotic function that scales the normalized sum of the excursions from objectives to yield a range from 0 to 100.

$$F3 = \left(\frac{\text{NSE}}{0.01\text{NSE} + 0.01} \right)$$

Table 8.2 WQI categorization according to quality

WQI	Descriptor words	Category	Drinkability	Treatment nature
91–100	Excellent	A	Yes	No
75–90	Good	B	Yes	Preliminary
51–74	Medium	C	No	Primary
26–50	Bad	D	No	Secondary
0–25	Very bad	E	No	Secondary/Advanced

Step 5—Calculation of WQI

$$WQI = 100 - \left(\frac{\sqrt{F1^2 + F2^2 + F3^2}}{1.732} \right)$$

According to above equations, it is possible to derive a unique index as a percentage for its quality of potability. At the same time, a level of its quality should be implied by adhering to a range. Hence, to achieve that, a range has been specified based on the quality of water as depicted in Table 8.2. At least the users to allow for usage of a scheme, it should be equal or more than 75% of WQI (CCME 2008). Although it is drinkable, it may cause health issues if any preliminary treatments are not adhered. Only schemes with excellent water quality index do not need a prior treatment train (Alison 2010).

8.2.4   Limitations of Current WQI

- For each and every element, no attention is paid-off
- Loss of information on interactions between variables. (Debel et al. 2005)
- Same weight is given to every parameter (Mnisi 2010)
- $F1$ not working appropriately when too few variables are considered or when too much covariance exists among them. (Khan et al. 2003)
- Difficulty of understanding the method of calculation to general public.

8.3   Results and Discussions

WQI was calculated adhering to above equations as specified in the above section. For each sample, it has tailored 15 water quality parameters and calculated five independent parameters for each sample as shown in WQI calculation. The results and corresponding discussions are shown below. Note: All blue dots indicated in all graphs represent the calculated WQI of a scheme and the yellow horizontal line indicated in all graph shows the drinkable WQI (75%) (less than 75% is not allowed for drinking).According to Fig. 8.1, only 2% out of total 172 samples are excellent as of Table 8.2. 32% are good. 47% are medium quality while 19% are bad. No any single sample with very bad water quality. At the same time, 34% samples can be used for drinking purposed as per the guidelines (75%) which has been depicted in yellow line in the graph (Fig. 8.2.) itself, as of Canadian water quality index guidelines.

Among all the schemes, 111 schemes are stream-based schemes. Their water quality variation has been graphically represented in Fig. 8.2. According to Table 8.2 categorization, no any single sample is with excellent water quality. Out of all 111 stream schemes, 26% are good quality, while 48% and 26% medium and bad quality, respectively. No any sample with very bad quality. But according to guidelines, only 26% schemes can be used for potability.

56 schemes from 172 schemes are shallow well-based schemes. Their variability of WQI index is shown in Fig. 8.3. Among them, there are three schemes with

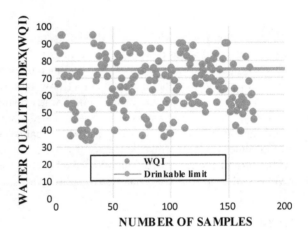

Fig. 8.1 WQI of total sources

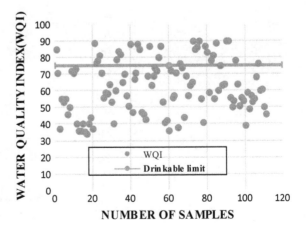

Fig. 8.2 WQI of stream sources

Fig. 8.3 WQI of shallow
well sources

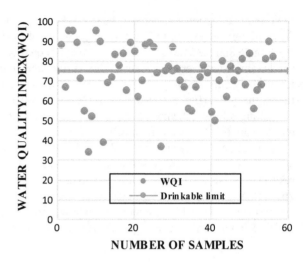

excellent water quality while 5, 43, 45, and 7% out of shallow well schemes, good, medium, and bad water quality index, respectively. No scheme is there with very bad water quality index. Only 48% are with the quality of drinking. There are only five deep well schemes for all 172 schemes (for water quality variation see Fig. 8.4). Among them, no any scheme with excellent water quality. 40% are good schemes by quality. Another 40% are medium water quality schemes. While 20% are bad in quality and no any scheme with very bad water quality. Among these samples, only 40% schemes are with drinkable quality. The following table depicts comparison of maximum and minimum water quality with other countries as according to literature reviews with respect to Canadian water quality indices.

Fig. 8.4 WQI of deep well
sources

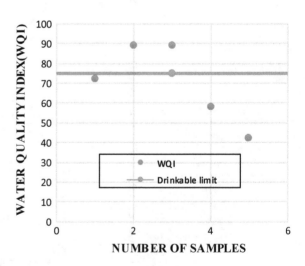

When we analyze the maximum and minimum water quality indexes of South Asian countries such as India, Nepal, as in Table 8.3, it is obvious that maximum WQI is 87 and 91 for India and Nepal, respectively. If we compare it with Sri Lanka (Galle), maximum WQI is 95. Through that, it is clear that WQI of Galle is high. This is also very obvious when we compare the maximum with South African countries such as Malawi and South Africa as of Table 8.3. When we compare the minimum, we can observe that WQI of Galle is categorized under "bad" criteria. When compared with South Asian Countries like India and Nepal, their minimum WQI is categorized under "very bad" category. Same is for Malawi and South Africa. Therefore, it is obvious that quality of water in Sri Lanka is good with respect to other countries. The results of calculated WQI of all 172 CMWSS's are shown in Fig. 8.1. There are three sources of water supply to these CMWSS. They are stream, shallow well, and deep well. According to Fig. 8.1, only 2% out of total 172 samples are excellent as of Table 8.1 while 32% are good. In addition to that 47% are medium quality while 19% are with bad water quality. At the same time, 34% samples can be used for drinking purposed as per the guidelines (75%) which has been depicted in yellow line in the graph (Fig. 8.2.) itself, as of Canadian water quality index guidelines (Fig. 8.5).

Following GIS map shows (Fig. 8.6) how the distribution of samples and their level of quality.

Table 8.3 Comparison of WQI with other countries

Country	Maximum	Minimum
India	87	21
Nepal	91	23
Malawi	83	18
South Africa	87	18
Sri Lanka (Galle)	95	34

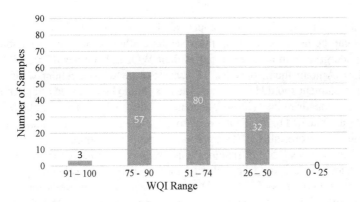

Fig. 8.5 Results of WQI

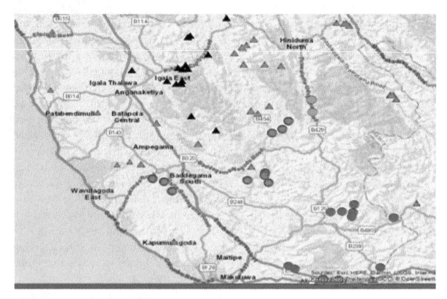

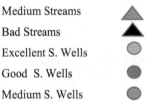

Medium Streams

Bad Streams

Excellent S. Wells

Good S. Wells

Medium S. Wells

Fig. 8.6 GIS location of sampling points and quality variation

8.4 Conclusion

The results of this study indicate that out of all the schemes, only 34% (58 schemes)of schemes can be used for drinking purposes (even though they are drinkable, they should undergo preliminary treatment as their WQI falls between 75 and 90) (see Table 8.2). Among them, only 2% (3 schemes) are with excellent water quality where no treatment should be adhered. The rest should be followed after a treatment procedure before drinking. The 66% schemes should be developed to potable level as only 34% are reached to drinkable level, by adhering at least to a primary treatment technique. Therefore, all the stream sources, deep well sources, and majority of shallow well sources (95% of schemes) have been contaminated. Therefore, a prior treatment technique should be followed in each and every scheme (preliminary, primary or secondary as of Table 8.2) in order to enhance the water quality. If not, it would create huge social and health problems. Therefore, CMWSS water has been contaminated to a considerable extent. In addition to that, when we analyze the schemes under sources, they all show their mean WQI value in the range of medium water quality index. Although CMWSS water is contaminated, Sri Lankan potable water gained through CMWSS is not contaminated when compared with other countries.

Acknowledgements Authors would like to acknowledge the financial support provided by the Sothern Provincial Council, Ministry of agriculture and water supply. And to district office of Department of National Community Water Supply located at Galle, Irrigation department of Southern Province for granting the co-operation to gain details of these schemes. A huge support was given by Department of Civil and Environmental Engineering by providing well equipped laboratory facilities and Ms. Nimalashanthie, technical officer of the laboratory gave remarkable contribution in testing processes.

References

Alison K (2010) Development and use of water quality indices (WQI) to assess the impact of bmp implementation on water quality in the cool run tributary of the white clay creek watershed, pp 50–51. University of Delaware, Asiaweek, 31 July

Amarasiri S (2008) Caring for water. Sri Lanka Nature forum, Nugegoda

CCME (2008) Canadian environmental quality guidelines for the protection of aquatic life, CCME water quality index: technical report, 1.0, 2001, Canada

Community Water Supply Department (2014) Annual Report 2013, Colombo, Sri Lanka

Community Water Supply Department (2016) Annual Report 2015, Colombo, Sri Lanka

Cude C (2001) Oregon water quality index: a tool for evaluating water quality management effectiveness. J Am Water Res Assoc 37(1):125–137

Debel SP, Figueroa R, Urrutia R, Barra R, Niell X (2005) Evaluation of water quality in the Chilla´n River (Central Chile) using physicochemical parameters and a modified water quality index. Environ Monit Assess 110:301–322

Health UNICEF state of world's children, Table 3: 2009. Available online: http://www.unicef.org/sowcc)housing.net.htm. Accessed on 28 April 2018

Hofkes EH, Huisman L (1983) International reference centre for community water supply and sanitation. Review Publications, London, UK

House MA (1990) Water quality indices as indicators of ecosystem change environmental monitoring and assessment. J Environ Sci 15:255–263

Khan F, Husain T, Lumb A (2003) Water quality evaluation and trend analysis in selected watersheds of the Atlantic Region of Canada. Environ Monit Assess 88:221–242

Kumar M, Patel AK, Das A, Kumar P, Goswami R, Deka JP, Das N (2017) Hydrogeochemical controls on mobilization of arsenic and associated health risk in Nagaon district of the central Brahmaputra Plain. India Environ Geochem Health 39(1):161–178

Kumar M, Chaminda T, Honda R, Furumai H (2019) Vulnerability of urban waters to emerging contaminants in India and Sri Lanka: resilience framework and strategy. APN Sci Bull 9(1). https://doi:10.30852/sb.2019.799

Maggie AM, Menachem E (2007) Water and sanitation in developing countries: including health in the equation. Environ Sci Technol from https://pubs.acs.org. Accessed on 28 Nov 2018

Mimrose D, Gunawardena N, Nayakakorala H (2011) Assessment of sustainability of community water supply projects in Kandy district. Tropical Agric Res 23(1):51–60

Mnisi LN (2010) Assessment of the state of the water quality of the Lusushwana River, Swaziland, using selected water qualityindices. M.Sc. thesis, University of Zimbabwe, Harare

Montgomery et al. (2009) Sustainability of community managed water supply schemes in Ethiopia, Chubley Publications. In: The 12th world lake conference, pp 567–575

Murcott S (2001) Clean water for 1.7 Billion people? In: paper presented to the Development by Design, Workshop, MIT, Cambridge, Mass, USA

NWSDB (2011) Annual Report 2011. Colombo, Sri Lanka

Patel AK, Das N, Kumar M (2019) Multilayer arsenic mobilization and multimetal co-enrichment in the alluvium (Brahmaputra) plains of India: a tale of redox domination along the depth. Chemosphere 224:140–150

Shweta T, Bhavtosh S, Prashant S, Rajendra D (2013) Water quality assessment in terms of water quality index. Am J Water Res 1(3):34–38. https://doi.org/10.12691/ajwr-1-3-3

Southern Provincial Council (2016) Annual Report 2016. Galle, Sri Lanka

Swami PK, Tragi A (2007) Improve method for aggregation of water quality sub-indices. J Environ Eng 133(2):220–225

Terrado M, Barcelo D, Tauler R, Borrell E, Campos SD (2010) Surface-water-quality indices for the analysis of data generated by automated sampling networks. Trends Anal Chem 29(1):40–52

WHO (2004) Meeting the MDG drinking water and sanitation target: a mid-term assessment of progress. WHO, Geneva

Yogendra K, Puttaiah ET (2008) Determination of water quality index and suitability of an urban water body in Shimoga Town, Karnataka. In: Proceedings of Taal2007: the 12th world lake conference, pp 342–346

Chapter 9
Scenario of Worldwide Preponderance of Contaminants of Emerging Concern in the Hydrosphere

Kiran Patni, Chitra Pande, and Tanuj Joshi

9.1 Introduction

Humans always tend to deteriorate the provided natural sources. Organic contaminants introduced in the environment are an emerging issue from a long time. They are getting introduced in the hydrosphere which significantly affects both human health and the environment. Hydrosphere consists of total amount of water present in the planet. Hydrosphere includes water that is present on the surface of the earth, underground and in the air. In the early 1800s, a new class of pollutants called emerging contaminants (ECs) also known as contaminants of emerging concern (CECs), and emerging organic contaminants came into light in water and aquatic environment (Miraji et al. 2016; Kumar et al. 2019a, b; Dey et al. 2019). Various definitions have been proposed to define emerging contaminants. ECs or CECs may be defined as the substances which are released in the environment, but no regulations are established for their environmental monitoring (Thomaidis et al. 2012). According to Alexandros, ECs may be defined as the recently discovered group of unregulated contaminants that are present in groundwater and surface water (Stefanakis and Becker 2015). They are known as emerging contaminants because previously they were unrecognized and lacked standard guidelines for their monitoring, but now due

Objectives This book chapter consists of description of various CECs in aquatic environment, their classification and various conventional and advanced techniques available for the removal of these contaminants.

K. Patni (✉)
Graphic Era Hill University, Bhimtal, Uttarakhand, India
e-mail: 93kiranpatni@gmail.com

C. Pande
Kumaun University, DSB Campus, Nainital, India

T. Joshi
Faculty, Department of Pharmaceutical Sciences, Bhimtal, Uttarakhand, India

© Springer Nature Singapore Pte Ltd. 2020
M. Kumar et al. (eds.), *Resilience, Response, and Risk in Water Systems*,
Springer Transactions in Civil and Environmental Engineering,
https://doi.org/10.1007/978-981-15-4668-6_9

151

to their health effects, they are gaining scientific attention (Nosek et al. 2014). They form a newly discovered class of chemicals present in groundwater and surface water. These emerging contaminants include compounds that are being used in our daily life and are involved with various industrial activities. These emerging contaminants are also known as micropollutants, and the reason behind their presence can be anthropogenic as well as natural. They are present in water sources at trace levels ranging from nanograms per liter (ng/L) to micrograms per liter (μg/L) (Luo et al. 2014). All the chemical pollutants like Dechlorane plus (DP), hexabromocyclododecanes (HBCDs), phthalate esters, pyrethroids, etc., were being used as the replacement of toxic chemicals in the following manner: DP was being used as a substitute of mirex which is a persistent organic pollutant (Xian et al. 2011), HBCDs were being used as a halogenated flame retardant and as an alternative of polybrominated diphenyl ethers (PBDEs). Nowadays, short-chain chlorinated paraffin (SCPPs) is gaining a lot of interest among scientists, and they are working on the persistence of SCPPs in the environment and its accumulation in humans (Zeng et al. 2011). Contaminants of emerging concern are not newly developed chemicals, they are dwelling in the environment for decades, but their presence is being investigated recently after noticing their health hazards.

The number of contaminating pollutants is increasing continuously which include industrial compounds, pharmaceuticals, personal care product, antibiotics, hormones, biocides, alkylphenols, plasticizers, plant protection product, perfluorinated compounds, nanomaterials, pesticides, flame retardant (Montagner et al. 2019). Many natural water sources, e.g., river, lakes, reservoir, contain these contaminants worldwide (Lai et al. 2016; Mukherjee et al. 2020; Singh et al. 2020; Wanda et al. 2017; Jaimes et al. 2018). These emerging contaminants get introduced in the water through discharge of wastewater and surface water including urban stormwater runoff (Kolpin et al. 2004; Fairbairn et al. 2018), agricultural runoff, streams, rivers (Kolpin et al. 2002; Lee et al. 2011), lakes (Ferrey et al. 2015), source drinking water and in shallow groundwater (Furlong et al. 2017). Their presence can be a cause of concern if used for drinking purposes (Riva et al. 2018). Sometimes, the transformation products (TPs) are more hazardous in comparison with their parent ECs (Richardson and Ternes 2018).

There are various conventional methods available for the removal of CECs from the water like coagulation, sedimentation, sand filtration, chlorination and advanced treatment processes like ozonation, activated carbon and ultra-membrane filtration (Lv et al. 2016). In addition to this, it is also important to evaluate the concentration of these emerging contaminants in water as they are being a serious issue concerning health and safety. Aquatic life is found to be the most affected due to these contaminants, so there is an urge for immediate research in the field of ECs. The USA, China, Canada, Spain, Germany, Japan, Africa, India, Brazil, Sweden, Norway and Switzerland are some countries that are involved in research on ECs (Bao et al. 2014).

9.2 Occurrence and Fate of CECs

The CECs in our environment can occur as pharmaceutical drugs, herbicides, pesticides, insecticides, etc. Antibiotics, analgesics, anti-epileptic, anti-inflammatory, betablockers, fragrances, barbiturates, diuretics, lipid-lowering agents, etc., are some classes of drugs that comprise emerging contaminants. Also, various insecticides, pesticides, herbicides, etc., which are frequently used in agricultural practices, enter the water bodies and form the CECs. CECs can enter the water bodies through diffuse sources and point sources of pollution. Examples of point sources of pollution that can lead to the occurrence of CECs in water bodies include municipal sewage treatment plants, septic tanks, industrial effluents, etc. Diffuse sources of pollution include urban and stormwater runoff, runoff from agricultural manures, runoff from leakage of urban sewage. It has been found that pharmaceutical products find their presence in various water bodies throughout the world. Like in Germany, effluents from sewage treatment units and rivers have been found to contain carbamazepine, diclofenac, naproxen, etc. Also, CECs like carbamazepine have been found in rivers of Madrid, Spain. In sludge of sewage, synthetic musk has been detected in countries like Germany, United Kingdom, China, Switzerland, Spain, Hong Kong, etc. It is found that the concentration of CECs varies to different extents in different countries. This probably depends upon the extent of use of various CECs and the efficiency of the technology of the particular country in removing CECs from waste material and sewage. The fate of a CECs can depend upon various factors. Factors like physicochemical properties (water solubility) of a particular CEC and also environmental conditions will dictate the dissipation of a particular CEC. The duration for which a particular CEC will persist in the subsurface and groundwater will depend upon various factors. Some of these factors are groundwater residence time, properties of the contaminant, redox conditions, etc. Certain mechanisms that operate to control the levels of CECs in nature are ion exchange in the aquifers and soils, sorption and degradation by microbes (Thomaidis et al. 2012).

9.2.1 Classification of Contaminants of Emerging Concern

There is a wide range of contaminants which are called as emerging. Some of them are represented below in Table 9.1 (with their molecular formula and possible sources of contamination) and Fig. 9.1 (representing structures of some of the CECs).

9.2.1.1 Endocrine-Disrupting Chemicals (EDCs)

EDCs are a group of natural or synthetic compounds that interfere with the functioning of the hormone system resulting in unnatural responses in the receiving organism. It may also be classified as a group of endocrine disruptors that alter the

Table 9.1 Representation of various contaminants of emerging concern (https://pubchem.ncbi. nlm.nih.gov)

S. No.	Contaminants of emerging concern	Examples	Molecular formula	Source
1	Pharmaceuticals and personal care products (PPCs)	Bisphenol A	$C_{15}H_{16}O_2$	Release of antibiotic to the environment, shampoos, soaps, deodorants, cosmetics
		Triclosan	$C_{12}H_7Cl_3O_2$	
		Triclocarban	$C_{13}H_9Cl_3N_2O$	
2	Perfluorochemicals	Perfluoromethanesulfonic acid	$C_8HF_{17}O_3S$	Used for the preparation of heat, oil, grease, water-resistant products
		Perfluorooctanoic acid	$C_8HF_{15}O_2$	
3	Siloxanes	Polydimethylsiloxanes	$C_8H_{24}O_2Si_3$	Used in paints, cosmetics, medical products
4	Quaternary ammonium compound	Benzyldimethyltetradecylammonium chloride	$C_{23}H42ClN$	Disinfectants, fabric softner, surfactants, antistatics
5	Artificial sweetners	Sucralose	$C_{12}H_{19}Cl_3O_8$	Domestic wastewater, groundwater
		Acesulfame	$C_4H_5NO_4S$	
		Saccharin	$C_7H_5NO_3S$	
6	Anticorrosives	Benzotriazoles	$C_6H_5N_3$	Used as corrosion inhibitors, herbicides, antialgal agent, slimicides in paper and pulp industry
		Benzothiazoles	C_7H_5NS	
7	Polybrominated-diphenyl ethers	2,3,4,5,6 penta bromo diphenyl ethers	$C_{12}H_5Br_5O$	Used in dielectric fluids, engine oil additives, electroplating masking compounds, wood preservatives, lubricants and for dye production

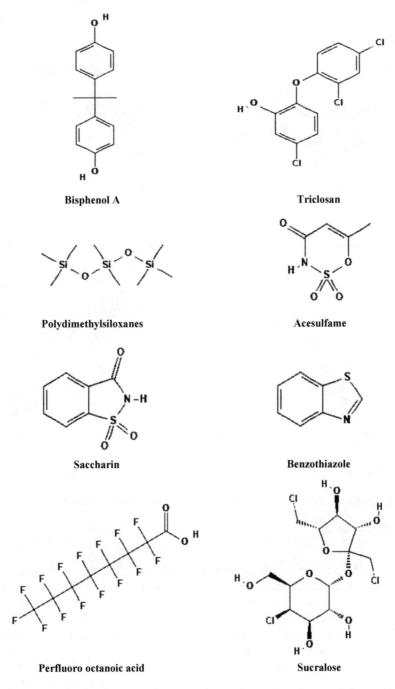

Fig. 9.1 Structures of various emerging contaminants. *Source* https://www.pubchem.ncbi.nlm. nih.gov

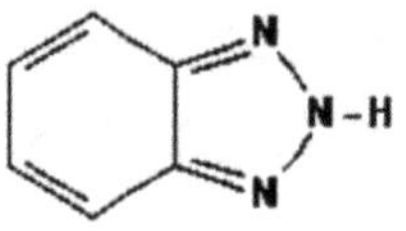

Benzotriazoles 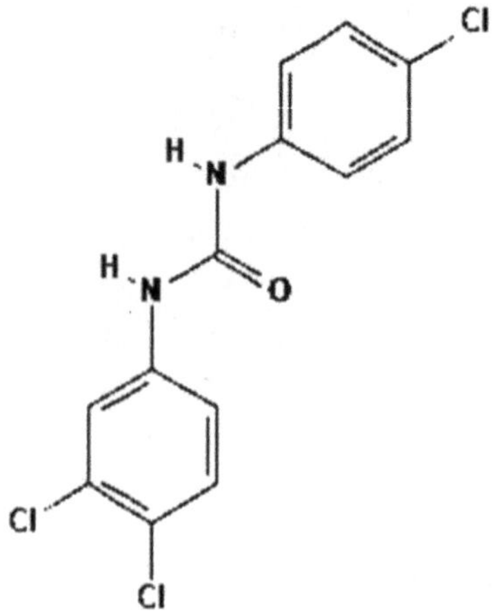   **Triclocarbon**

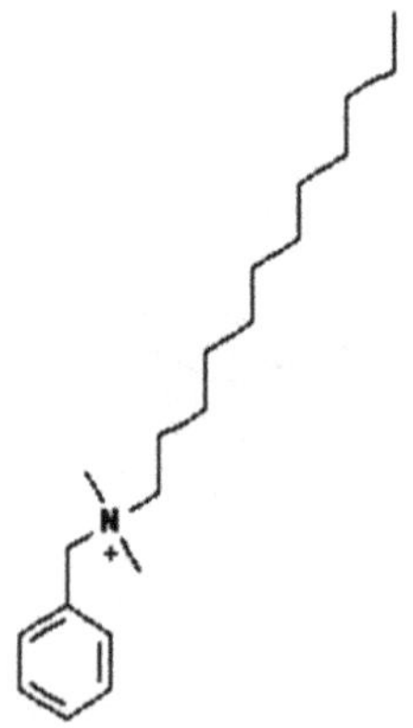

Benzyldodecyldimethylammonium 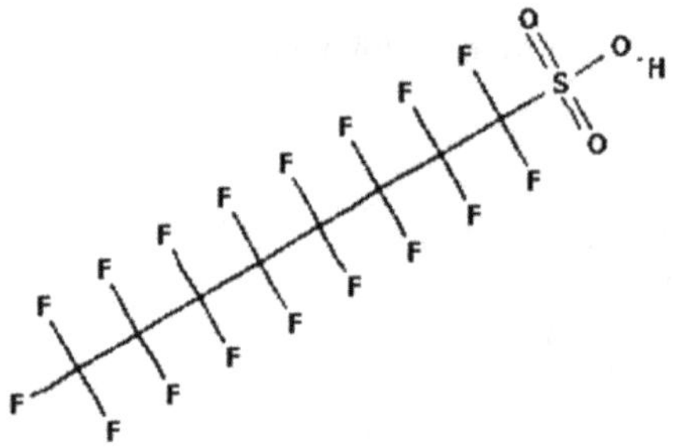   **Perfluorooctanesulfonic acid**

Fig. 9.1 (continued)

functions of the endocrine system resulting in adverse health effects (Damstra et al. 2002). EDCs can be introduced through sewage effluents and industrial wastewater in the mainstream and affect aquatic life as well as humans and animals. Initially, EDCs were studied for their influence on estrogenic hormones (Gómez et al. 2013), but later research showed that they disrupt thyroid hormones (TH) and androgenic hormones (AH) as well (Gong et al. 2014). Some irreversible changes like mental retardation, neurological deficits and testicular cancers are observed as a result of TH or AH disruption. Some chemicals that interact with the endogenous system cause serious health risks and classified as emerging contaminants are polycyclic aromatic hydrocarbons (PAHs), pesticides, alkylphenol, phthalate esters, 17 β estradiol (E2, natural estrogenic steroid), 17 α-ethynylestradiol (EE2, synthetic steroid), 4-tert-octylphenol (4-t-OP), 4-n-nonylphenol (4-n-NP), bisphenol A (BPA, an industrial chemical), nonylphenol and octylphenol (Shi et al. 2016; Ying and Kookana 2003). Infertility, impairments in pregnancy, birth defects, ovarian failure and growth retardation can also be caused due to EDCs. At present, EDCs are being considered as a possible reason for the degradation of aquatic life (Wang and Zhou 2013). A lot of literature is available which indicates that endocrine disruption in aquatic life occurs at a higher rate in comparison with the telluric lives since water body forms the most common sink for the disposal of almost all domestic wastes, industrial effluents, etc. (Jafari et al. 2009). Seven hundred and eighty-five species including mammals, seabirds, fish, crustaceans and gastropods have become extinct or are at the edge of extinction (Stork 2010) over the last 100 years and the reasons suggested for this extinction are overexploitation, climate change, loss of habitat, pollution, etc. Studies suggest that EDCs are one of the reasons for the decrease in wildlife in recent years (Millsa and Chichester 2005).

9.2.1.2 Pharmaceuticals and Personal Care Products (PPCPs)

The aquatic occurrence of PPCPs is getting more attention nowadays. Chemicals used in these products cause a harmful effect on the public. Pharmaceuticals have gained attention as EC due to their pharmacological activities. Many countries are being aware of these ECs and adopting a precautionary approach by dealing with ECs at their point sources like hospitals and WWTP. ECs enter the aquatic environment through the discharge of effluents released from the wastewater treatment plant (WWTP). PPCPs are such products that are resistant to degradation, persist in the aqueous system, environment and cause harmful effects (Kümmerer 2009). Personal care products (PCPs) consist of a wide range of products like soaps, shampoos, lipsticks, lotions, creams, cologne, toiletries, etc. (Sang and Leung 2016). PCPs may get introduced in the aquatic environment through showers, cleaning, washing machines, use of sunscreen/lotions, etc. (Rodil and Moeder 2008; Tsui et al. 2014). PCPs are prevalent and are considered as 'pseudo-persistent' because of their constant presence and inefficient removal from water (Blüthgen et al. 2014). Octocrylene (OC) which acts as a UV-filter is a main component of sunscreen to protect the skin from harmful UV-radiation. OC is used in other skincare products also and acts as a stabilizer for other UV-filters in different formulations or in plastics. OC is non-degradable, photostable and a lipophilic neutral compound (Zhu et al. 2016). Other factors like population, climate, dilution potential of water, availability of manufacturing sites, anthropogenic activities in that particular region are also responsible for the accumulation of chemicals in the environment and aquatic life. Most of the antibiotics are not metabolized completely and get excreted in urine and feces. They then enter sewage treatment plant (STP)/WWTP from where after degradation, they finally get introduced in surface/groundwater (Behera et al. 2011). Due to all these reasons, STP/WWTP effluents require efficient monitoring regarding antibiotics as they are being used excessively for treating the infection as well as for promoting fruit growth (Samaraweera et al. 2019).

9.2.1.3 Artificial Sweetner

Artificial sweeteners are used as a substitute for sugar all over the world in food, beverages, drugs, etc. Artificial sweeteners are regarded as water contaminants. Two artificial sweeteners acesulfame (ACE) and sucralose (SUC) are found to be present in the aquatic environment in higher concentrations in comparison with other waste specific anthropogenic organic chemicals (Lange et al. 2012). Due to their use as food additives, artificial sweeteners are extensively tested for adverse health effects in humans. Few studies are available regarding the ecotoxicological impact of SUC, which was the first artificial sweetener detected in the environment (Lange et al. 2012).

9.2.1.4 Perfluorooctane Sulfonate/Perfluorooctanoic Acid (PFOS/PFOA)

Perfluorooctane sulfonate and perfluorooctanoic acid are globally present compounds (Squadrone et al. 2015). Due to their chemical stability, surface tension lowering properties, they find use in coating, fire-fighting, foams; as water repellent agents in leather; paper and textiles (Van Asselt et al. 2013). Various health problems like hepatotoxicity, developmental toxicity, neurobehavioral toxicity, immunotoxicity, reproductive toxicity, lung toxicity, etc., were encountered due to the continuous use of PFOS and PFOA (EFSA 2011).

9.2.1.5 Benzotriazoles and Naphthenic Acid

Benzotriazoles are used as anti-corrosives and widely used in engine coolants, air-craft deicers and antifreeze liquid. It can cause harmful effect on endocrine system. Neurotoxicity in fish was also found due to benzotriazoles (Casado et al. 2014).

9.2.1.6 Algal Toxins

Blue–green algal (cyanobacterial) blooms can cause the production of toxins that pollute the drinking water. Some of the most common species of algae involved in this process are *Anabaena bergii*, *Aphanizomenon ovalisporum*, *Microcystis aeruginosa*, *Aphanizomenon flosaquae*, *Umezakia natans*, and *Raphidiopsis curvata* (Falconer and Humpage 2006). In America, Europe and Australia, algal toxins have shown hazardous effects on humans (Falconer and Humpage 2005). *Microcystis aeruginosa* acts as a hepatotoxin that can cause liver damage, and *Anabaena flos aquae* produces a neurotoxin that can attack the central nervous system (Hal et al. 2007).

9.2.1.7 Perchlorate

Perchlorate is a highly stable anion that resides in nature due to anthropogenic activities as well as natural reasons. It is a strong oxidizing agent used for various purposes like missile fuel, fireworks, vehicle airbags and fertilizers (Steinmaus 2016). Perchlorate can be found in water, soil and plants. Perchlorate is soluble in water and shows high mobility in soil (Steinmaus 2016). The higher concentration of perchlorate may result in thyroid disorder (Cal Baier-Anderson and Anderson 2006).

9.2.1.8 Herbicides

Herbicides are used widely for the protection of crops by killing weeds. Herbicides can be classified as organic and inorganic herbicides. Chloroacetanilide and chloroacetamide are widely used herbicides. The primary degradation of these herbicides includes metabolism in the soil, and the prime metabolites obtained are ethane sulfonic acid (ESA) and oxalic acid (OXA), which are water-soluble (Vargo 2013). Although many of the available herbicides are not toxic, yet they can be converted to toxic products, which can produce various carcinogens, teratogens, phytotoxins and insecticidal or fungicidal products. There are many herbicides that pass through farms and agricultural lands to water bodies in the nearby area. Some of these herbicides like oryzalin, ronstar, roundup and trifluralin can be harmful to the aquatic life found in these water bodies (Rashid et al. 2010).

9.3 Methods of Removal of CECs

The presence of the contaminants of emerging concern has been a barrier in the field of water pollution control, so the proper disposal and treatment of these contaminants become very necessary. There are various methods proposed for the removal of these contaminants viz. biological method, physical and chemical separation, coagulation and sorption, chemical oxidation, etc.

9.3.1 Biological Treatment for the Removal of CECs

Out of the total available techniques present for the removal of CECs, biological treatment method is one of the effective, sustainable and economical technique. It is one of the methods, which is widely used for the removal of CECs from wastewater (Ahmed et al. 2016). Biological treatment involves the removal of CECs through biodegradation method in which high molecular weight CECs are degraded into small molecules and biomineralized to small inorganic molecules like water and carbon dioxide with the help of bacteria, algae and fungi (Rodríguez et al. 2014). This process can be divided into conventional and non-conventional methods. Conventional methods for the removal of CECs consist of biological activated carbon, biological nitrification and denitrification, microalgae/fungi-based treatment and activated sludge process. Non-conventional methods consist of biosorption, membrane bioreactor (MBR) and constructed wetlands which are discussed below.

9.3.2   Conventional Biological Treatment Methods for the Removal of CECs

9.3.2.1   Activated Sludge

Activated sludge is one of the most used treatments for the removal of CECs, in which bacteria and protozoa are used for treating sewage and wastewater. It is a process in which biomass produced in wastewater by the growth of microorganisms in aeration tanks takes place in the presence of dissolved oxygen (Buttiglieri and Knepper 2008). This method is generally designed to remove the pathogens, organic and inorganic contaminants. This process has lower capital cost in comparison with advanced oxidation processes (AOPs) and is more ecological than chlorination process (Luo et al. 2014). This process is helpful in the removal of almost 102 target contaminants which include EDCs, pesticides, beta blockers, personal care products (78–90%, except for celestolide which is degraded up to 60% only), surfactants and pharmaceuticals (65–100%). Some beta blockers viz. atenolol, metoprolol were not removed efficiently through this method. For the better removal of contaminants, activated sludge method can be coupled with ozonation or MBR also.

9.3.2.2   Biological Activated Carbon (BAC)

Biological activated carbon is developed on the basis of activated carbon process. Mostly, BAC process is coupled with some other oxidation processes like ozonation in order to obtain better removal of the contaminants (Jin et al. 2013). At present, BAC has become a widely used treatment for the removal of contaminants from industrial wastewater as well as for wastewater reclamation. It is generally applied after ozonation process for further removal of contaminants (Kalkan et al. 2011), and it was found to be more effective in the removal of some pesticides, betablockers and pharmaceuticals when used after the ozonation process, but some other EDCs (E3, bisphenol A, octylphenol) were not removed efficiently (Gerrity et al. 2011).

9.3.3 Non-conventional Biological Treatment Methods for the Removal of CECs

9.3.3.1 Biosorption

This method is used for the removal of contaminants from water through the united action of adsorption and biological destruction. For adsorption, activated carbon is used, but sometimes, other solids like gravel, sand, clay, porous organic adsorbants are also used (Pidlisnyuk et al. 2003). This method is effective in 100% removal of some pharmaceuticals like ibuprofen, naprox and gemifibrozil. Along with this, some other contaminants like 17 β-estradiol- 17 α-acetate, pentachlorophenol, 4-tert-octylphenol and triclosan were also removed very efficiently (Banihashemi and Droste 2014). Estrogens can be removed effectively by the combination of biosorption and biodegradation interaction due to low Henry's law coefficient, low biodegradation and high octanol-water partition coefficients (K_{ow}) (Kumar et al. 2009).

9.3.3.2 Membrane Bioreactor (MBR)

MBR is a widely used technology for removal of contaminants from municipal and industrial wastewater treatment plant. It can effectively remove a large number of pollutants which are resistant to activated sludge process and constructed wetland (Radjenović et al. 2009). Even, when higher removal of effluents is required, some amount of activated carbon can also be added which can remove the contaminants more efficiently (Li et al. 2015). Various pesticides, beta blockers, PCPs and EDCs can be removed through this technology. MBR is a better technology than conventional sludge process as it can remove high amount of EDCs from water (Nguyen et al. 2013).

9.3.3.3 Constructed Wetlands (CWs)

Constructed wetland is an engineered land-based treatment method which is an integrated combination of biological, physicochemical and chemical interactions (Töre et al. 2012). CWs can be classified as subsurface/surface flow (SFWC), horizontal flow (HFWC) and vertical flow (VFWC) on the basis of the wastewater flow management, and hybrid CWs can also be formed by mixing of these different systems (Rodríguez et al. 2014). CWs are highly effective in the removal of pharmaceuticals and personal care products. It can easily remove pesticides, herbicides, beta blockers, NSAIDs, diuretics, etc. Some EDCs like E1, E2, EE2, bisphenol A, phthalates can also be removed successfully (75–100%) through this method (Matamoros et al. 2008). Although out of total available treatment methods, this method for the removal of contaminants is only partially explored. It is a very environmental friendly, low operating cost and green method available for the removal of contaminants (Töre et al. 2012).

9.3.4 Chemical Treatment for the Removal of CECs

All the biological treatments are not effective for the removal of CECs from water, so a wide range of chemical treatments is used for that purpose. Chemical treatment for the removal of CECs can be broadly classified into conventional chemical treatment methods and advanced oxidation processes (AOP). Conventional methods consist of photolysis, chlorination, Fenton process, and AOPs consist of ozonation, heterogeneous photocatalysis (UV/TiO$_2$), photo-Fenton process (UV/H$_2$O$_2$), electro-Fenton process, etc.

9.3.5 Conventional Chemical Treatment Methods for the Removal of CECs

9.3.5.1 Photolysis

Photolysis is one of the conventional methods, used for the decomposition of CECs in water. In this process, removal of CECs using UV remains a commonly used technique. Basically, there are two types of photolysis viz. direct photolysis in which there occur the direct absorption of photons, resulting in the degradation of CECs and another one is indirect photolysis in which photosensitizers are used for processing the reaction. This process is helpful in the complete removal of some pharmaceuticals like diclofenac, iopamidol, ketoprofen, mefenamic acid, oxytetracycline and tetracycline. Although UV photolysis was found to be less effective in the removal of beta blockers (Rodríguez et al. 2008), some pesticides were removed efficiently (80–100%) through this method (Liu et al. 2009; Nguyen et al. 2013).

9.3.5.2 Chlorination

Chlorination is one of the conventional treatment method used to reduce the pathogenic content of biologically treated wastewater. Endocrine-disrupting chemicals (EDCs) and non-steroidal inflammatory drugs (NSAIDs) are characterized as CECs, and they cannot be removed completely through biological wastewater treatment methods (Noutsopoulos et al. 2014). So, this method is used for their removal. It was also found that 17 β Estradiol was 100% removed within 10 min using chlorination method (Belgiorno et al. 2007). But on comparing chlorination process with ozonation, it was found that rate constant value for ozonation was around three orders of magnitude higher than that of chlorination during the removal of some ECs viz. amitriptyline hydrochloride, methyl salicylate, etc. (Real et al. 2014). Along with this, it is also found that during wastewater treatment through this process, some sub-products are formed (Utrilla et al. 2013).

9.3.5.3   Fenton Process

It is an oxidation process in which hydrogen peroxide reacts in the presence of iron to produce hydroxyl radicals. pH plays a very important role in Fenton process as it has been observed that maximum degradation occurs at pH range 2–4 (Petrovic and Barcelo 2007).

Fenton process is represented below:

$$Fe^{+2} + H_2O_2 = Fe^{+3}OH^. + OH^- \tag{1}$$

$$Fe^{+3} + H_2O_2 = Fe^{+2} + HO_2^. + H^+ \tag{2}$$

Although Fe^{+2} can be regenerated from Fe^{+3}, but reaction (2) is much slower than reaction (1), due to which Fe^{+3} accumulates in solution and forms precipitate of $Fe(OH)_3$. Along with this, this process requires significant amount of reagents due to which it becomes costly, and there occurs the unintended consumption of OH and ferrous ions. The removal of CECs using Fenton process was not found to be satisfactory in comparison with other oxidation processes (Ahmed et al. 2015).

9.3.6   Advanced Oxidation Process for the Removal of CECs

9.3.6.1   Ozonation

Ozonation is one of the AO processes, which can go through two mechanisms direct and indirect for the removal of CECs. In direct mechanism, there occurs the reaction with ozone directly, and in indirect mechanism, hydroxyl radical (OH$^.$) obtained from ozone in aqueous solutions participates in the reaction (Rizzo et al. 2019; Utrilla et al. 2013). Ozone selectively reacts with olefins and aromatic ring containing emerging contaminants (Acero et al. 2015). Ozonation is effective in the removal of ECs viz. E1, E3, E2, EE2, bisphenol A and nonylphenol, atrazine, chlorfenvinphos, alachlor, diuron, isobroturum with 90% or higher removal efficiency (Esplugas et al. 2007). The drawback of this process is that it requires high amount of energy, formation of by-products and interference of radical scavengers (Luo et al. 2014).

9.3.6.2 Heterogeneous Photocatalysis (UV/TiO₂)

In photolysis process, the rate of degradation of CECs is quite low, so this advance process is used for the better results. In this process, heterogeneous catalyst is used for the removal of CECs, which is activated in the presence of light (Macwan et al. 2011). TiO_2-based materials are widely used as catalyst for this process which may be due to its photo stability, economic, particle size and inert nature (Gaya and Abdullah 2008). UV/TiO_2 process is effective in degradation of E1, E2, EE2, E3, bisphenol A, progesterone up to 100%. Along with this, this method can also be used for the degradation of pharmaceuticals (mainly analgesics) and pesticides (Gaya and Abdullah 2008).

9.3.6.3 Photo-Fenton Process (UV/H₂O₂)

This method is also widely used for the removal of CECs from wastewater. In this method, UV light is used for the formation of radicals, and the formation occurs by the reaction of hydrogen peroxide in the presence of ferrous ion, but this process can also be possible in the absence of UV light, by using sunlight only. Photo-Fenton process is a pH dependent method, and generally, this reaction takes place in acidic medium (optimum pH $=$ 2.8). Ferrous ion in acidic medium forms $[Fe(OH)_2]^{+2}$, which by the absorption of $h\nu$, go through photoreduction and form OH and Fe^{+2} (Eq. 3).

$$[Fe(OH)]^{+2} + h\nu = Fe^{+2} + OH \qquad (3)$$

Now, this Fe^{+2} ion can further react with H_2O_2 and form OH again (Will et al. 2004). In this process, amount of Fe^{+2} is increased continuously (Tamimi et al. 2008), and during this process, pH reaches to near neutral. The oxidized ligand can further involve in various reactions for the degradation of CECs (Cruz et al. 2012). H_2O_2 and iron concentrations, pH and organic/inorganic content in wastewater are some factors that govern the efficiency of Photo-Fenton process. This process is effective in the removal of various types of pharmaceuticals (except penicillin). In the degradation of some anti-inflammatory pharmaceuticals, viz. antipyrine, 4AA, 4AAA, 4FAA, 4MAA and metronidazole, this process is more effective in comparison with other processes (Tijani et al. 2013). Various pesticides including atrazine, diuron, mecoprop and terbutryn were also oxidized efficiently with the help of photo-Fenton process (Klamerth et al. 2013).

9.3.6.4 Electro-Fenton Process

It is an advanced process of classical Fenton process which has been developed for the better removal of CECs. In this process, H_2O_2 is generated electrochemically in a controlled way (Roth et al. 2016). This method is useful in the removal of iopromide, atenolol, metoprolol, propranolol, triclosan, triclocarban and some antibiotics also (Ganzenko et al. 2014; Estrada et al. 2012). This process of CECs removal is very effective as it avoids the cost of reagent and formation of sludge, etc. (Ganzenko et al. 2014).

9.3.7   Physical Treatment for the Removal of CECs

There are various physical techniques like membrane process, carbon adsorption, mineral surface adsorption and ion exchange, which can be used for the treatment of CECs in water.

9.3.8   Membrane Process

This process is one of the assuring methods among the physical methods of removal of contaminants from surface water and wastewater. Ultrafiltration (UF) and micro-filtration (MF) are used for the advanced treatment of urban wastewater to remove the total suspended solids (TSS) and microorganism. Nano-filtration (NF) and RO membranes are used for treating that water which is already filtered and has low TSS concentration. Various studies show that the pharmaceutical compound, carbamazepine can be removed by 60–90%, 32–40% and >97% by using NF membranes, and in the case of RO membranes, the rejection of carbamazepine was reported to be >99%. Other than carbamazepine, other CECs like diclofenac, E2 a natural steroid hormone, NDMA (undesired by-product of oxidation and disinfection processes) can also be removed using NF and RO process (Rizzo et al. 2019).

9.3.9   Activated Carbon Adsorption

In this process, there occurs the transfer of CECs from liquid phase to the solid phase. Activated carbon (AC) can be applied in two forms namely powdered activated carbon (PAC) and granular activated carbon (GAC), which will be decided on the basis of the nature of adsorbate and absorbant both (Luo et al. 2014). Along with this, the effectiveness of the process also depends upon adsorbate solubility, the adsorbate and adsorbent hydrophobicity. This process is very useful for the removal of pharmaceuticals, and the advantage of this method is that it does not form toxic and pharmacologically active products (Utrilla et al. 2013).

9.4 Conclusion

Emerging contaminants are present in our environment due to various activities of humans. ECs or CECs have been defined as those contaminants for which no strict regulations have been imposed. Thus, there is a constant need to monitor ECs or CECs. With increase in globalization and industrialization, their number and concentration are increasing day by day. However, concern regarding emerging contaminants is only being developed lately. Emerging contaminants have been present in industries, wastewater treatment plants, cosmetics, pharmaceuticals, etc., but it is only recently that scientists and common people are realizing their non-degradable and long-lasting nature in the environment. The above-mentioned problem combined with their harmful effects on human and animal health has emerged as a threat for both humans and animals. Many CECs like PPCPs, POPs, disinfectants, insecticides, PFOS, PFOA, algal toxins, etc., are harmful to health. Detection of these CECs from time to time is the need of the hour. Many countries around the world are suffering from problem related to CECs. Anthropogenic activities are the key factors responsible for occurrence of these contaminants in the environment and water bodies. Various conventional and advanced techniques are available for the detection of these emerging contaminants. The ECs can be dealt with by biological, chemical and physical processes. Sedimentation, membrane filtration, chemical treatments are some important methods for the removal of ECs from our environment. The threat of ECs can be fought by avoiding/limiting the use of harmful chemicals in industries. Also, the threat can be tackled by using bio-degradable substances obtained from nature in the manufacturing process and developing newer technologies to remove emerging contaminants from our environment. It is very important for humans to identify the substances, drugs and chemicals having the characteristics of ECs or CECs. After identification of ECs or CECs, governments around the world should take quick actions to set up regulations which will help in the removal of these contaminants so that the environment in which mankind resides remains safe and pure. Extensive and exhaustive research needs to be carried out by various countries around the world for developing newer methods for eradication and removal of ECs and CECs from our environment.

References

Acero JL, Benitez FJ, Real FJ, Rodriguez E (2015) Erlimination of selected emerging contaminants by the combination of membrane filtration and chemical oxidation processes. Water Air Soil Pollut 226:1–14
Ahmed MB, Zhou JL, Ngo HH, Guo W (2015) Adsorptive removal of antibiotics from water and wastewater: progress and challenges. Sci Total Environ 532:112–126
Ahmed MB, Zhou JL, Ngo HH, Guo W, Thomaidis NS, Xu J (2016) Progress in the biological and chemical treatment technologies for emerging contaminant removal from wastewater: a critical review. J Hazard Mater 323:274–298

Banihashemi B, Droste RL (2014) Sorption–desorption and biosorption of bisphenol A, triclosan, and 17α-ethinylestradiol to sewage sludge. Sci Total Environ 487:813–821

Bao LJ, Wei YL, Yao Y, Ruan QQ, Zeng EY (2014) Global trends of research on emerging contaminants in the environment and humans: a literature assimilation. Environ Sci Pollut Res 22:1635–1643

Behera SK, Kim HW, Oh JE, Park HS (2011) Occurrence and removal of antibiotics, hormones and several other pharmaceuticals in wastewater treatment plants of the largest industrial city of Korea. Sci Total Environ 409:4351–4360

Belgiorno V, Rizzo L, Fatta D, Rocca CD, Lofrano G, Nikolaou A, Meric S (2007) Review on endocrine disrupting-emerging compounds in urban wastewater: occurrence and removal by photocatalysis and ultrasonic irradiation for wastewater reuse. Desalination 215:166–176

Blüthgen N, Meili N, Chew G, Odermatt A, Fent K (2014) Accumulation and effects of the UV-filter octocrylene in adult and embryonic zebrafish (Danio rerio). Sci Total Environ 476–477:207–217

Buttiglieri G, Knepper T (2008) Removal of emerging contaminants in wastewater treatment: conventional activated sludge treatment. In: Barcelo MPD (ed) Emerging contaminants from industrial and municipal waste. Springer, Berlin, pp 1–35

Cal Baier-Anderson BC, Anderson CB (2006) Estimates of exposures to perchlorate from consumption of human milk, dairy milk, and water, and comparison to current reference dose. J Toxicol Environ Health 69:319–330

Casado J, Nescatelli R, Rodríguez I, Ramil M, Marini F, Cela R (2014) Determination of benzotriazoles in water samples by concurrent derivatization–dispersive liquid–liquid microextraction followed by gas chromatography–mass spectrometry. J Chromatogr A 1336:1–9

Cruz ND, Giménez J, Esplugas S, Grandjean D, Alencastro LD, Pulgarín C (2012) Degradation of 32 emergent contaminants by UV and neutral photo-fenton in domestic wastewater effluent previously treated by activated sludge. Water Res 46:1947–1957

Damstra T, Barlow S, Bergman A, Kavlock R, Van der Kraak G (2002) Global assessment of the state-of-the-science of endocrine disruptors. WHO, 180

Dey S, Bano F, Malik A (2019) Pharmaceuticals and personal care product (PPCP) contamination—a global discharge inventory. In: Pharmaceuticals and personal care products: waste management and treatment technology. Elsevier, Amsterdam, pp 1–26

EFSA (2011) Results of the monitoring of perfluoroalkylated substances in food in the period 2000–2009. EFSA J 9:34

Esplugas S, Bila DM, Krause LG, Dezotti M (2007) Ozonation and advanced oxidation technologies to remove endocrine disrupting chemicals (EDCs) and pharmaceuticals and personal care products (PPCPs) in water effluents. J Hazard Mater 149:631–642

Estrada AL, Li YY, Wang A (2012) Biodegradability enhancement of wastewater containing cefalexin by means of the electro-Fenton oxidation process. J Hazard Mater 227:41–48

Fairbairn DJ, Elliott SM, Kiesling RL, Schoenfuss HL, Ferrey ML, Westerhoff BM (2018) Contaminants of emerging concern in urban stormwater: spatiotemporal patterns and removal by iron-enhanced sand filters (IESFs). Water Res 145:332–345

Falconer IR, Humpage AR (2005) Health risk assessment of cyanobacterial (blue-green algal) toxins in drinking water. Int J Environ Res Public Health 2:43–50

Falconer IR, Humpage AR (2006) Cyanobacterial (blue-green algal) toxins in water supplies: cylindrospermopsins. Environ Toxicol 21:299–304

Ferrey ML, Heiskary S, Grace R, Hamilton M, Lueck A (2015) Pharmaceuticals and other anthropogenic tracers in surface water: a randomized survey of 50 minnesota lakes. Environ Toxicol Chem 34:2475–2488

Furlong ET, Batt AL, Glassmeyer ST, Noriega MC, Kolpin DW, Mash H, Schenck KM (2017) Nationwide reconnaissance of contaminants of emerging concern in source and treated drinking waters of the United States: pharmaceuticals. Sci Total Environ 579:1629–1642

Ganzenko O, Huguenot D, Hullebusch ED, Esposito G, Oturan MA (2014) Electrochemical advanced oxidation and biological processes for wastewater treatment: a review of the combined approaches. Environ Sci Pollut Res 21:8493–8524

Gaya UI, Abdullah AH (2008) Heterogeneous photocatalytic degradation of organic contaminants over titanium dioxide: a review of fundamentals, progress and problems. J Photochem Photobiol C 9:1–12

Gerrity D, Gamage S, Holady JC, Mawhinney DB, Quiñones O, Trenholm RA, Snyder SA (2011) Pilot-scale evaluation of ozone and biological activated carbon for trace organic contaminant mitigation and disinfection. Water Res 45:2155–2165

Gómez CM, Lamoree M, Hamers T, Velzen M, Kamstra J, Fernández B, Vethaak A (2013) Integrated chemical and biological analysis to explain estrogenic potency in bile extracts of red mullet (*Mullus barbatus*). Aquat Toxicol 134–135:1–10

Gong Y, Tian H, Wang L, Yu S, Ru S (2014) An integrated approach combining chemical analysis and an in vivo bioassay to assess the estrogenic potency of a municipal solid waste landfill leachate in Qingdao. PLoS ONE 9:1–9

Hal T, Hart J, Croll B, Gregory R (2007) Laboratory-scale investigations of algal toxin removal by water treatment. Water Environ J 14:143–149

https://pubchem.ncbi.nlm.nih.gov

Jafari AJ, Abasabad RP, Salehzadeh A (2009) Endocrine disrupting contaminants in water resources and sewage in Hamadan city of Iran. Iran J Environ Health Sci Eng 6:89–96

Jaimes JA, Postigo C, Alemán RM, Aceña J, Barceló D, Alda ML (2018) Study of pharmaceuticals in surface and wastewater from Cuernavaca, Morelos, Mexico: occurrence and environmental risk assessment. Sci Total Environ 613–614:1263–1274

Jin P, Jin X, Wang X, Feng Y, Wang XC (2013) Biological activated carbon treatment process for advanced water and wastewater treatment. Biomass Now Cultivation Utilization 153–192

Kalkan Ç, Yapsakli K, Mertoglu B, Tufan D, Saatci A (2011) Evaluation of biological activated carbon (BAC) process in wastewater treatment secondary effluent for reclamation purposes. Desalination 265:266–273

Klamerth N, Malato S, Agüera A, Alba AF (2013) Photo-Fenton and modified photo-Fenton at neutral pH for the treatment of emerging contaminants in wastewater treatment plant effluents: a comparison. Water Res 47:833–840

Kolpin D, Furlong ET, Meyer MT, Thurman EM, Zaugg S, Barber LB, Buxton H (2002) Pharmaceuticals, hormones, and other organic wastewater contaminants in U.S. streams, 1999–2000: a national reconnaissance. Environ Sci Technol 36:1202–1211

Kolpin DW, Skopec M, Meyer MT, Furlong ET, Zaugg SD (2004) Urban contribution of pharmaceuticals and other organic wastewater contaminants to streams during differing flow conditions. Sci Total Environ 328:119–130

Kumar A, Mohan S, Sarma PN (2009) Sorptive removal of endocrine-disruptive compound (estriol, E3) from aqueous phase by batch and column studies: kinetic and mechanistic evaluation. J Hazard Mater 164:820–828

Kumar M, Ram B, Honda R, Poopipattana C, Canh VD, Chaminda T, Furumai H (2019a) Concurrence of antibiotic resistant bacteria (ARB), viruses, pharmaceuticals and personal care products (PPCPs) in ambient waters of Guwahati, India: urban vulnerability and resilience perspective. Sci Total Environ 693: 133640. https://doi.org/10.1016/j.scitotenv.2019.133640

Kumar M, Chaminda T, Honda R, Furumai H (2019b) Vulnerability of urban waters to emerging contaminants in India and Sri Lanka: resilience framework and strategy. APN Sci Bull 9(1). https://doi.org/10.30852/sb.2019.799

Kümmerer K (2009) The presence of pharmaceuticals in the environment due to human use—present knowledge and future challenges. J Environ Manag 90:2354–2366

Lai WP, Lin YC, Tung HH, Lo SL, Lin AY (2016) Occurrence of pharmaceuticals and perfluorinated compounds and evaluation of the availability of reclaimed water in Kinmen. Emerg Contam 2:1–10

Lange FT, Scheurer M, Brauch HJ (2012) Artificial sweeteners—a recently recognized class of emerging environmental contaminants: a review. Anal Bioanal Chem 403:2503–2518

Lee KE, Langer SK, Menheer MA, Foreman WT, Furlong ET, Smith SG (2011) Chemicals of emerging concern in water and bottom sediment in Great Lakes areas of concern, 2010 to 2011—collection methods, analyses methods, quality assurance, and data. Report, U.S. Geological Survey

Li C, Cabassud C, Guigui C (2015) Evaluation of membrane bioreactor on removal of pharmaceutical micropollutants: a review. Desalin Water Treat 55:845–858

Liu ZH, Kanjo Y, Mizutani S (2009) Removal mechanisms for endocrine disrupting compounds (EDCs) in wastewater treatment—physical means, biodegradation, and chemical advanced oxidation: a review. Sci Total Environ 407:731–748

Luo Y, Guo W, Ngo HH, Nghiem LD, Hai FI, Zhang J, Wang XC (2014) A review on the occurrence of micropollutants in the aquatic environment and their fate and removal during wastewater treatment. Sci Total Environ 473–474:619–641

Lv X, Xiao S, Zhang G, Jiang P, Tang F (2016) Occurrence and removal of phenolic endocrine disrupting chemicals in the water treatment processes. Report 6:1–10

Macwan D, Dave PN, Chaturvedi S (2011) A review on nano-TiO2 sol–gel type syntheses and its applications. J Mater Sci 46:3669–3686

Matamoros V, García J, Bayona JM (2008) Organic micropollutant removal in a full-scale surface flow constructed wetland fed with secondary effluent. Water Res 42:653–660

Millsa LJ, Chichester C (2005) Review of evidence: Are endocrine-disrupting chemicals in the aquatic environment impacting fish populations? Sci Total Environ 343:1–34

Miraji H, Othman OC, Ngassapa FN, Mureithi E (2016) Research trends in emerging contaminants on the aquatic environments of Tanzania. Scientifica 2016:1–6

Montagner CC, Sodré FF, Acayaba RD, Vidal C, Campestrini I, Locatelli MA, Jardim WF (2019) Ten years-snapshot of the occurrence of emerging contaminants in drinking, surface and ground waters and wastewaters from São Paulo State, Brazil. J Braz Chem Soc 30:614–632

Mukherjee S, Patel AKR, Kumar M (2020) Water scarcity and land degradation nexus in the era of anthropocene: some reformations to encounter the environmental challenges for advanced water management systems meeting the sustainable development. In: Kumar M, Snow D, Honda R (eds) Emerging issues in the water environment during anthropocene: a South East Asian perspective. Springer Nature. ISBN 978-93-81891-41-4

Nguyen LN, Hai FI, Kang J, Price WE, Nghiem LD (2013) Removal of emerging trace organic contaminants by MBR-based hybrid treatment processes. Int Biodeterior Biodegradation 85:474–482

Nosek K, Styszko K, Golas J (2014) Combined method of solid-phase extraction and GC-MS for determination of acidic, neutral, and basic emerging contaminants in wastewater (Poland). Int J Environ Anal Chem 94:1–14

Noutsopoulos C, Koumaki E, Mamais D, Nika MC, Bletsou AA, Thomaidis NS (2014) Removal of endocrine disruptors and non-steroidal anti-inflammatory drugs through wastewater chlorination: the effect of pH, total suspended solids and humic acids and identification of degradation by-products. Chemosphere 119:5109–5114

Petrovic M, Barcelo D (2007) LC-MS for identifying photodegradation products of pharmaceuticals in the environment. Trends Anal Chem 26:486–493

Pidlisnyuk VV, Marutovsky RM, Radeke KH, Klimenko NA (2003) Biosorption processes for natural and waste water treatment—part II: experimental studies and theoretical model of a biosorption fixed bed. Eng Life Sci 3:439–445

Radjenović J, Petrović M, Barceló D (2009) Fate and distribution of pharmaceuticals in wastewater and sewage sludge of the conventional activated sludge (CAS) and advanced membrane bioreactor (MBR) treatment. Water Res 43:831–841

Rashid B, Husnain T, Riazuddin S (2010) Herbicides and pesticides as potential pollutants: a global problem. In: Plant adaptation and phytoremediation. Springer, Berlin, pp 427–447

Real FJ, Benitez JF, Acero JL, Casas F (2014) Comparison between chlorination and ozonation treatments for the elimination of the emerging contaminants amitriptyline hydrochloride, methyl salicylate and 2-phenoxyethanol in surface waters and secondary effluents. J Chem Technol Biotechnol 90:1400–1407

Richardson SD, Ternes TA (2018) Water analysis: emerging contaminants and current issues. Anal Chem 90:398–428

Riva F, Castiglioni S, Fattore E, Manenti A, Davoli E (2018) Monitoring emerging contaminants in the drinking water of Milan and assessment of the human risk. Int J Hyg Environ Health 221:451–457

Rizzo L, Malato S, Antakyali D, Beretsou VG, Đolić MB, Gernjak W, Kassinos DF (2019) Consolidated vs new advanced treatment methods for the removal of contaminants of emerging concern from urban wastewater. Sci Total Environ 655:986–1008

Rodil R, Moeder M (2008) Development of a simultaneous pressurised-liquid extraction and cleanup procedure for the determination of UV filters in sediments. Anal Chim Acta 612:152–159

Rodríguez A, Rosal R, Perdigón-Melón JA, Mezcua M, Agüera A, Hernando MD, Letón P, Fernández-Alba AR, García-Calvo E (2008) Ozone-based technologies in water and wastewater treatment. In: The handbook of environmental chemistry, vol 5, pp 127–175

Rodríguez AG, Matamoros V, Fontàs C, Salvadó V (2014) The ability of biologically based wastewater treatment systems to remove emerging organic contaminants—a review. Environ Sci Pollut Res 21:11708–11728

Roth H, Gendel Y, Buzatu P, David O, Wessling M (2016) Tubular carbon nanotube-based gas diffusion electrode removes persistent organic pollutants by a cyclic adsorption—electro-Fenton process. J Hazard Mater 307:1–6

Samaraweera DN, Liu X, Zhong G, Priyadarshana T, Malik RN, Zhang G, Peng X (2019) Antibiotics in two municipal sewage treatment plants in Sri Lanka: occurrence, consumption and removal efficiency. Emerg Contam 5:272–278

Sang Z, Leung KS (2016) Environmental occurrence and ecological risk assessment of organic UV filters in marine organisms from Hong Kong coastal waters. Sci Total Environ 566–567:489–498

Shi W, Deng D, Wang Y, Hu G, Guo J, Zhang X, Wang Z (2016) Causes of endocrine disrupting potencies in surface water in East China. Chemosphere 216:1435–1442

Singh A, Patel AK, Kumar M (2020) Mitigating the risk of arsenic and fluoride contamination of groundwater through a multi-model framework of statistical assessment and natural remediation techniques. In: Kumar M, Snow D, Honda R (eds) Emerging issues in the water environment during anthropocene: a South East Asian perspective. Springer Nature. ISBN 978-93-81891-41-4

Squadrone S, Ciccotelli V, Prearo M, Favaro L, Scanzio T, Foglini C, Abete M (2015) Perfluorooctane sulfonate (PFOS) and perfluorooctanoic acid (PFOA): emerging contaminants of increasing concern in fish from Lake Varese, Italy. Environ Monit Assess 187:438

Stefanakis AI, Becker JA (2015) A review of emerging contaminants in water: classification, sources, and potential risks. In: Impact of water pollution on human health and environmental sustainability. IGI Global Disseminator of Knowledge, pp 55–80

Steinmaus CM (2016) Perchlorate in water supplies: sources, exposures, and health effects. Curr Environ Health Rep 3:136–143

Stork NE (2010) Re-assessing current extinction rates. Biodivers Conserv 19:357–371

Tamimi M, Qourzal S, Barka N, Assabbane A, Ait-Ichou Y (2008) Methomyl degradation in aqueous solutions by Fenton's reagent and the photo-Fenton system. Sep Purif Technol 61:103–108

Thomaidis N, Asimakopoulos A, Bletsou A (2012) Emerging contaminants: a tutorial mini-review. Glob NEST J 14:72–79

Tijani JO, Fatoba OO, Petrik LF (2013) A review of pharmaceuticals and endocrine-disrupting compounds: sources, effects, removal, and detections. Water Air Soil Pollut 224:1–29

Töre GY, Meriç S, Lofrano G, De Feo G (2012) Removal of trace pollutants from wastewater in constructed wetlands. In: Lofrano G (ed) Emerging compounds removal from wastewater. Springer, Berlin, pp 39–58

Tsui MM, Leung H, Wai TC, Yamashita N, Taniyasu S, Liu W, Murphy MB (2014) Occurrence, distribution and ecological risk assessment of multiple classes of UV filters in surface waters from different countries. Water Res 67:55–65

Utrilla JR, Polo MS, García MÁ, Joya GP (2013) Pharmaceuticals as emerging contaminants and their removal from water. A review. Chemosphere 93:1268–1287

Van Asselt ED, Kowalczyk J, Van Eijkeren JCH, Zeilmaker MJ, Ehlers S, Fürst P, Lahrssen-Wiederholt M, Van der Fels-Klerx HJ (2013) Transfer of perfluorooctane sulfonic acid (PFOS) from contaminated feed. Food Chem 141:1489–1495

Vargo JD (2013) Determination of chloroacetanilide and chloroacetamide herbicides and their polar degradation products in water by LC/MS/MS. In: Analytical chemistry. ACS Publications, Washington, D.C., pp 238–255

Wanda EM, Nyoni H, Mamba BB, Msagati TA (2017) Occurrence of emerging micropollutants in water systems in Gauteng, Mpumalanga, and North West provinces, South Africa. Int J Environ Res Public Health 14:79

Wang Y, Zhou J (2013) Endocrine disrupting chemicals in aquatic environments: a potential reason for organism extinction? Aquat Ecosyst Health Manage 16:88–93

Will IB, Moraes JE, Teixeira AC, Guardani R, Nascimento CA (2004) Photo-Fenton degradation of wastewater containing organic compounds in solar reactors. Sep Purif Technol 34:51–57

Xian Q, Siddique S, Li T, Feng YL, Takser L, Zhu J (2011) Sources and environmental behavior of dechlorane plus—a review. Environ Int 37:1273–1284

Ying GG, Kookana RS (2003) Degradation of five selected endocrine-disrupting chemicals in seawater and marine sediment. Environ Sci Technol 37:1256–1260

Zeng L, Wang T, Wang P, Liu Q, Han S, Yuan B, Jiang G (2011) Distribution and trophic transfer of short-chain chlorinated paraffins in an aquatic ecosystem receiving effluents from a sewage treatment plant. Environ Sci Technol 45:5529–5535

Zhu Y, Price OR, Kilgallon J, Rendal C, Tao S, Jones KC, Sweetman AJ (2016) A multimedia fate model to support chemical management in China: a case study for selected trace organics. Environ Sci Technol 50:7001–7009

Chapter 10
Water Crisis in the Asian Countries: Status and Future Trends

Vandana Shan, S. K. Singh, and A. K. Haritash

10.1 Introduction

Water is the most important natural resource required for the economic development of a nation. Safe and clean water is the basic necessity of humankind for their rational use. In general, the global water crisis is majorly related to decrease in the availability of water quantity due to irrational use by the human population. Not only the freshwater accessibility but water quality is also affected due to various natural degradation and anthropogenic activities subjected to frequent floods, drought and uneven rainfall pattern and dumping of effluent from industries, untreated sewage waste, agrochemicals run-off from agricultural fields into water bodies, respectively (Abbaspour 2011; Mukherjee et al. 2020). Six transboundary rivers, namely Mekong, Irrawaddy, The Red, Salween Ganges-Brahmaputra-Meghna and Indus Rivers and their tributaries are responsible for higher socioeconomic growth, food production, environmental stability and sustainable development in various Asian countries (Ismail 2016). But in recent years, conditions have become worse as we stepped into twenty-first century. Water scarcity with respect to quantity and water pollution with respect to quality has become major issues in developing as well as developed nations. The current situation in Asian developing countries is highly variable and majorly affected by socioeconomic and physical conditions along with various phases of civilisation and industrialisation. Human Development Report (2006) of UNDP stated that water crisis is the most serious problem for human in Asian countries due to increased demand for domestic, industrial, irrigational as well as various economic and developmental activities (Srivastava et al. 2008) which exerted multiple pressure on the water resources of the region. Water-related issues in Asian countries are serious and diverse. It was stated that an increased number of East, West and South Asian countries will be facing a water deficit situation by 2050 (Falkenmark et al. 2009;

V. Shan (✉) · S. K. Singh · A. K. Haritash
Department of Environmental Engineering, Delhi Technological University, Delhi, India
e-mail: vandanashan@dce.ac.in

© Springer Nature Singapore Pte Ltd. 2020
M. Kumar et al. (eds.), *Resilience, Response, and Risk in Water Systems*,
Springer Transactions in Civil and Environmental Engineering,
https://doi.org/10.1007/978-981-15-4668-6_10

Singh et al. 2020). Expanding population and unsustainable use of natural resources, especially water, are causes of great concern in Asian countries. Around 31% of the earth's total arable land supports nearly 56% of the world's population only in Asia. Rapid urbanisation has also robbed many of the Asian countries' environmental wealth by deteriorating the quality of soil, air and water. Accessibility of clean and safe water is essential for human health and development. Both are affected if the water is polluted and unfit for consumption. Also, polluted water poses a major threat to human health (UNEP 2016). More than 50% of the population in most of the Asian countries, viz. China, Malaysia, Indonesia, Iran, Brunei, Mongolia, Singapore, North Korea, Japan, Pakistan, Kazakhstan, South Korea, Philippines and Turkmenistan will be urban by 2025. The estimated urban population proportion percentage projected to be 52, 53 and 45 in East Asia, South-east Asia, and South and Central Asia, respectively (EWCRP 2002). Not only the problems of surface water pollution but depletion of groundwater reserves due to excessive utilisation for irrigation in many Asian countries is of great concern threatening current and future generations. According to current situation it was estimated that by 2080, the percentage of the world population have to live forcefully in water-scarce regions will increase from 28 to 50.

10.2   Global Water Distribution on Earth Surface

Water in lakes, rivers, streams, reservoirs and groundwater is the principal source of water for human use (Srivastava et al. 2008). The total volume of water on earth surface is 1.39 billion cubic kilometres (Table 10.1). Around 96.5% of earth water is saline and constituted mainly by global oceans. In case of freshwater content and its availability, a very small amount of freshwater is available for human consumption on the earth surface as approximately 1.7% freshwater is locked up in glaciers, polar ice caps and snow and only 1.7% is stored in groundwater, lakes, ponds, streams, rivers and soil (Table 10.1). About 1% of total water exists in water vapour on the earth's atmosphere (Graham et al. 2010).

10.3   Global Water Consumption

It was estimated that a large amount of freshwater is available on the earth surface to support 9 billion population and more. But most of it is in inaccessible form. Due to the unequal distribution of water in various geographical regions and unequal consumption of it has created water-scarce conditions in major regions of the world. Global water consumption is maximum for Asia and least for Australia and Oceania. More than half of the world's water consumption is in Asian countries which support more than 50% of the world population. Around 70% of freshwater is consumed in agriculture, after that in industries and rest is used for various developmental

Table 10.1 Worldwide distribution of water content on earth surface (Gleick 1996)

Estimate of worldwide water distribution	Volume (10^3 km^3)	Total water percentage	Freshwater percentage
Oceans, seas and bays	1,338,000	96.5	–
Ice caps, glaciers and permanent snow	24,064	1.74	68.7
Groundwater	23,400	1.7	–
Fresh	(10,530)	(0.76)	30.1
Saline	(12,870)	(0.94)	–
Soil moisture	16.5	0.001	0.05
Ground ice and permafrost	300	0.022	0.86
Lakes	176.4	0.013	–
Fresh	(91.0)	(0.007)	0.26
Saline	(85.4)	(0.006)	–
Atmosphere	12.9	0.001	0.04
Swamp water	11.47	0.0008	0.03
Rivers	2.12	0.0002	0.006
Biological water	1.12	0.0001	0.003
Total	1,385,984	100.0	100.0

activities, domestic use and human consumption. Consumption of water is 10 times more in developed countries than developing nation but overall consumption is more in developing nation due to large population and extensive economical activities (UNWWD 2015) (Table 10.2).

Table 10.2 Global distribution of water consumption

Geographic location	Percentage
Asia	55
North America	19
Europe	9.2
Africa	4.7
South America	3.3
Rest of the world	8.8

10.4   Global Water Stress

Being a relative concept, water scarcity can appear at any stage of water supply or demand. Water scarcity leads to unavailability or shortage of freshwater resources which leads to unsatisfied demand, economic competition for water quality and quantity, fast abstraction of groundwater, conflicts between states at national and international level and irreversible impacts on the environment and its segments. According to hydrologists (Falkenmark et al. 1989), population–water equation is an indicator of water stress conditions in a particular region or country. Population in a particular region having annual water supply level below 1700 m^3 per person experiences water stress and faces water scarcity if annual water supply falls below 1000 m^3 per person and absolute water scarcity is due to annual water supply drop below 500 m^3 per person. In general, periodic or limited water shortage levels are expected to be at levels between 1700 and 1000 m^3 per person per year. Those regions which have water level below 1000 m^3 per person per year have to face water scarcity (Falkenmark and Lindh 1976). It was stated that about one-third of the population in developing nations will have to face severe water scarcity by 2025 (Seckler et al. 1998; Patel et al. 2019a, b) (Fig. 10.1).

Physical water scarcity is related to the quantity and economic water scarcity is related to quality. Physical water scarcity in most of the regions is due to lack of enough water resources and economic water scarcity is due to lack of access to safe water. Water scarcity is a global issue. Around 1.2 billion people of the world population live in the regions of physical scarcity and about 500 million and more are approaching along with this harsh situation. About 1.6 billion people are facing economic water scarcity even if the water is present locally (HDR 2006; UNDP 2006). Around 700 million populations in 43 countries of the world were reported to live below 1700 m^3 per person (WWDR 2012). One-fourth of the population

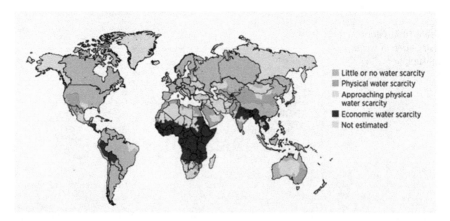

Fig. 10.1 Physical and economical water scarcity at global level (World Water Development Report 4. World Water Assessment Programme (WWAP), March 2012a)

in China, India and Sub-Saharan Africa countries are living under water-stressed conditions, and the number is continuously increasing. Middle East is known as the world's most water-stressed region having an average value of 1200 m^3 of water per person (WWDR 2012). Water availability per person, per year, is decreasing continuously likely to cause water crisis in coming years as shown in Fig. 10.2.

10.5   Current Scenario in Asian Countries

Demographic and economic growth in most Asian countries is responsible for increased water withdrawal and water utilisation in the last 20–30 years (Kundzewicz et al. 2007). In the year 2000, Industrial, agricultural and domestic water consumption was about 18%, 70% and 13% of total water withdrawal by continents, respectively, while in Asia, its consumption varies by 85% in agriculture, 7% in industries and 8% in domestic use (Gleick et al. 2006; Wada et al. 2011; Kumar et al. 2009). A number of Asian countries have expanding economies which require large water supply for energy generation and infrastructure development. A reliable supply of freshwater is a basic need in various industrial processes which is required for the economic growth of the region. Increased communities' wealth further triggers in the requirement for protection of ecosystem (AWDO 2013; Patel et al. 2019a, b).

10.5.1   Southeast Asia

Myanmar, Laos, Cambodia, Malaysia, Thailand, Vietnam, Indonesia, Philippines and Singapore, these nine countries are situated in South-east Asia. The Mekong and Salween Rivers fed almost all of these countries situated near and in their basin and could be used for economic and social development of the region. This region is facing a tremendous pressure of population growth along with urbanisation. According to current estimates, it was found that out of 22 of 34 Indian cities face shortage of water daily. Geographical and economic inequalities in South-east Asian zone have made most of these countries to face the problems of unavailability of safe and clean water and sanitation. World Bank Report stated that in Indonesia, around 80% population is unable to access safe water and in Jakarta, 4.5 million of population have facilities of piped water and 35% of population use water from artesian wells, 15% consume bottled water and 8% population is still dependent on rainwater (Arshad 2016). Governments are planning to overcome current situations by 40% by 2019. In 2011, Thailand faces a great impact on its economy and experiences extensive flooding which made government look into the situation of water scarcity. Thailand is the country where increased economic growth is enabling a great impact on water quality management and environmental conservation. To maintain a proper balance between the use of various water resources between industrial, domestic, energy, agricultural sectors in developed economies in this region is highly needed.

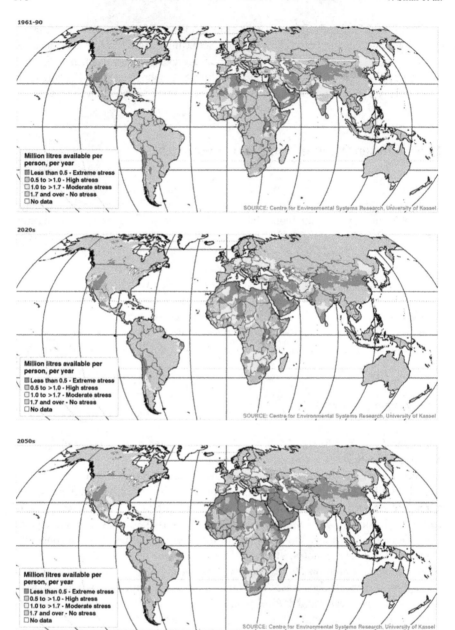

Fig. 10.2 Water availability scenario based on temperature, population and industrialisation (WWAP 2012a, b)

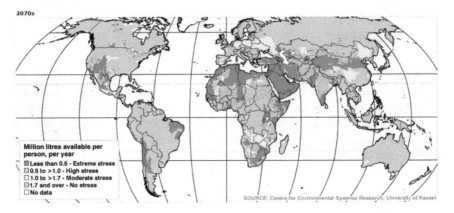

Fig. 10.2   (continued)

10.5.2   East Asia

In East Asia, major consumer of water is agriculture sector, while industries and municipal water supply accounts 22% and 14%, respectively. Water plays an important role in economic growth and poverty reduction in a country. So considering the importance of water, Government of Republic of China State Council has started water sector investment programme which deals in improved quality of water in lakes and rivers, increased water productivity and its use efficiency and total national water abstraction at 700 km^3 by 2030 (CSC 2012). Due to increase in demand in thermal energy generation, demand for water has also been doubled improved water use efficiency.

10.5.3   Central and West Asia

Mostly, the Central Asia regions producing cotton and wheat are deficient in agriculture and food security. Since the last five decades, the ecosystems of Aral Sea have been depleted due to various agricultural activities. Lack of appropriate technologies and nonavailability of financial facilities resulted in deterioration of the irrigational system. Moreover, the development of irrigation canals crossing international boundaries increases new limitations related to the management of irrigational system along with water allocation to a different region. Apart from this scanty rainfall, conventional irrigation methods and saline groundwater resources add up to the agricultural production of the Central Asia Region (Asian Water Development Outlook 2013).

10.6   The Major Contributing Factor for Water Stress Conditions

Asian zone is a highly populous region which is undergoing urbanisation, energy production, agricultural and industrial advancement with a fast pace. All of these are highly needed aspects for the social and economic growth of the region. But these aspects may act as a driving force which may lead to unfavourable situations of water stress and decreased region's capacity to maintain a proper balance in socioeconomic need of water development.

10.6.1   Increased Population and Pollution

The main contributing factor of water scarcity in developing nations is the exponential growth rate of the human population which extensively and unsustainably use renewable freshwater resources (Marcoux 1996). It was roughly estimated that our world's population will shoot 10 billion people by 2050. Increased population is responsible for alleviating various environmental issues, ultimately exerted tremendous pressure on natural resources. Enhanced growth in population will demand additional land for cultivation, human settlements and various economic activities which lead to deforestation, forest clearing, overexploitation of water resources and wildlife habitat destruction. China and Vietnam are only two countries which affirm the slowing down in population growth and have positive impacts on the environment, and Singapore and Malaysia favour in population growth. No country is ready to slow down the process of urbanisation (World Bank 1989). Most of the Asian countries have sufficient natural resources, but the continuous annual increase in human population in various Asian countries like in China, Indonesia, Philippines and Vietnam is 13 million, 3 million, 1.4 million and 1.6 million, respectively (World Bank 1995; Yimin 1994; SRV 1995; Cruz and Cruz 1990). Extensive uses of fertilisers and pesticides have further worsened the quality of freshwater bodies. Eutrophication of lakes is the major problems encountered in many emerging economies. Some researchers have also reported that around 80% of untreated sewage is discharged in surface water bodies in developing countries (UNWWAP 2009) which affect not only the water ecosystems but also human health (Palaniappan et al. 2010). Also, arsenic, nitrate and fluoride pollution of groundwater has affected the health of a large human population and aquatic biota which are the dependents on these resources for their livelihood (Gun 2012; Das et al. 2016). Illegal dumping of untreated industrial waste has further deteriorated water quality and decreased the self-purification capacities of major rivers. River health index stated that around eighty per cent of rivers in Asian region is in very poor health conditions. Pollution from cities is only a part of the challenge to the security of the water environment. Most of the rivers in South Asia and Central and West Asia are reported rivers that are in very poor health condition where environmental water security is threatened in river basins of Azerbaijan, Thailand, Bangladesh, Sri Lanka and Pakistan.

10.6.2   Increased Water Consumption by the Agricultural Sector and Food Security

Increased population will lead to increased food demand which will be more than 50% by the current value. If the same trend in the current consumption of freshwater resources will continue then about two-thirds of the global population would have to live in water-stressed conditions by 2025. India and China are two major countries which support more than 60% of the total Asian population. Agriculture is a major factor in water scarcity. A huge amount of water is required for the production of good amount of crop. More than 70% water is used in agricultural production (FAO 2011; Oude Essink et al. 2010). To feed increasing population, farmers of next generations have to face challenges to double the food production to feed increasing population by using half the water available, without use of fossil fuels, costly pesticides, chemicals and fertilisers, less available arable land, infertile or less fertile soil and under the pressure of changed climatic conditions. Today, one-fourth of total world land has been highly degraded compounding with mining, hazardous pollutants, recreation, urban shifting and rising sea levels. It was estimated that in last two decades, 15–80% degradation of global agricultural land has been noticed (Bindraban et al. 2012; Oldeman et al. 1991; Pimentel and Burgess 2013; Bai et al. 2008). And figure increased in present years by 25% for highly degraded, 44% slightly degraded and about 10% recovering from degradation (Bindraban et al. 2012; FAO 2011; Cribb 2010). As most of the developing countries in Asian region have insufficient or unsuitable agricultural technologies as compared to developed nations and are not helpful to overcome hunger and unsustainability due to high costs and less skill (Rekacewicz 2008). In last two decades, around an average estimate of 2000 ha land in more than 75 countries in arid and semi-arid regions have been degraded by salt, reported by UN University's Canadian-based Institute for Water, Environment and Health. Recently, about 20% of world's irrigated land area have affected several soil properties and food products due to less rainfall, poor rainwater percolation through the soil, unscientific irrigation practices without drainage system in dry areas (Sinha 2014). Researchers gave a rough estimate of around 50% of irrigated land in parts of Central Asia which are waterlogged and affected by high salt content (Qadir et al. 2009). About 88% of water withdrawals from the Brahmaputra, Indus and Meghna River Basin and used for irrigation and often the excess run-off water contaminated with agricultural chemicals and pesticides from irrigated land return flows to the river systems and affect water quality (Babel and Wahid 2008).

At the global level, around 1989, due to various anthropogenic activities, 15% of the land area was degraded globally. In this region, around 50% of the countries are near to the global average, but most of them are highly degraded. Most affected regions are China, DPR Korea or Vietnam, and Thailand. Only Laos is markedly better off than average from this standpoint. 18% of total Asian land proportion is degraded till now (Beck et al. 2016).

10.6.3 Urbanisation, Industrialisation and Various Developmental Activities

Rapidly expanding cities are experiencing increased in water scarcity (Vörösmarty et al. 2000). Problems related to water pollution and waterborne diseases are escalating from 1995 to 2025 when the urban population doubled to 5 million from 1995 to 2025. Driving force behind water demand and supply are fast-growing population and economic growth with a small impact of climate change on natural resources (Foster 2010). Asia is one of the most populous and fast-growing regions in the world with an annual urban population growth of 2.3%, which is more than the annual global average urban population growth of 2%. Most of the largest megacities are situated in Asia having more than 10 million urban populations in more than 10 metropolitan cities. And the figure is estimated to increase in 2015 and in 2022 urban population is expected to increase as compared to rural population (ADB 2008). Cambodia, Myanmar and Nepal are the countries which have the highest urban population growth rate compounded with various problems related to developmental and management activities. Rapid industrial and economic growth in South Asian regions has a negative impact on the river health. Each year toxic and harmful chemical waste, organic solvents, dyes and other substances are estimated to be discharged in major river basins (Babel and Wahid 2008).

10.6.4 Abstraction Pressure

Excessive and unsustainable withdrawal of a large amount of groundwater has an adverse impact on the aquifers and coastal regions where water table is near-surface and is badly affected by seawater or brackish water intrusion which degrades the quality of water in the nearby region and in the absence of surface water availability pressure shifts towards groundwater resources. In most Asian developing countries, excessive groundwater abstraction causes groundwater depletion and increased salination due to intrusion of seawater in nations near coastal regions (FAO 2011).

Most of the Asian countries are situated in arid and semi-arid zones where the groundwater table is very low. But to meet the food requirements for a large section of the population, these resources are exploited in an unsustainable manner which further makes the situation worse as most regions have non-renewable groundwater resources. Also, the regions having shallow groundwater experience the severe pollution due to the disposal of untreated urban and industrial effluent discharge which

Table 10.3 Maximum annual abstraction of groundwater in various countries (as per 2010) (Margat 2008)	S. No.	Country	Abstraction (km^3/year)
	1	India	251
	2	China	112
	3	USA	112
	4	Pakistan	64
	5	Iran	60
	6	Bangladesh	35
	7	Mexico	29
	8	Saudi Arabia	23
	9	Indonesia	14
	10	Italy	14

ultimately degrades the quality of freshwater and may cause disasters over the long term. Maximum annual groundwater abstraction of the top ten countries are given in Table 10.3 (Margat 2008). Maximum annual abstraction is by India following after China. Also among 10 countries, seven are Asian countries which indicate maximum groundwater extraction (Table 10.3).

In Saudi Arabia, aquifers systems are facing highly declined groundwater levels (Brown 2011; Van der Gun and Ahmed 1995), and also highland basins of Yemen are reported with the same problem. Also, Iran's mountain basins experience a constant or steady decline in groundwater levels (Motagh et al. 2008) of various Indian states, under Indus basin, especially Rajasthan, Gujarat, Punjab, Haryana and Delhi (Jia and You 2010) suffer extensive abstraction of aquifer systems. The aquifer in North China has experienced a severe drop in groundwater levels (Kendy et al. 2004; Liu et al. 2001; Ismail 2016). Apart from these recorded studies, there are also a huge number of aquifers located in various regions in the world where the groundwater table is lowering or are still declining which exert a negative impact on the environment and social well-being (Rodell et al. 2009).

10.6.5 *Hydro-Morphological Alteration*

Hydro-morphological alteration in water bodies is due to flow alteration or regulation by constructing dams on rivers and its tributaries for hydel power and water supply in cities and extraction of excess water for irrigational, domestic and industrial use. Also, canal constructions and deforestation in river basin cause change in water flow direction and habitat destruction of various living species. Pollution and overharvesting also affect the hydro-morphological characteristics of water resources and make them unfit for human consumption and breeding place of various disease-causing microorganisms (Dudgeon 2000). According to Fu et al. (1998), annual deforestation rate in Asia is estimated in a range between 0.9 and 2.1% and when coupled with high and low river discharge during wet and dry seasons, respectively, and influences the floodplain regions and their inhabitants (Fu et al. 1998). A number of Asian countries like Bangladesh, Bhutan, India, Nepal and Pakistan are located near major six transboundary rivers and their basins, viz. the Red, Mekong, Salween, Irrawaddy, Ganges-Brahmaputra-Meghna and Indus River. Also, the river basins are extensively used for various water infrastructure projects like canal and reservoir construction, hydroelectric dam formation on rivers and its tributaries. Due to overexploitation and unsustainable use of water resources, a number of south Asian countries are facing a tremendous pressure of water stress and facing a permanent state of water stress conditions (Chua 2015).

10.6.6 Environmental Degradation and Climate Change

Asia is one among seven continents which have a large region in an arid and semi-arid zone, facing the strongest climatic impacts on the relative change in water scarcity. But during the last two or three decades, adverse impacts on environmental changes were also experienced by Europe, North America and South-east Asia (Schlosser 2014). The major impact of climatic change has been reflected in Asian countries. As most of these are developing nations and indulge in urban, social and economic development. There is extensive use of water resources in various developmental activities which lead to negative pressure on the environment and various natural resources like soil, air and water and adversely affect quality and accessibility in Asia. Most of the developing Asian regions are characterised by rapidly growing economies associated with increased problems related to safe and clean water, sanitation facilities, uniform water supply, depletion in groundwater levels, flooding and water pollution (Dudgeon 2000). Water stress or severe drought is expected to be increased in many other parts of Asia due to climate change (Ban Ki-moon). It is expected that by 2050, there will be increased in earth temperature from 1.6 to 6 °C in all around the world. Around 7% of the world population will have to face a decrease in one-fifth part of renewable water resources. Increased temperature will increase high crop water demand due to excessive evaporation and transpiration from surface water resources and plant surfaces respectively. Also, water scarcity will produce drought conditions and will affect agricultural productivity (Bandyopadhyay 2007). The study indicates that the vast stretches of ice mass of Hindukush mountain range are susceptible to changed climatic conditions which are the main region for storing water and providing continuous river flows during dry periods. Thus, due to changed climatic conditions, are affecting the energy, food and water supply of this region (Rasul 2014).

10.6.7   Water Sharing Conflict Within the Countries

Freshwater resources in Asian countries are facing drastic changes in water quantity and quality due to accelerated growth in the human population increased urbanisation and industrialisation in the last few decades. Growing pressure on the water resources has attracted developing countries towards this global issue (Arfanuzzaman 2017). Also, South Asian region is known as the most water-stressed region in the world with an annual growing population of 25 million. Due to the increase in human resources, the demand for water, food and land has increased which subsequently reduced the per capita water availability by 70% since 1950 (Singh 2008). Three major Himalayan rivers, the Indus, the Ganges and the Brahmaputra, are among the largest rivers of the world. These three rivers are the backbone of South Asian countries in terms of infrastructure and economic prosperity. Transboundary river basins interconnect South Asian riparian countries and their states in a complex network of economic, political and environmental interdependence (Ismail 2016). Only 80% of annual water flow in IGBM river basin system occurs in monsoon (June to September) and the remaining 20% occurring during the rest of the dry months period (Singh 2008). This type of alternative scene of water excess and scarcity is the main reason for conflicts over water sharing between countries. Upstream countries which are highly dependent on water flow from nearby countries are at higher risks. Also, the countries with fast-growing populations and resource consumption are extremely protective with respect to their water share. Most of the South Asian nations, including two major populous countries, China and India, are going to plan diversion of water flow in some of these transboundary rivers and their tributaries. Also, downstream countries near to transboundary rivers receive reduce water availability due to infrastructure development and have to manage wisely to avoid tension (Kameri-Mbote 2007). According to UNESCO, a number of conflicts have occurred in Middle East interstates due to Tigris and Euphrates rivers, among Iraq, Syria and Turkey; and a serious situation among Jordan, Israel, Lebanon and the State of Palestine have stated due to the Jordan River. Nile river water is a major cause of conflicts between Egypt, Ethiopia and Sudan in Africa (Ismail 2016). As well as the Aral Sea conflict among Kazakhstan, Uzbekistan, Turkmenistan, Tajikistan and Kyrgyzstan in Asia. For India and Bangladesh region, Farakka barrage continues to be a source of tension between India and Pakistan, and Indus River is also another source of tensions. To resolve issues between India and Pakistan, the Indus Water Treaty is playing a major role since it was signed in 196 (Appelegreen 1992). In India, Bangladesh and Pakistan, it is difficult to assess water resources at a basin-wide level due to fragmented water information. Lack of proper information related to river basins leads a number of difficulties to develop treaties on use of shared water resources, which further creates the regional tensions in the future and induced forced migration due to complex water scarcity issues.

10.7   Indian Scenario

Developed and developing countries both are facing water scarcity in their regions. Only developed countries have the potential to adapt to water scarcity due to their economic potential. But due to the high population, lack of infrastructure, technologically backward and lack of water management policies in developing countries made them face water stress conditions. Developing countries are the centre zone where current and potential water conflict in the world takes place. UN climate report stated that the Himalayan glaciers that feed Asia's biggest rivers—Ganges, Indus, Brahmaputra, Yangtze, Mekong, Salween and Yellow—could disappear by 2035 if the temperature of earth surface continues to rise as a present (Bagla 2009). It was later revealed that the source used by the UN climate report actually stated in 2350, not 2035. Increasing demand in freshwater resources due to fast-growing population in India is a great challenge to handle the current situation. Also the centre and state government are not doing justice at their part in the proper management of water quantity and quality in India. Also, the government is unsuccessful and helpless in implementing existing laws and regulations. Also, the public and industries are violating these rules without taking care of the protection and preservation of existing water resources. In the last 20–30 years, India is facing critical water stress conditions which negatively affect the health of both human as well as the environment. Most of Indian rivers are in very worse situation as compared to past. Increased pollution due to organic and hazardous waste in the surface water is deteriorating the water quality with a large extent. Apart from this, the nation is facing rising interstate water disputes for water share which are highly intensive and widespread and creating most complex situation for management authorities. In most Indian states, Andhra Pradesh, Tamil Nadu, Gujarat, Karnataka and Maharashtra, groundwater is declining at an alarming rate due to overextraction and poor water infiltration. According to the current estimate, annual groundwater consumption in India is 230–250 km^3 which account around one-fourth of the world's groundwater consumption. Due to maximum groundwater consumption in agriculture (70%) and domestic use (85%), India places next to China and USA combined. According to the NASA report in 2009, Indus basin, which mainly includes the Indian granaries states, Punjab and Haryana, was stated second most stressed aquifer all over the world with groundwater depletion rate by one metre in every three years. If the present situations remain to continue, more than 50% of aquifers in India will be in a highly critical situation by 2030 which will result in one-fourth of agricultural production at higher risk and lead economic downfall. Also, less advanced agricultural techniques and unsustainable utilisation of groundwater are creating a big problem in developing nation India as more than 85%. In India, major reasons for water scarcity are unequal access and inequitable distribution of water in different sectors, viz. domestic, agricultural, industrial and energy, unsustainable and excessive use of surface as well as groundwater resources, seasonal and annual variations in rainfall pattern, overexploitation of groundwater than surface water in agricultural fields, degradation of surface and groundwater quality due to discharge of untreated sewage waste, industrial wastes and agricultural waste which make water in most of rivers for unfit for human consumption.

10.8 Current Status and Future Trends for Water Management in Asian Countries

10.8.1 National Level

Water resources are at greater risk in Asian countries as there a large gap between water supply and demand. Almost half of the world population is unable to access safe and clean water supply and sanitation. Population in the rural area is suffering more as compared to urban areas, also declines in groundwater level and is dropping drastically in most Asian parts and rivers in these regions which are running short of water which is also a major issue of concern. Due to a wide pattern of economies and water environment, management challenges are also highly variable in Asia, the world's fastest growing zone. Only 4.6% of global annual renewable water resources (Hirji et al. 2017) are unequally distributed between South Asian countries (Afghanistan, Bangladesh, Bhutan, India, Nepal, Pakistan, Sri Lanka) and their river basins. Growth in Nepal and Afghanistan is slow as compared to India, Bangladesh and Bhutan which are facing a lack of proper water supply and water scarcity in large areas, including major cities.

Improved water management, viz. proper water supply, better monitoring strategies for surface and groundwater, recycling and reuse of water, sustainable use of surface as well as groundwater is the prime concern for the government and management authorities because new water storage infrastructure is costly (Thomas 2014). By 2030, water demand is expected to be increased twice as compared to currently available water supply in Asian countries which affect their growth and development. Both water demand and water supply management across the region is the main component for various water management strategies. Main features for water resource management at national level in Asian countries should be proper management of watersheds and groundwater aquifers, surface water bodies like lakes and wetlands, protecting and maintaining water quality and managing watersheds, aquifers, lakes and wetlands, sustainable use of freshwater resources. Judiciously utilisation of existing water resources and improved planning and management in water resources incorporate better utilisation of existing water resources in regions. Also, enhancement of equitable and desirable water supply in different sectors, domestic, industrial, agriculture is highly needed to meet water and energy demand and their use for economic growth and development. A proper strategy between various elements, planning, development and management is highly required to overcome the water-related problems like floods, drought and siltation. Also, a great need for cooperative and mutual beneficial management strategies should be adopted for shared water between countries having transboundary river basins and aquifers among their neighbourhood states and nations. Water policies coherence and integration across the economic sector and administrative jurisdictions can help out the water problems efficiently and funded in investments in water supply, flood management and overcoming drought problems. Also, technological advancement in the various sector can also improve the management of water resources, for example, Drip or sprinkler irrigation systems, rainwater harvesting, replacement of open canals by underground pipes, change in crop pattern, soil management, IS technology in water resources and land use pattern management (Suhardiman et al. 2017).

10.8.2 International Level

10.8.2.1 Water Sharing Plans Across National Boundaries

South Asian zone is characterised by a highly interconnected network of rivers and aquifers. There is a high need for cooperation among these riparian countries is required for water use agreements of some of the rivers. The situation of water sharing is highly complex due to asymmetric equations between countries sharing transboundary river water, depending upon economic, political and geographical locations. The most significant challenge to achieve such cooperation is based on country priorities. Indo-Gangetic and Brahmaputra basins in Nepal, India and Bangladesh are the main lifelines of people residing in these regions alone. South Asian Association for Regional Cooperation (SAARC) and the South Asia Cooperative Environment Programme (SACEP) exist in regional political systems. But due to lack in progress, they lead only region-wide adaptation initiatives (Suhardiman et al. 2017). In South Asia management of transboundary water is governed by bilateral treaties which are signed by Nepal, Bangladesh, Pakistan with India (situated in upper riparian zone in most cases, except with Nepal). Among these treaties, some have worked and rest have not. And also all of them are have some controversy and misgivings at different stages (FAO 1994).

10.8.2.2 Basin-Level Planning and Management

There is an extremely high need for the introduction of basin planning in south Asian countries like Nepal, Pakistan and India. Basin-level water resources planning and management practices in water resources instruments have already been adopted by a number of Asian countries, Afghanistan, Bhutan, India and Nepal. Main functions of these authorities in these regions are—developing rules for water allocation, monitoring of water use and enforcing action if there is a violation of rules in water use permits, issuing water use permit depending upon water availability, allocating responsibilities to the sub-basin level during emergency. Also, water infrastructure development responsibilities come under the supervision of basin authorities in some countries.

10.8.2.3   Water Resource Management

For overcoming the decline in the water table in growing economies, various methods for groundwater storage and management is implemented, which includes management of green cover in recharge zone, maintaining river flows over recharge beds, land use management in recharge zones, using aquifer recharge methods (MAR) to store water during water shortage conditions. Awareness among land users to avoid contamination of aquifers, basic education and behaviour change among groundwater users and the development and enforcement of water quality standards is necessary for groundwater protection. Rural and urban communities are largely affected by the saltwater intrusion into coastal aquifers which further affect the groundwater quality in these regions. The strong initiative should be taken for operations and management of existing coastal water supply planning in urban and rural areas, developing cost-effective protective measures and proper monitoring of saltwater intrusion along the large stretch of coastal regions of South Asian countries, Bangladesh, Pakistan, Sri Lanka and India. Equal distribution of water supply in different sectors in the urban region is a prime objective of the government. For this purpose, various policies and frameworks are incorporated in the protection of urban water utilities, urban planning, sewage management and stormwater management.

10.8.2.4   Policies and Legal Framework

A country required comprehensive national policy, for water distribution in different sectors, utilisation and its conservation. Also, these effective policies and plans are needed to mitigate various problems in the transboundary river basin. Implementations of policies and rules by government must effectively regulate the water consumption by the population (World Bank 1993). Sustainable water utilising methods measures for improving efficiency levels in water utilisation (domestic, agricultural or industrial) are extremely important. In semi-arid and arid regions due to a shortage of seasonal water, it is highly required to conserve water. National water conservation authorities and planners should introduce sustainable cropping systems, water-efficient technologies, appropriate cost recovery level and work on reduction in distribution loss, recycling wastewater and increasing public awareness. An effective legal framework of rules and regulations should incorporate to address various issues like water inventory, planning, use, quality and protection. There should be some act which covers the protection and preservation of water resources. New laws should be implemented for the development of water resources.

10.8.2.5   Community Participation

Various awareness programmes and special practices are highly needed to aware general public, communities, various authorities in different sectors about the harmful and long-lasting impacts of water scarcity and required adaptation strategies which can further overcome these impacts. Adequate staff with advanced scientific techniques can bring improvement in the field of water resource management and highly skilled resource person. Also, a better understanding of various water institutions at the national level is required which distribute the responsibility and provide resources at the local level. LAPAs in Nepal, while needing further strengthening, provide a model for community-level adaptive action.

10.9 Conclusion

Freshwater availability is facing rising challenges across the world due to its overconsumption in developing as well as developed nations. Due to unsustainable use and poor management of water in developing nations, this problem is becoming more complex. If proper attention is not given towards this problem, it might become a major source of dispute and war between various countries. Not only the water is utilised according to proper need and demands but also its quality is deteriorating due to increased pollution. There is a great need for integrated water resource management. Effective system related to measurement and control at the local, national and global levels is required to protect this basic commodity. Proper cooperation and coordination between government, private sector and community are required to work on this problem to find out a fruitful solution. A great need for global governance and new policies should be adopted to tackle this situation. To abate pressure on various natural resources, international discussions and treaties will be helpful.

References

Abbaspour S (2011) Water quality in developing countries, South Asia, South Africa, water quality management and activities that cause water pollution. In: International conference on environmental and agriculture engineering, vol 15, pp 94–102

ADB (2008) Managing Asian cities, sustainable and inclusive urban solutions. Manila. Available at http://www.adb.org/publications/managing-asian-cities?ref=themes/urban-development/publications

Appelegreen KW (1992) Management of transboundary water resources. Nat Resour Forum 21(2):92

Arfanuzzaman M (2017) Economics of transboundary water: an evaluation of a glacier and snowpack-dependent river basin of the Hindu Kush Himalayan region. Water Policy Uncorrected Proof (1–19)

Arshad A (2016) 140 m in South East-Asia do not have clean water, e-paper. The Straits Times, Asia. https://www.straitstimes.com/asia/se-asia/140m-in-se-asia-do-not-have-clean-water

Asian Water Development Outlook (2013) Measuring water security in Asia and the Pacific. Asian Development Bank. ISBN: 978-92-9092-988-8 (Print), 978-92-9092-989-5

Babel MS, Wahid SM (2008) Freshwater under threat—South Asia: vulnerability assessment of freshwater resources to environmental change. United Nations Environment Programme, Nairobi and Asian Institute of Technology, Bangkok

Bagla P (2009) Himalayan glaciers melting deadline 'a mistake'. http://news.bbc.co.uk/2/hi/south_asia/8387737.STM.BBC

Bai ZG, Dent DL, Olsson L, Schaepman ME (2008) Proxy global assessment of land degradation. Soil Use Manag 24:223–234

Bandyopadhyay J (2007) Water systems management in South Asia: the need for a research framework. Econ Polit Wkly 42(10):863–873

Beck A, Haerlin B, Richter L (2016) Agriculture at a crossroads: finding and recommendations for future farming. The Foundation on Future Farming, Berlin, Germany. https://www.globalagriculture.org/fileadmin/files/weltagrarbericht/EnglishBrochure/BrochureIAASTD_en_web_small.pdf

Bindraban PS et al (2012) Assessing the impact of soil degradation on food production. Curr Opin Environ Sustain 4:478–488

Brown LR (2011) World on the Edge. Routledge, London. https://doi.org/10.4324/9781849775205

China State Council (CSC) (2012) Regulation on implementing the strictest water resources management system, Beijing

Chua J (2015) Solving Asia's water crisis. EcoBuisness magazine. http://www.eco-business.com/news/solving-asias-water-crisis/

Cosgrove WJ, Rijsberman FR (2014) World water vision: making water everybody's business. Earthscan Publications, London, UK

Cribb J (2010) The coming famine: risks and solutions for global food security. https://www.sciencealert.com/the-coming-famine-risks-and-solutions-for-global-food-security. Coping with water scarcity. An action framework for agriculture and food stress (PDF). Food and Agriculture Organization of the United Nations, 2012. Retrieved 31 December 2017

Cruz W, Cruz MC (1990) Population pressure and deforestation in the Philippines. ASEAN Econ Bull 7(2):200–212

Das N, Deka JP, Shim J, Patel AK, Kumar A, Sarma KP, Kumar M (2016) Effect of river proximity on the arsenic and fluoride distribution in the aquifers of Brahmaputra floodplains, Assam, Northeast India. J Groundwater Sustain Dev 2–3:130–142

Dudgeon D (2000) Large-scale hydrological changes in tropical Asia: prospects for riverine biodiversity. The construction of large dams will have an impact on the biodiversity of tropical Asian rivers and their associated wetlands. Bioscience 50(9):793–806. https://doi.org/10.1641/0006-3568

EN/AR/FR/RU/ES REPORT from UN Environment Programme (2016) A snapshot of the world's water quality: towards a global assessment

EWCRP (2002) Population, natural resources, and environment. The future of population of Asia. https://www.eastwestcenter.org/publications/future-population-asia

Falkenmark M, Lindh G (1976) Quoted in UNEP/WMO. Climate change 2001: Working Group II: impacts, adaptation and vulnerability

Falkenmark M, Lundqvist J, Widstrand C (1989) Macro-scale water scarcity requires micro-scale approaches: aspects of vulnerability in semi-arid development. Nat Resour Forum 13(4):258–267

Falkenmark M, Rockström J, Karlberg L (2009) Present and future water requirements for feeding humanity. Food Secur 1:59–69. https://doi.org/10.1007/s12571-008-0003-x

FAO (2011) The State of the World's Land and Water Resources for Food and Agriculture (SOLAW)—managing systems at risk. Food and Agriculture Organization of the United Nations, Rome, Italy; Earthscan, London, UK. Available online: http://www.fao.org/docrep/017/i1688e/i1688e.pdf

Food and Agriculture Organization legislative study preparing national regulation for water resources management: principles and practices (1994). FAO Publication, Rome, pp 49–51

Foster W (2010) Exploring alternative futures of the world water system. Building a second generation of world water scenarios driving force: Technology

Fu CB, Kim JW, Zhao ZC (1998) Preliminary assessment of impacts on global change in Asia. In: Galloway JN, Melillo JM (eds) Asian change in the context of global climate change. Cambridge University Press, Cambridge (UK), pp 308–341

Gleick PH (1996) Water resources. In: Schneider SH (ed) Encyclopedia of climate and weather, vol 2. Oxford University Press, New York, pp 817–823

Gleick PH, Cain N, Haasa D, Hunt C, Kiparsky M, Moench M, Srinivasan V, Wolf G (2006) The biennial report on freshwater resources. The world's water 2006–2007

Graham S, Parkinson C, Chahine M (2010) The water cycle. Earth Observatory. https://earthobservatory.nasa.gov/Features/Water/printall.php

Gun J (2012) Groundwater and global change: trends, opportunities and challenges. United Nations World Water Assessment Programme, United Nations Educational, Scientific and Cultural Organization, Paris, France

Hirji R, Nicol A, Davis R (2017) South Asia climate change risks in water management climate risks and solutions: adaptation frameworks for water resources planning. Development and Management in South Asia, Report No: AUS14873

Human Development Report (2006) Beyond scarcity—power, poverty and the global water crisis. Palgrave Macmillan, Basingstoke, United Kingdom. http://hdr.undp.org/en/content/human-development-report-2006

Ismail H (2016) Water scarcity, migration and regional security in South Asia. Global Food and Water Research Programme

Jia Y, You J (2010) Sustainable groundwater management in the North China Plain: main issues, practices and foresight. Extended abstracts No. 517, 38th IAH Congress, University of Silesia, Krakow, Poland, pp 855–862

Kameri-Mbote PG (2007) Water, conflict, and cooperation: lessons from the Nile River Basin. Woodrow Wilson International Center for Scholars

Kendy L, Zhang Y, Liu C, Wang J, Steenhuis T (2004) Groundwater recharge from irrigated cropland in the North China Plain: a case study of Luancheng County, Hebei Province, 1949–2000. Hydrol Process 18:2289–2302

Kumar M, Furumai H, Kurisu F, Kasuga I (2009) Understanding the partitioning processes of mobile lead in soakaway sediments using sequential extraction and isotope analysis. Water Sci Technol 60(8):2085–2091

Kundzewicz ZW, Mata LJ, Arnell NW, Döll P, Kabat P, Jiménez B, Miller KA, Oki T, Sen Z, Shiklomanov IA (2007) Freshwater resources and their management. In: Parry ML, Canziani OF, Palutikof JP, van der Linden PJ, Hanson CE (eds) Climate change 2008: impacts, adaptation and vulnerability. Contribution of Working Group II to the fourth assessment report of the Intergovernmental Panel on Climate Change

Liu C, Yu J, Kendy E (2001) Groundwater exploitation and its impact on the environment in the North China Plain. Water Int 26:265–272

Marcoux A (1996) Population change-natural resources-environment linkages in Central and South Asia

Margat J (2008) Groundwater Around the World. Orleans / Paris, BGRM / UNESCO

Motagh M, Walter TR, Sharifi MA, Fielding E, Schenk A, Anderssohn J, Zschau J (2008) Land subsidence in Iran caused by widespread water reservoir overexploitation. Geophys Res Lett 35(16)

Mukherjee S, Patel AKR, Kumar M (2020) Water scarcity and land degradation nexus in the era of anthropocene: some reformations to encounter the environmental challenges for advanced water management systems meeting the sustainable development. In: Kumar M, Snow D, Honda R (eds) Emerging issues in the water environment during anthropocene: a South East Asian perspective. Springer Nature. ISBN 978-93-81891-41-4

Mulvey S (2009) Averting a perfect storm of shortages. BBC News. http://news.bbc.co.uk/2/hi/8213884.stm

Oldeman LR, Hakkeling RTA, Sombroek WG (1991) World map of the status of human-induced soil degradation: an explanatory note, second revised edition. ISRIC/UNEP. http://www.isric.org/sites/default/files/ExplanNote_1.pdf

Oude Essink GH, Van Baaren ES, De Louw PG (2010) Effects of climate change on coastal groundwater systems: a modelling study in the Netherlands. Water Resour Res 46(10)

Palaniappan M, Gleick PH, Allen L, Cohen MJ, Christian J, Smith C (2010) Clearing the waters: a focus on water quality solutions. In: Ross N (ed) United Nations environment programme. UNON Publishing Services Section, Nairobi

Patel AK, Das N, Goswami R, Kumar M (2019a) Arsenic mobility and potential co-leaching of fluoride from the sediments of three tributaries of the Upper Brahmaputra floodplain, Lakhimpur, Assam, India. J Geochem Explor 203:45–58

Patel AK, Das N, Kumar M (2019b) Multilayer arsenic mobilization and multimetal co-enrichment in the alluvium (Brahmaputra) plains of India: a tale of redox domination along the depth. Chemosphere 224:140–150

Pimentel P, Burgess M (2013) Soil erosion threatens food production. Agriculture 3:443–463

Qadir M, Noble AD, Qureshi AS, Gupta RK, Yuldashev T, Karimov A (2009) Salt-induced land and water degradation in the aral sea basin: a challenge to sustainable agriculture in Central Asia. Nat Resour Forum 33:134–149

Rasul G (2014) Food, water and energy security in South Asia: a nexus perspective from the Hindu Kush Himalayan region. Environ Sci Policy 39:35–48

Rekacewicz P (2008) Global soil degradation. UNEP/GRID-Arendal—from collection. IAASTD—International Assessment of Agricultural Science and Technology for Development. http://www.grida.no/graphicslib/detail/global-soil-degradation_9aa7

Rodell M, Velicogna I, Famiglietti JS (2009) Satellite-based estimates of groundwater depletion in India. Nature 460(7258):999

Schlosser CA et al (2014) The future of global water stress: an integrated assessment. Joint program report series, 30

Seckler D, Molden D, Barker R (1998) Water scarcity in the twenty-first century. IWMI water brief. International Water Management Institute, Colombo, Sri Lanka

Singh R (2008) Trans-boundary water politics and conflicts in South Asia: towards 'Water for Peace'. A report of Centre for Democracy and Social Action. Centre For Democracy And Social Action (CDSA), New Delhi, India

Singh A, Patel AK, Kumar M (2020) Mitigating the risk of arsenic and fluoride contamination of groundwater through a multi-model framework of statistical assessment and natural remediation techniques. In: Kumar M, Snow D, Honda R (eds) Emerging issues in the water environment during anthropocene: a South East Asian perspective. Springer Nature. ISBN 978-93-81891-41-4

Sinha K (2014) Salt invasion in Indo-Gangetic basin has led to 40% increase in human health problems. UN, TNN. Updated: Oct 28, 2014, 13:29 IST

Srivastava J, Gupta A, Chandra H (2008) Managing water quality with aquatic macrophytes. Rev Environ Sci Biotechnol 7:255–266

SRV (1995) Viet Nam national environmental action plan, final draft. SRV/IDRC

Suhardiman D, Silva SD, Arulingam I, Rodrigo S, Nicol A (2017) Review of water and climate adaptation financing and institutional frameworks in South Asia. Background paper 3

The World Bank (1993) Water resources management: a World Bank policy paper. The World Bank, Washington, p 14

Thomas S (2014) The global water scenario: human and socio-economic impacts and solutions, the future of science. In: 10th world conference—'the eradication of hunger', Venice, Italy

UNDP (2006) Coping with water scarcity. Challenge of the twenty-first century

UNWWAP (2009) The 3rd United Nations world water development report: water in a changing world (WWDR-3) and facing the challenges. New York

UNWWD (2015) Water for a sustainable world. World water development report. Vanishing Himalayan glaciers threaten a billion. http://www.planetark.com/dailynewsstory.cfm/newsid/42387/st

Van der Gun JA, Ahmed AA (1995) The water resources of Yemen: a summary and digest of available information. Delft and Sana'a, WRAY-Project

Vörösmarty CJ, Green P, Salisbury J, Lammers RB (2000) Global water resources: vulnerability from climate change and population growth. Science 289:284–288. The American Association for the Advancement of Science, Washington, DC

Wada Y, Van Beek LP, Viviroli D, Dürr HH, Weingartner R, Bierkens MF (2011) Global monthly water stress: 2. Water demand and severity of water stress. Water Resour Res 47. https://doi.org/10.1029/2010wr009792

World Bank (1989) Indonesia—forest, land and water: issues in sustainable development, Washington

World Bank (1995) Vietnam. Environmental program and policy priorities for a socialist economy in transition, vol 2. Washington

WWDR (2012) Managing Water under Uncertainty and Risks. The United Nations World Water Development Report 4., vol I. Water. World Water Assessment Programme, UN

WWAP (World Water Assessment Programme) (2012a) World water development report 4

WWAP (2012b) The United Nations world water development report 4: managing water under uncertainty and risk. UNESCO, Paris

Yimin S (1994) An initial look into China's population, environment and sustainable development, Cairo

Chapter 11
Water Supply, Urbanization and Climate Change

Venkata Sandeep, Ashwini Khandekar, and Manish Kumar

11.1 Introduction

Urbanization refers to the gradual rise in the population of urban areas, the shift of population from rural areas to urban areas (Accra 2000). The higher proportion of urbanization is not occurring only in developed nations but also in developing nations (Salerno et al. 2018; Kumar et al. 2019a, b). Besides economic growth, there are many environmental impacts and implications regard to urbanization on water supply. These implications are getting exacerbated due to climate change.

The implications of urbanization are insufficient supply of clean water, excessive usage of energy, industrialization, contaminants of freshwater from factories, household effluents, land use change due to urbanization has changed the flow patterns of wastewater and stormwater which in turn has an impact on rivers (Hassan Rashid et al. 2018). When there is an insufficient water supply, people opt for private wells as a source of water. Private wells make the particular house less vulnerable to the water scarcity, but on a larger scale, the whole area where such thing is practised, on a long time it makes the region more vulnerable to water crisis (Srinivasan et al. 2013).

The three main causes of urban flooding are climate change, urbanization and lack of urban planning. As more evaporation rate and extreme precipitation caused by the climate change, it further leads to urban flooding. In case of urbanization, impervious surface areas lead to water run-off and decreasing number of wetlands, ponds and lakes. Improper urban planning depicts improper drainage system designs which cause the stormwater run-off the ground and make the water more polluted (Nguyen et al. 2019). Climate change and urbanization are key factors affecting the future of water quality in urbanized catchments (Salerno et al. 2018).

V. Sandeep · A. Khandekar · M. Kumar (✉)
Discipline of Earth Sciences, Indian Institute of Technology Gandhinagar, Gandhinagar, Gujarat 382355, India
e-mail: manish.kumar@iitgn.ac.in

© Springer Nature Singapore Pte Ltd. 2020
M. Kumar et al. (eds.), *Resilience, Response, and Risk in Water Systems*,
Springer Transactions in Civil and Environmental Engineering,
https://doi.org/10.1007/978-981-15-4668-6_11

From (Astaraie-imani et al. 2012) the case study, it clearly demonstrates that climate change combined with increasing urbanisation is likely to lead to worsening river water quality in terms of both frequency and magnitude of breaching threshold **dissolved oxygen** and ammonium concentrations.

There are very few reviews which give overall glimpse of urbanization and climate change implications on water supply. This chapter discusses the detailed review of the implications of urbanization and climate change on water supply and the adaptation strategies or responses to improve water resilience in the face of climate change and urbanization.

11.2 Water Supply

11.2.1 Introduction

The main forces that affect the global water scenarios are a rapid increase in the population, economic growth, demographic change and environmental quality. Along with the rapid increase in population, climate change also introduces additional stress to the hydrological cycle resulting in an alteration in the water resources systems (Manju and Sagar 2017; Singh et al. 2020). The increase in the flood events in the country causes deaths and need to relocate people to safer places; whereas on the other side, many regions of the country face water scarcity and need to go kilometres in search of water (UNWWD Report 2006). The United Nations Development Programme's Sixth Sustainable Development Goal mentions that one of the key elements for attaining Sustainable Development is the availability of clean water (Sustainable Development Goals 2016).

The water shortage can be overcome by using the water in a sustainable manner as well as developing new technologies which can provide safe drinking water and also treat the wastewater and reuse it for other purposes. India gets nearly 75% of total annual rainfall in monsoon, out of which 48% represents surface water. The entire India receives less than 4000 billion cubic metres (bcm) of rainfall annually, including snowfall (Manju and Sagar 2017). The main dependency for water supply in urban and especially rural areas of India is groundwater, which ultimately depends on the amount of precipitation occurring.

Due to urbanization, there is a high influx of people coming from rural areas to urban and so the demand for water also increases. In shortage of water availability, many developing countries have to provide intermittent supply resulting in less per capita supply. According to United Nations World Water Development Report, in 2006, 54% of the population of the world had a piped connection to their dwelling, plot or yard, and 33% population used other improved sources of drinking water. The remaining 13% (884 million people) rely on some unimproved sources.

11.2.2 Chennai Water Crisis 2019: A Case Study

According to the 2011 census, the household water supply in the city of Chennai was 55 litres per capita per day (lpcpd), which was much less than the Ministry of Urban Development Benchmark of 135 lpcpd, making Chennai a water-deficient city. Even then, the urban and industrial growth did not stop. Real estate had been growing since then. Its municipal corporation boundary had expanded from 175 to 426 km^2 in 2011, and the Chennai Metropolitan Area is now being considered for expansion from 1189 km^2 to more than 8878 km^2. Despite the advancement in the prediction of weather systems, it is still difficult to predict weak or strong monsoon precisely. In 2015, the heavy flood was faced by the city and Chembarambakkam Lake was overflowing while in 2019, the lake had dried up (Dhanapal 2019). The approximate contribution to the city water supply is 51% from surface reservoirs, 1% from well fields, 17% from the desalination plant, 19% from Veeranam lake and 11% from Krishna River (Paul and Elango 2018; Kumar et al. 2019a, b). Three rivers Coovum, Adayar and the Kotalaiyar supply Chennai with its water. Of these, the Coovum and Adayar are highly contaminated by domestic waste and effluent and are very saline due to saltwater intrusion from the Bay of Bengal. This is one of the reasons for the poor drinking water supply. In addition, it is seen from the local water supply data that even with an annual rainfall of between 700 and 1100 mm, Chennai has an absolute water shortage (Sethuram and Cooper 2017). After the severe drought in 2000, it was made mandatory in Chennai to have rainwater harvesting system and it worked well initially, but later on, it suffered problems of poor maintenance and many of the rainwater harvesting systems are now dysfunctional (Guntoju et al. 2019; Saikia et al. 2017). According to a study by Paul and Elango (2018), using WEAP model, the gap between demand and supply in the city of Chennai is increasing from the year 2016 and it will approximately double by 2030. It also shows that the demand would increase by 20 MLD per year, but the supply cannot meet the demand and cannot increase beyond 800 MLD due to the absence of any other sources of water. Thus, proper management is required to save the city from any further crisis. One of the solutions for such a huge demand of water can be solved by setting up the desalination plant. CMWSSB (Chennai Metro Water Supply and Sewerage Board) is operating two desalination plants successfully and has also initiated the process of constructing a new desalination plant at Pattipulam. But the establishment of desalination plants is not a sustainable solution to it as it consumes a lot of energy and creates a lot of water wastage.

11.2.3 Desalination Plants as an Alternative in Fighting Water Crisis

Desalination means removing the salts from the water in order to make it drinkable. The seawater or saline water is processed in order to bring the TDS within the limits set by WHO or the limits set by national bodies. In the thermal process, the seawater or the saline water is heated and vapourized. This vapour is condensed and distilled water is formed. There are various methods such as multi-stage flash (MSF) distillation, vapour compression (VC), solar stills and multiple effect distillation (MED). The membrane processes include reverse osmosis and electrodialysis. In these, the semi-permeable membranes are used which act as selective barriers to separate salts and impurities. In electrodialysis, the salt ions are separated using selective ion exchange membranes, and in RO, the water is passed through a semi-permeable membrane by high pressure, and thus, water is filtered (Manju and Sagar 2017). The demerits of these processes are that it consumes a high amount of energy and also has a huge carbon footprint (Abdelmoez et al. 2014). Solar stills are good alternatives to reduce the carbon emissions but they can be established where the demand of water is less. Again, the maintenance cost is high and there are additional costs if the water has to be supplied for long distance. 0.86 kWh/m^3 energy is used to desalinate the water with salinity 34,500 ppm, but the plants use 5–26 times more than this depending on the type of process used. In isolated places, where there is no good accessibility to electricity, plants operated by renewable energy can be established. These plants can be run using solar, wind or tidal energy to produce electricity for desalination plants (Manju and Sagar 2017) (Fig. 11.1).

Establishing desalination plants seems to be a good option in water crisis conditions, but it has a severe impact on the ecological environment. Many marine animals get entrapped and killed, while the water is being drawn. Also, it produces air pollution and a large quantity of waste streams containing the cleaning waste, heavy metals from corrosion and brine solution. It is essential to dilute and treat this waste before disposing into the seawater (Manju and Sagar 2017; Kim et al. 2011).

There have been many cases of water crisis faced in different countries. Here are some identified case studies (Table 11.1).

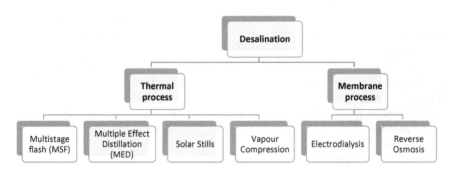

Fig. 11.1 Classification of the desalination processes of water

Table 11.1 Extreme water crisis and faced in various parts of the world and its effects

Place	Period	Population affected	Reason	Effects	Mitigation	References
Chennai, India	2019		Drought	Drying up of all the reservoirs which supply water Water brought to city by train Supply of water by private tankers	Desalination plants Rainwater harvesting	Dhanapal (2019), Guntoju et al. (2019)
Cape Town, South Africa	2017–2018	3.8 million	Constant drought since 2015	Declaration of day zero where no water would be supplied	Increased tariffs and penalties for high volume users Voluntary water restrictions Campaigns to reduce water consumption Releasing critical water shortages disaster plan	Enqvist and Ziervogel (2019), Miller (2019)
Most parts of Australia, especially East Coast	1996–2009		Below average rainfall	Fall in crop production, water level got down	Recycled water scheme, desalination plant	Caball and Malekpour (2019)
Sao Paulo, Brazil	2014–2015				Reduction in cost for those who lowered their water consumption Pressure reduction in water supply pipelines	Millington (2018)
California, USA	2012–2016		Drought	Death of 102 million forest trees Agricultural and economic losses	Groundwater sustainability legislation State urban conservation mandates	Lund et al. (2018)

(continued)

Table 11.1 (continued)

Place	Period	Population affected	Reason	Effects	Mitigation	References
Istanbul, Turkey	2006–2007		Lowest rainfall since last 50 years	Cotton production decreased from 2.2 to 1.093 Gton	Grants to the farmers who are affected by drought Protection of water sources	Kayam and Cetin (2012)
Kenya and Uganda	2015–2016	6.4 lakh (Kenya) 3.93 lakh (Uganda)	Drought	Herd size sensibly reduced, below average crop production	Increasing water harvesting capacity, Improving access to water	FAO (2016)
Yunnan Province, China	2012	6.3 million	Drought	Lack of access to drinking water to 2.4 million people, reduced income for farmers and agri-business	Effective monitoring, early warning systems to help decision-makers get timely information	Horn et al. (2008)

11.2.4   Water Supply System in India and the Problems Faced

Urban water supplies consist of a grid network or reticulate system in accordance with the existing street pattern of the city. In India, the majority of the cities have intermittent water supply which means the water supply is just for few hours daily. According to the research carried out (Vairavamoorthy et al. 2001), the case study revealed the design engineers stated that although it was known, while designing that a system would operate intermittently, it was designed as a continuous system. They acknowledged that this was not the correct approach, but argued that they did not have any alternative as there were no proper guidelines and design tools were developed specifically for intermittent systems. The problem faced with intermittent supply system is that the pressure maintained is not sufficient and equal at all the places. As the supply is for limited hours, people try to collect as much water as they can and store it for further use. So, the residents of the areas where the pressure of water is high get more amount of water and others get less resulting in unequal water supply even though the supply hours are the same. Also, in the time of absence of water supply, there are chances of the addition of contaminants into the pipelines resulting in the degradation of water quality as well as an increase in chances of occurrence of disease.

In some of the areas of India, especially the rural ones, there is serious health threat as many people still practise open defecation. In a case study conducted in four villages in Bihar (Vairavamoorthy et al. 2008), sanitary checks were carried out if the toilet was at least 10 m away from the water supply line (which is recommended) or not. Also, 150 supplies were sampled for TTC thermotolerant (faecal) coliforms and TLF tryptophan like fluorescence. It was found that 18% of the samples were contaminated by TTCs and 91% of those were within 10 m of water supply. Also, 58% of TLF were above the detection limit. Thus, such kind of places tends to have more serious threat as they leach into the underground water supply nearby.

To avoid the contamination in water, the water must be disinfected to make it safe for drinking. But there are numerous reasons for which the water supplies are not disinfected, which includes the intentional absence of disinfection due to consumer resistance for the taste that disinfected water has (Diergaardt and Lemmer 1995).

Another reason for unequal or less supply of water is leakages in the pipelines or illegal connections. Illegal connections also result in less pressure. When the supply is not enough, people have to rely on some other sources of water like groundwater and they have to bear the additional cost of installing the bore wells. It is estimated that in India, more than 60% of irrigated agriculture and 85% of drinking water supplies are dependent on groundwater (The World Bank 2012). Also, water costing is not equal in India. The state governments are responsible for levying the tariffs. This results in unequal tax being paid by the residents of different cities. To meet the increasing demand of water in water-stressed cities of India, energy-efficient measured must be adopted by the local governments. Energy efficiency lowers down the cost in a longer time and results in a cleaner environment (Ray 2018).

11.2.5   Solutions for Sustainable Water Supply and Use

- 80% of the wastewater goes into the waterways without adequate treatment (UNDP SDG:6, Facts and Figures). So, one must put the use of technologies that reuse the wastewater (Fig. 11.2).
- Maintenance such as leak detection and repair, maintaining a minimum pressure in the system, adequate pricing through setting tariffs, careful billing of users and monitoring and evaluation of services provided.
- Installing water metres. A large amount of water gets lost and resulting in loss of revenue. The metres should be automated and timely repaired. This would help in keeping a check on the amount of water consumed.
- Providing nature-based solutions to the problems.

11.3   Urbanization

It refers to the gradual rise in the population of urban areas and shift of population from rural areas to urban areas. Around half of the human population lives in the

Individual level	Societal/community level	Governing body
• Reusing the water(grey water) • Harnessing the rain water • Wise usage of water	• Efforts to conserve natural bodies. • Awareness campaigns	• Revising or forming new laws and policies • Integrated water resource planninng and management • Levying water tariffs • Funding for related technologies • Regular maintainance of water supply systems.

Fig. 11.2 Different levels in fighting water crisis

towns and approximately 60% of the population will be the urban residents within two decades, globally. Every second, the urban population rises by 2 people.

- Globally, 141 million urban residents do not have access to improved drinking water.
- Improved sanitation facilities were not in access by the one out of four city dwellers, 794 million in total.
- 43% of South-central Asian urban dwellers and 62% of the Sub-Saharan Africa town population resides in slums.
- In the cities like Accra which is in Ghana, the urban poor pay high about 12 times more than richer neighbours, since they often depend upon non-public vendors (Accra 2000).

11.3.1 Casual Factors Behind Urbanization

The foremost reasons for the rising urban population are migration of people from rural areas to urban areas, including international migration to smaller extent, enlargement or reclassification of existing boundaries of the city to incorporate more population that were previously classified as being resident outside the city limits. These are assessed to contribute about 60% of the region's urban growth, while natural increase counts for some 40% (Ichimura 2003). The higher proportion of urbanization is not occurring only in developed nations but also in developing nations (Salerno et al. 2018). The driving forces for this rural to urban migration are services and opportunities obtainable in cities.

11.3.2   Risks of Water System in Face of Urbanization

Due to urbanization, population and construction of buildings, road enlargement or reconstruction increase and that will eventually lead to urban flooding as more impervious surfaces get raised. Low infiltration rate is causing due to water quality is affected by a wide range of natural and human influences and the main source of drinking water is from rivers. As the urban flooding water is mixing with the sewage systems in urban areas, it leads to water run-off to rivers (Garg 2016).

11.3.3   Urban Water Management Strategies for Water Resilience

Management of water is the primary feature regards to the sustainable development of urban areas. Due to fast urbanization and dangerous weather, problems such as water shortage, more urban floods, groundwater overexploitation, wasting of rainwater sources and water pollution occur.

The three main causes of urban flooding are climate change, urbanization and lack of urban planning. As more evaporation rate and extreme precipitation caused by the climate change, it further leads to urban flooding. In case of urbanization, impervious surface areas lead to water run-off and decreasing number of wetlands, ponds and lakes. Improper urban planning depicts improper drainage system designs which cause the stormwater run-off the ground and make the water more polluted (Nguyen et al. 2019; Shafique et al. 2018).

11.3.4   Resourcing Rainwater

As population and industrialization are rising, water resources are being expunged. In order to encounter this problem, harvesting rainwater is the better solution not only to decrease the water demand and to attenuate the water run-off. To guarantee the Sponge City effectively resources like the rainwater, it is essential to know the hydrological characteristics of that area such as water surface run-off, flow time, speed, discharge and peak time to well connect between natural water networks and drainage systems. In this manner, the water storage capacity of urban infrastructure can be improved (Nguyen et al. 2019).

Millions of gallons of freshwater fall on the earth each year through precipitation. It is a solution to the problem of rising water demand, and it is for reducing water run-off (Simhan 2018). Municipal water supply is significantly replaced by rainwater as an alternative source. From Takagi et al. (2019), it is identified that by introducing a standard size water storage tank, toilet water demand of 90% of four number families can be satisfied by means of rainwater in the study area named Galle, Sri Lanka. By

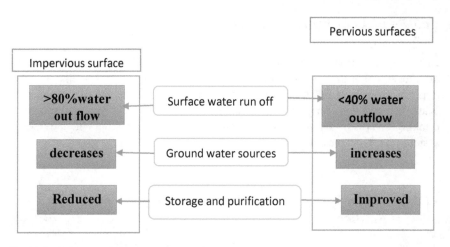

Fig. 11.3 Differences between pervious and impervious surfaces

resourcing rainwater, a tariff which is charged by the municipal corporation can be avoided and along with that government itself will deliver the subsidy for the construction of rainwater storage. It is the best method to follow if the water supply is in adequate to satisfy the demand. Monthly and annual water demand was assessed for flushing and laundry. Tank sizes are based on demands for harvesting rainwater (Olowoiya and Omotayo 2010).

11.3.5   Changing Impervious Surfaces to Pervious Surfaces

As the rise of buildings and pavements due to population growth, stormwater runs off the ground and with that the occurrence of urban flooding causes. The technological remedy to address this issue is permeable pavement which is the utilization of permeable materials to construct ground pavement. By means of this technology, water run-off has been reduced, groundwater is purified for urban and noise reduction and aid rainwater infiltration, etc. Figure 11.3 shows the differences between pervious and impervious surfaces (Nguyen et al. 2019).

11.3.6   Green Infrastructures

The objective of green infrastructures is to alleviate the passage of stormwater into the systems of drainage(Salerno et al. 2018). The portion of roof surface areas is 40–50% out of total impervious surface areas in the urban areas. It can keep more water than a normal roof to avoid stormwater run-off. The results show that average run-off retention on green roof was 10–60% in all rain events (Shafique et al. 2018).

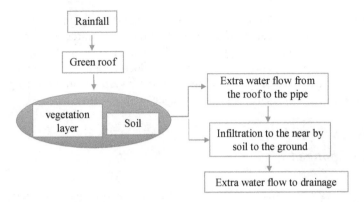

Fig. 11.4 Rainfall run-off management process at the site

The antecedent dry weather period (ADWP) is also considered a noticeable factor which affects the hydrologic performance of the green roof. Poor construction of this has an important effect on water bodies. However, additional investigations should concentrate on the green roof construction materials, especially the substrate layer and on the maintenance/management problems of extensive green roofs. Figure 11.4 shows the rainfall run-off management process at the site (Shafique et al. 2018). They serve to mitigate stormwater run-off; attenuate energy utilization; alleviate the urban heat island effect; enhance quality of air and water; improve wildlife habitat and plant life; and boost recreational activities through green areas (Nguyen et al. 2019).

11.3.7 Wastewater Treatment

As surface water and groundwater are closely related, water run-off leads to the contamination of both surface water and groundwater. The cost for the cleaning of groundwater is very huge if it is polluted. So, by means of wastewater treatment we can mitigate this issue. The techniques to achieve this treatment are ozonation, coagulation–precipitation, biological treatment and activated carbon, reverse osmosis ("Lenntech" 2019).

11.3.8 Ecological Water Management

Rising trend of population and industrial activities have endangered the water quality in many areas. This new strategy guarantees that the water environment is reinstated ecologically through a waterfront design with healthy landscapes for wildlife and people and self-purification system. Self-purification comprises biological, chemical and physical processes. Among all biological process is eco-friendly and the factors that influence self-purification of water are hydrodynamic force, soil and the third one is plants that eliminate heavy metal pollutants, nitrogen and phosphorous and the last one is microorganisms can help in the degradation of contaminants. The objective of waterfront design is to integrate water system and cities to grow the macro- and micro-urban environments (Nguyen et al. 2019).

11.3.9 Sponge City

China is taking consequences extremely. One repercussion of our warming world is growing often and more extreme flooding. In order to encounter these problems, a new urban water management strategy called Sponge City has been applied by china in the year 2013 (Nguyen et al. 2019). Sponge City has been found to be of huge value for urban resilience. The objective of this initiative is that 80% of urban areas in China will reuse 70% of their rainwater by 2020 (World Economic Forum 2019). The four principles of Sponge City are resourcing rainwater, ecological water management, urban permeable pavement and green infrastructure.

11.3.10 Barriers and Opportunities of Sponge Cities Adoption

Public acceptance is one of the robust barriers in its approval. And insufficient subsidies are provided by the central government because of the requirement of more and long-term funding. So, funding from non-government organizations is very crucial. In this case study, china central government and policy makers identified it in the long term. So, they come with public and private partnership in this strategy to make it feasible. The important technical troubles are deficient in a suitable simulation model that incorporates the pertinent factors based on evidence for designs of Sponge City implementation (Nguyen et al. 2019). The 30 towns comprised in the initiative have received more than $12 billion in sponge projects funding. However, the national government can provide funding in between 15 and 20%. And the remaining amount of funding is from the private investors and local governments (World Economic Forum 2019). To develop resilient, sustainable, healthier cities in the era of rapid urbanization and climate change, urban policies and strategies will play a significant role on promoting this concept (Nguyen et al. 2019).

Building code has been initiated in UK that assesses the sustainability of a new home against the nine classes of sustainable design. This type of labelling targets introducing evident criteria for buildings energy efficiency rating, delivering market signals which support the building stock transformation towards high efficiency ("Adaptation of urban planning: Water and energy" 2015). Table 11.2 shows the

Table 11.2 Implications and responses of water supply in case of urbanization

Implications	Responses	References
Energy consumption increases with urbanization	Installation of more efficient appliances, i.e. water-efficient fixtures Adaptation of water-saving measures	Topcu and Girgin (2016), "Adaptation of urban planning: Water and energy" (2015)
Deterioration of river water quality	Primary water treatment plants Strict laws for effluent to be discharged	Astaraie-imani et al. (2012), Kambole (2003)
Increase in water demand	Recycling and reuse Rainwater harvesting Dual flush toilets Water leakage reduction programmes Installing water metres	Simhan (2018), Mukheibir (2007, 2008)
Decrease in groundwater level	Technological use in water transfer, Induced recharge and collecting more water	Giordano (2009)
Surface water-run off	Wastewater treatment	Lenntech (2019)
Urban flooding	Urban planning Government subsidies to implement mitigation strategies	Nguyen et al. (2019)
Low infiltration rate	Green infrastructures Construction of pervious areas	Nguyen et al. (2019)

implications and responses of water domain in the face of urbanization.

11.4   Climate Change

11.4.1   How Is Climate Change Impacting Water Cycle?

The water cycle designates the movement of water from the surface. Water evaporates from the land, sea and condenses into snow or rain in clouds which ultimately comes down to earth via rain and snow. Climate change strengthens this cycle as the temperature of air rises, more water gets evaporated and hot air can grip more water vapour which further leads to heavy forceful rainstorms, causes problems such as dangerous flooding in coastal groups around the earth.

11.4.2 Effects of Climate Change on Available Water

Climate change effects on available water are much hard to forecast because many are combined. If the temperature rises, there will be more transpiration from plants and evaporation from the soil and water flow will be weakened in rivers or seep into underground aquifers. If precipitation is highly severe, more proportion of water will flow into grounds as floods or deeper groundwater gets infiltrated through the soil. Variations in the carbon dioxide concentrations, rainfall, temperature will influence land use and plant cover which will in turn considerably affect the water behaviour when it falls as rain (Muller 2007).

Singh and Kumar (2015) assessed the climate change thresholds for water resources vulnerability throughout India. Enumerating the reliance on future water availability on changing climate is critical for water resources management and planning in water-stressed countries like India.

And the impacts due to anthropogenic activities also been considered—changes in land use, for instance, the water availability will also be affected by cropping systems. Hence, it is clear that the forecasting of stream flows and groundwater regimes under climate change scenarios is an ambitious undertaking (Muller 2007).

11.4.3 Climate Change Implications on Water Supply: Drought and Flood

Around 40% of the world's population lives within 100 km of coast (Sedac 2017). Out of which many are water-stressed and experiencing fast population growth. These areas are facing the problem of sea-level rise may become unsuitable for living in, shifting populations and forcing them to secure the water resources due to saltwater intrusion. This is of specific concern for small low-lying islands and very low land nations like Bangladesh. Groundwater levels may rise, where long-term rainfall rises, alleviating the natural purification processes efficiency, rising infectious disease risks and toxic chemicals exposure. Climate variability is already a danger to sanitation and water supplies. Due to climate change, floods and droughts are occurring frequently in many places. Floods can have disastrous results for basic water infrastructure. It even takes years to repair the infrastructure which is affected by flood (World Health Organization 2009; Kayam and Cetin 2012).

There are three types of drought. One is meteorological drought which is in relation to rainfall, estimating degree of dryness (by comparing with local average) and dry period duration. This type of drought is more specific to a region as average rainfall might change spatially. Second type is hydrological drought which affects streamflow, soil moisture, levels of reservoir and lakes, groundwater discharge by reduced precipitation (UCUSA 2019). Thirdly, physiological drought articulated as water deficiency that adversely impacts the production of crops (Střelcová et al.

2009). US drought monitor has been producing weekly maps for the conditions of drought since 1999 to mitigate the effects (UCUSA 2019).

The indirect effects of climate change on water supply and sanitation comprises energy interruption impacts, growing the unreliability of piped water and sewerage systems. Droughts can occur randomly, worldwide and become more frequent with climate change. Falling groundwater tables and decreased surface water flows can lead to desiccation of wells, increasing distances that necessarily be travelled to collect water, and rising pollution of the water source as temperature increase leads to lesser levels of dissolved oxygen in the water. Henceforth, the insects, fish and other types of aquatic animals that depend on oxygen would be in stress. Other reasons for alleviation in water quality due to climate change are increased run-off as more precipitation happens ("Water and Climate change" 2018) In response, drilling rigs—which are utilized to rise access—may be renewed or replaced out of service wells, slowing progress in extending access. Climate change is not occurring separately. Challenges like other sector's water demands may worsen its impacts. These effects were alleviated by rising the wastewater reuse for agriculture (World Health Organization 2009). Single measures like leakage minimization, management of demand have significant potential in contributing to both adaptation of technologies and mitigation of adverse effects to increase resilience. In response to the multiple adverse effects, improved planning events and the development and use of technologies will give support.

To determine the water availability, the adaptive capacity of sanitation services and water supply, vulnerability, management and technology interact with local conditions such as demand. For both sanitation and water supply, conferring to their climate change resilience technologies were classified, considering both vulnerability and adaptive capacity. This classification was founded on published literature information, a web-facilitated questionnaire survey and a series of semi-structured interviews. More investigation is necessary to further enhance these categories and consider multiple source use (World Health Organization 2009).

11.4.4 Climate Change Implications on Lakes and Rivers

Climate variability and rising global temperatures will increase the harshness and occurrence of dangerous storm events. Due to rising temperature, there is an exponential rise in the atmospheric capacity to hold moisture which leads to higher capacity in the events of precipitation and floods. Ambient air temperature may cause fluctuation in water temperature thereby causing a variation in different water parameter values which in turn affects the bio/physico-chemical equilibrium of aquatic environment such as metabolic activities of aquatic species and determines the solubility, availability and toxicity of certain bioactive compounds.

Across the nation, evaporation from lakes and reservoirs gets increased due to hotter temperatures, balancing increases in precipitation in some regions and magnifying cuts in western areas. Reservoirs on the **Colorado River** previously lose

1.8 million acre-feet of water due to evaporation is about 13% in river's annual flow in an average year ("The impacts of climate change on rivers" 2017).

Rising temperature and warm weather conditions due to climate change increase the nutrient load in rivers and lakes (Kumar and Taneja 2018). Nutrient load changes are linked with transport rates of sediments and surface run-off, which are affected by climate change (Todd 2015). Climate change will increase mosquito-virus borne disease due to increased intensity of close contact between humans and spoiled water. Climate change enhances genetic changes in bacteria and allows them to live in water (Kumar and Taneja 2018).

Figure 11.5 shows the linkages between climate change, water supply and urbanization implications.

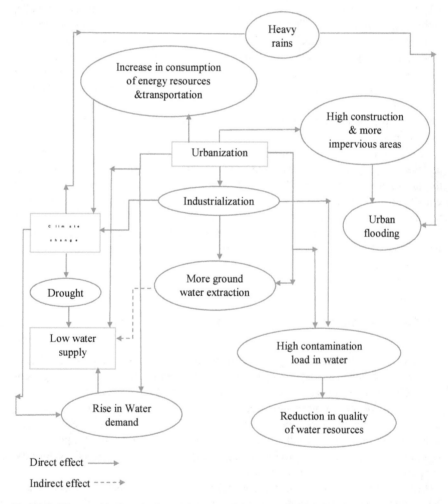

Direct effect ──────►

Indirect effect - - - - ►

Fig. 11.5 Direct and indirect indicators between climate change, urbanization, water supply

Table 11.3 Water technology resilience to climate change by 2030

Group 1: Potentially resilient to all expected climate changes	Utility piped water supply Tube wells
Group 2: Potentially resilient to most of expected climate changes	Small piped systems Protected springs
Group 3: Potentially resilient to limited climate changes only	Dug wells Rainwater gathering
Technologies characterized by Joint Monitoring Programme (JMP) as not improved drinking water sources	Unprotected dug wells Unprotected springs Carts with small tank or drum Surface water (rivers, dams, lakes, ponds, streams, canals, irrigation channels)

Source World Health Organization (2009)

11.4.5 *Climate Change Adaptation Strategies*

Ziervogel et al. (2010) focussed on the facilitation of adaptation of climate change within urban water sector in the area of Cape Town. This study suggests government support and commitment is needed in support of water supply adaptation with respect to climate change implications. Important factors that inhibit the implementation of strategies for climate change implications are lack of local capacity, running cost of most of the strategies, low financial resource base to cover the capital (Mukheibir 2007).

The adaptation strategies or responses for climate change impacts in supply side and demand side are education programmes on water conservation measures, use of greywater, utilizing saline water in toilets, local water resource management and monitoring, artificial groundwater recharge, household-level rainwater harvesting, dual flush toilets, water leakage mitigation programmes at distribution and household level, regional water resource planning, conjunctive use of surface and groundwater, initiation of tariff structures for alleviating water demand, groundwater desalination and changes in agricultural management practices (Mukheibir 2007, 2008).

From Bertule (2018), climate change adaptation strategies in addition to the above ones are shifting the time of use, water metering, water re-allocation, water savings requirements in building codes, natural wetlands, solar water distillation, water recycling and reuse and soil moisture conservation techniques. Table 11.3 shows the water technology resilience to climate change from World Health Organization.

11.5 Conclusion

Rapid urbanisation is seen not only in developed countries but also in developing countries. Urbanisation indirectly contributes to climate change. Major results of climate change are droughts and floods which lead to extra stress over water utilization and water supply issues. Many of the cities are facing issues related to the water crisis.

Insufficient rainfall or drought is the main reasons for it. Responses for the sustainable use of water can be rainwater harvesting, reusing the water, good water policy planning, water tariffs and incentives. Technological advancements like establishing desalination plants can be an alternative but they have ecological impacts and other limitations. Climate change also leads to deteriorating water quality. Rising temperature and warm weather conditions due to climate change increase the nutrient load in rivers and lakes. Climate change will increase mosquito-virus borne disease due to increased intensity of close contact between humans and spoiled water and it enhances genetic changes in bacteria and allows them to live in water. Overutilization of water, depleting groundwater, low infiltration rates of lands which lead to urban flooding are some of the impacts of urbanisation. These problems can be handled by proper planning by the government, smart planning of cities, technological use and public awareness. The responses identified and enumerated from the literature to mitigate the impacts of climate change are education programmes on water conservation measures, use of greywater, utilizing saline water in toilets, local water resource management and monitoring, artificial groundwater recharge, household-level rainwater harvesting, dual flush toilets, water leakage mitigation programmes at distribution and household level, regional water resource planning, conjunctive use of surface and groundwater and changes in agricultural management practices.

References

Abdelmoez W, Mahmoud MS, Farrag TE (2014) Water desalination using humidification/dehumidification (HDH) technique powered by solar energy: a detailed review. Desalin Water Treat 52(25–27):4622–4640

Accra I (2000) Water and urbanisation main challenges, 1–4

Adaptation of urban planning: water and energy (2015). Retrieved August 24, 2019, from https://climate-adapt.eea.europa.eu/metadata/adaptation-options/adaptation-of-urban-planning-water-and-energy

Astaraie-imani M, Kapelan Z, Fu G, Butler D (2012) Assessing the combined effects of urbanisation and climate change on the river water quality in an integrated urban wastewater system in the UK. J Environ Manage 112:1–9. https://doi.org/10.1016/j.jenvman.2012.06.039

Bertule M (2018) Climate change adaptation technologies for water. Retrieved from https://www.ctc-n.org/sites/www.ctc-n.org/files/resources/water_adaptation_technologies_0.pdf

Caball R, Malekpour S (2019) Decision making under crisis: lessons from the millennium drought in Australia. Int J Disaster Risk Reduction 1(34):387–396

Dhanapal G (2019) Chennai's water crisis: a lesson for Indian cities. https://www.downtoearth.org.in/blog/water/chennai-s-water-crisis-a-lesson-for-indian-cities-65606

Diergaardt GF, Lemmer TN (1995) Alternative disinfection methods for small water supply schemes with chlorination problems. Water Supply 13(2):309–312

Enqvist JP, Ziervogel G (2019) Water governance and justice in cape town: an overview. Wiley Interdisc Rev Water e1354

FAO (2016) Africa: the case of the 2015–2016 El Nino phenomenon, August, 8–12. Retrieved from file:///D:/water supply/FAO.pdf

Garg A (2016) Water quality deterioration factors. Retrieved August 20, 2019, from https://www.slideshare.net/AnchalGarg8/water-quality-deterioration-factors-70146986

Giordano M (2009) Global groundwater? issues and solutions. Annu Rev Environ Resour 21(34):153–178

Groundwater: a valuable but diminishing resource World Bank article India. https://www.worldbank.org/en/news/feature/2012/03/06/india-groundwater-critical-diminishing

Guntoju SS, Alam MF, Sikka A (2019) Chennai water crisis: a wake-up call for Indian cities. https://www.downtoearth.org.in/blog/water/chennai-water-crisis-a-wake-up-call-for-indian-cities-66024

Hassan Rashid MAU, Manzoor MM, Mukhtar S (2018) Urbanization and its effects on water resources: an exploratory analysis. Asian J Water Environ Pollut 15(1):67–74. https://doi.org/10.3233/AJW-180007

Horn T, Carolina N, States U (2008) Drought: a slow, creeping natural disaster

Ichimura M (2003) Urbanization, urban environment and land use: challenges and opportunities. An issue paper, January, 1–14

Kambole MS (2003) Managing the water quality of the Kafue River. Phys Chem Earth Parts A/B/C 28(20–27):1105–1109

Kayam Y, Cetin O (2012) The impacts of drought and mitigation strategies in Turkey. In 5th International Scientific Conference on Water, Climate and Environment. BALWOIS

Kim J, Kuwahara Y, Kumar M (2011) A DEM-based evaluation of potential flood risk to enhance decision support system for safe evacuation. Nat Hazards 59:1561–1572

Kumar M, Taneja P (2018) Implications of climate change on water supply. IIT Gandhinagar, Ahmedabad

Kumar M, Ram B, Honda R, Poopipattana C, Canh VD, Chaminda T, Furumai H (2019a) Concurrence of antibiotic resistant bacteria (ARB), viruses, pharmaceuticals and personal care products (PPCPs) in ambient waters of Guwahati, India: urban vulnerability and resilience perspective. Sci Total Environ 693:133640. https://doi.org/10.1016/j.scitotenv.2019.133640

Kumar M, Chaminda T, Honda R, Furumai H (2019b) Vulnerability of urban waters to emerging contaminants in India and Sri Lanka: resilience framework and strategy. APN Sci Bull 9(1). http://doi.org/10.30852/sb.2019.799

Lenntech (2019) Retrieved August 6, 2019, from https://www.lenntech.com/groundwater/reducing-contamination.htm

Lund J, Medellin-Azuara J, Durand J, Stone K (2018) Lessons from California's 2012–2016 drought. J Water Resour Plann Manage 144(10):04018067

Manju S, Sagar N (2017) Renewable energy integrated desalination: a sustainable solution to overcome future fresh-water scarcity in India. Renew Sustain Energy Rev 1(73):594–609

Miller J (2019) Managing urban water: the role of isotope hydrology and what the Cape Town water crisis taught us. Organismo Internacional de Energía Atómica Boletin 60(1):29–30

Millington N (2018) Producing water scarcity in São Paulo, Brazil: the 2014–2015 water crisis and the binding politics of infrastructure. Polit Geogr 1(65):26–34

Mukheibir P (2007) Qualitative assessment of municipal water resource management strategies under climate impacts: the case of the Northern Cape, South Africa. Water SA 33(4):575–581

Mukheibir P (2008) Water resources management strategies for adaptation to climate-induced impacts in South Africa. Water Resour Manage 22(9):1259–1276. https://doi.org/10.1007/s11269-007-9224-6

Muller M (2007) Adapting to climate change: water management for urban resilience. 19(1):99–113. https://doi.org/10.1177/0956247807076726

Nguyen TT, Ngo HH, Guo W, Wang XC, Ren N, Li G, Ding J, Liang H (2019) Implementation of a specific urban water management—sponge city. Sci Total Environ 652:147–162. https://doi.org/10.1016/j.scitotenv.2018.10.168

Olowoiya O, Omotayo A (2010) Assessing the potential for rainwater harvesting. 2129–2137. https://doi.org/10.1007/s11269-009-9542-y

Paul N, Elango L (2018) Predicting future water supply-demand gap with a new reservoir, desalination plant and waste water reuse by water evaluation and planning model for Chennai megacity, India. Groundwater Sustain Dev 1(7):8–19

Ray I (2018) Pay less for more: energy efficiency approach to municipal water supply in Indian cities. In: Low carbon pathways for growth in India. Springer, Singapore, pp 131–144

Saikia R, Goswami R, Bordoloi N, Pant KK, Kumar Manish, Kataki R (2017) Removal of arsenic and fluoride from aqueous solution by biomass based activated biochar: optimization through response surface methodology. J Environ Chem Eng 5:5528–5539

Salerno F, Viviano G, Tartari G (2018) Urbanization and climate change impacts on surface water quality: enhancing the resilience by reducing impervious surfaces. Water Res 144:491–502. https://doi.org/10.1016/j.watres.2018.07.058

Sedac (2017) People and oceans general. Retrieved from https://sedac.ciesin.columbia.edu/es/papers/Coastal_Zone_Pop_Method.pdf

Sethuram S, Cooper M (2017) Rivers and water security: supply adaptation strategies in the city of Chennai, India. In: Rivers and society. Routledge, London, pp 77–92

Shafique M, Kim R, Kyung-ho K (2018) Green roof for stormwater management in a highly urbanized area: the case of Seoul, Korea, 1–14. https://doi.org/10.3390/su10030584

Simhan R (2018, July 16) Tamilnadu leads in rain water harvesting. The Hindu Business Line. Retrieved from https://www.thehindubusinessline.com/specials/india-file/tamil-nadu-leads-in-rainwater-harvesting/article24436406.ece

Singh R, Kumar R (2015) Vulnerability of water availability in India due to climate change: a bottom-up probabilistic Budyko analysis. Geophys Res Lett 42(22):9799–9807. https://doi.org/10.1002/2015GL066363

Singh A, Patel AK, Kumar M (2020) Mitigating the risk of arsenic and fluoride contamination of groundwater through a multi-model framework of statistical assessment and natural remediation techniques. In: Kumar M, Snow D, Honda R (eds) Emerging issues in the water environment during anthropocene: a South East Asian perspective. Springer Nature. ISBN 978-93-81891-41-4

Srinivasan V, Seto KC, Emerson R, Gorelick SM (2013) The impact of urbanization on water vulnerability: a coupled human–environment system approach for Chennai, India. Global Environ Change 23(1):229–239

Střelcová K, Mátyás C, Kleidon A, Lapin M, Matejka F, Blaženec M, Škvarenina J, Holécy J (2009) Bioclimatology and natural hazards, pp 1–298. https://doi.org/10.1007/978-1-4020-8876-6

Sustainable Development Goals (2016) Clean water and sanitation | UNDP. https://www.undp.org/content/undp/en/home/sustainable-development-goals/goal-6-clean-water-and-sanitation.html

Takagi K, Otaki M, Otaki Y, Chaminda T (2019) Availability and public acceptability of residential rainwater use in Sri Lanka. J Cleaner Prod. https://doi.org/10.1016/j.jclepro.2019.06.263

The impacts of climate change on rivers (2017). Retrieved August 7, 2019, from https://www.americanrivers.org/threats-solutions/clean-water/impacts-rivers/

The United Nations World Water Development Report 3 (2006) Water in a changing world

Todd A (2015) Evaluating the impacts of climate change on catchment nitrogen transfer: a modelling study on Wensum River, UK, 1–32. Retrieved September 19, 2019, from https://www.semanticscholar.org/paper/evaluating-the-impacts-of-climate-change-on-%3a-a-on/83406b69bb860226de10a6e08c11e959c03f99a0

Topcu M, Girgin S (2016) The impact of urbanization on energy demand in the Middle East. J Int Glob Econ Stud 9:21–28

UCUSA (2019) Causes of drought. Retrieved August 26, 2019, from https://www.ucsusa.org/global-warming/science-and-impacts/impacts/causes-of-drought-climate-change-connection.html

Vairavamoorthy K, Akinpelu E, Lin Z, Ali M (2001) Design of sustainable water distribution systems in developing countries. In: Bridging the gap: meeting the world's water and environmental resources challenges, pp 1–10

Vairavamoorthy K, Gorantiwar SD, Pathirana A (2008) Managing urban water supplies in developing countries—climate change and water scarcity scenarios. Phys Chem Earth, Parts A/B/C 33(5):330–339

Water and climate change (2018). Retrieved September 20, 2019, from https://www.ucsusa.org/global-warming/science-and-impacts/impacts/water-and-climate-change.html

World Economic Forum (2019) China is fighting flooding with sponge cities. Retrieved from https://www.weforum.org/agenda/2017/11/china-is-building-30-sponge-cities-to-combat-climate-change

World Health Organization (2009) Summary and policy implications vision. Retrieved from papers2://publication/uuid/BDB05616-392B-40D4-B1C9-8C673E819AD3

Ziervogel G, Shale M, Du M (2010) Climate change adaptation in a developing country context: the case of urban water supply in Cape Town. Clim Dev 2(2):94–110. https://doi.org/10.3763/cdev.2010.0036

Chapter 12
Climate Change—Implication on Water Resources in South Asian Countries

Atul Srivastava, Anjali Singhal, and Pawan Kumar Jha

12.1 Introduction

South Asian countries extend between the tropics (Cancer and Capricorn). The geographical territory of South Asian countries from the northern Himalayan peaks (Bhutan and Nepal) to the Indian Ocean in the south (India and Sri Lanka) and Hindu Kush (Afghanistan) in the west to the vast delta of Sunderban (Bangladesh) in the east, (Adhikari 2014; Kumar et al. 2017; Das et al. 2015). Everyone needs water to survive. Water is used for irrigation of crops, fulfills the requirements of industries to generate energy, and consumption for household purposes. South Asian countries are highly populated and having large networks of rivers that flow down from the great range of mountains and the large North Indian Plain basin aquifers. North Indian Plain is the most fertile alluvial plain of South Asia region. It mainly supports agriculture activities and nourishes billions of people. The two largest economies of South Asia, India and Pakistan, are water-scarce and many areas, including major cities, lack piped water supply system.

In South Asia region, the gross domestic product (GDP) is 7.0% at the present time and will increase in the near future. For enhancing the gross domestic product growth of the region, the contribution from the export and agriculture sector should be increased to attain sustainable economical growth. Water availability is going to be a key factor for the GDP growth rate and it may suffer a decline due to water deficit in future decades (Artuc et al. 2019; Kumar et al. 2013). This water deficit condition may force migration of large populations from one geographic area to another leading to more socioeconomic problems in the region (United Nations-WATER). South Asian

A. Srivastava · P. K. Jha (✉)
Centre of Environmental Science, University of Allahabad, Prayagraj, India
e-mail: findpawan@gmail.com

A. Singhal
Department of Botany, University of Allahabad, Prayagraj, India

© Springer Nature Singapore Pte Ltd. 2020
M. Kumar et al. (eds.), *Resilience, Response, and Risk in Water Systems*,
Springer Transactions in Civil and Environmental Engineering,
https://doi.org/10.1007/978-981-15-4668-6_12

regions are prone to a vast variety of natural hazards, including floods, earthquakes, landslides, cyclones, and tsunamis, which put extra burden over the economy and natural resources present in the region (Mani et al. 2018; Seth et al. 2019).

12.1.1 Effect of Climate Change on Water Resources

The overall impact of global warming on water resources is shown as a flow diagram in Fig. 12.1. Global warming affects the melting of glaciers and polar ice which leads to increase in seawater level (IPCC 2014). An increase in the seawater level will affect the population residing in the coastal areas of South Asian countries. The large section of population residing in these areas will have to migrate to other regions which will create a socioeconomic crisis in the region (Ahmed and Suphachalasai 2014; Bush et al. 2011; Patel et al. 2019; Singh et al. 2020). The change in the precipitation pattern due to global warming can result into less water supply in certain part of South Asian region. This decline in water supply will result in low agricultural productivity, shortage of food grain in the region, and resultant population migration (Shah and Lele 2011; Hirji et al. 2017; Rahmasary et al. 2019; Mukherjee et al. 2020).

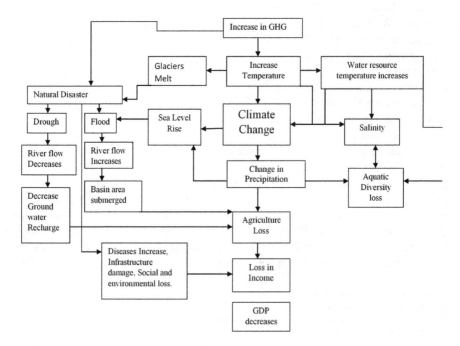

Fig. 12.1 Schematic diagram showing interlinked relation of climate change

The increase in temperature may result in higher precipitation in certain parts of South Asian region due to the increased supply of moisture from the evaporation and evapotranspiration processes. This excess precipitation when combined with the meltwater from glaciers can increase the overall water discharge and sediment carrying capacity of rivers present in the region (Asia-Pacific water forum, ADB 2013). The change in the precipitation pattern map in the South Asian region from 2003 to 2018 is given in Fig. 12.2. The global warming may result in the shrinkage of glaciers or its disappearance therefore reducing the flow of glacier-fed rivers during the dry season (Turer and Annamalai 2012; Hirji et al. 2017). The rate of urbanization is also increasing, and with the expansion of urban areas, more and more surface areas

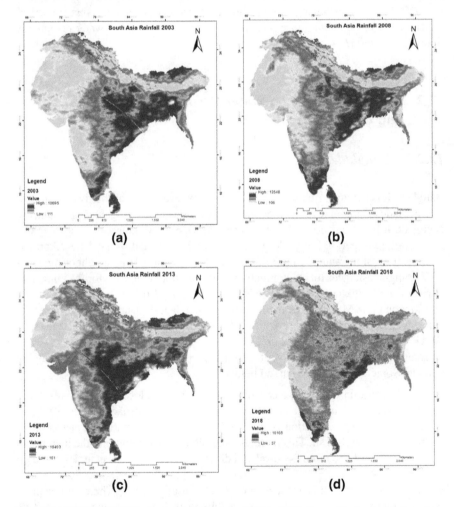

Fig. 12.2 PERSIANN-CCS, Annual Rainfall from 2003 to 2018. *Source* Centre for Hydrometeorology and Remote Sensing, University of California, Irvine-UCI

are getting impervious due to which we have increased in surface runoff and decrease in the groundwater recharge (Field 2014; Ray et al. 2015).

A recent regional study estimates that the storage volume in the first 200 depths of the Indo-Gangetic Basin (IGB) aquifer is almost 30,000 km^3. This is nearly 100 times the total constructed surface water storage (dams, reservoirs, and tanks) in the region (Ahmed et al. 2018). Also, its storage capacity is far more than the sum total of the annual flow of many large rivers such as Brahmaputra, Ganga, and Indus. The quality of IGB aquifer is deteriorating because of unscientific management water flow that leads to long-term losses in socioeconomic sector (Aryal et al. 2019; Alam and Huq 2019). The main objective of this paper to find out the answers to some of the questions like, what is the importance of water in the South Asian region and on its GDP? How climate change affects the water resources of South Asian countries? What are the present and future scenarios for the management of water resource in South Asia region?

12.2   Countries Profile on Water Resources, Climate Risks, and Uncertainty

12.2.1   Bangladesh

Bangladesh is having a rich network of rivers; however, most of the rivers that flow in Bangladesh originate from mountain ranges located in neighboring countries. Bangladesh is the last points for the rivers like Ganga, Brahmaputra, and Meghna. As a result, most part of Bangladesh is deltaic, floodplain, and wetland area. Approximately, 93% of the annual runoff of Bangladesh is coming from neighboring countries. Bangladesh faces lots of conflicts with its neighbor countries for the allocation of river water (Mukherjee et al. 2019; Ravenscroft et al. 2018). Ganga, Brahmaputra, and Meghna (GBM) are three main rivers of Bangladesh. During the rainy season (July–August), the combined peak flow of GBM varies from 80,000 to 140,000 m^3/s (WARPO 2001; Zevenbergen et al. 2018; Akanda 2019).

Bangladesh water resources will face more stress and problem due to changing climate. Some problem is discussed below:

1. The increase in the meltwater from glaciers due to global warming results in more water flow in Ganga and Brahmaputra rivers (Lacombe et al. 2019).
2. Increase in the precipitation as predicted by most of the climatic and meteorology models in Bangladesh. This increase precipitation will result into the increase in surface runoff and more flood-related disaster increase (Haigh et al. 2018; Singh et al. 2019).
3. Rise in sea level will increase incidences of coastal flooding. There will be more waterlogged in these rivers; thus, there will be prolonged riverine flooding. This is adversely affecting agricultural land and residential area leading to the greater socioeconomic burden on the country.

4. Climate change is a global-scale phenomenon that governs the rhythmic oscillation circulation pattern of the ocean system. Due to change in this pattern, catastrophic hazards like cyclone will affect the coastal areas of Bangladesh and most of the mangrove will be completely submerged after sea level rise (Kumar et al. 2018).

12.2.2   Bhutan

Bhutan is extended between the Tibet, India, and Southeast Asia. Bhutan is well known for its abundance of water resources, although the country facing water scarcity at the local level due to improper management of water resources. The areas that have reported water scarcity are Trashiyangtse, Trashigang, and Trongsa and also some parts of South Bhutan. Most of the rivers of Bhutan originated from a great and mid-Himalayan part. The Bhutan average water flow is 73 billion m^3/year (Amrith 2018). Irrigation and drinking water demands are mainly full filled by small rivers and canal, but in Bhutan, unsustainable consumption of river water makes them dry in some season. According to a recent report in the Great Himalayan range, 24 glaciers are melting at higher rate and poses flood hazard risk for Bhutan (Turner et al. 2017). The melting ice from retreating glaciers is increasing the volume of water in glacial lakes. Bhutan is also famous for tourism flash flood that will destroy the flora and fauna of the country and will affect the tourism and livelihood of people. Most of the electricity generations in Bhutan are based on hydropower, and any change in the water resource availability will affect the electricity generation (Mahanta et al. 2018; Sovacool et al. 2016).

12.2.3   India

India is the second highest populated country in the world. Most of the population is dependent on agriculture for livelihood. A vast area of agricultural lands is dependent on rainfall during monsoon season. Composite Water Management Index (CWMI) report states that 21 Indian cities (Delhi, Bengaluru, Chennai, Hyderabad, Maharashtra, Tamil Nadu, and others) are having declining groundwater level and will not have any groundwater left for use by 2020, affecting 100 million population residing in these cities. Due to current water scarcity, farmers commit suicide and lakhs of people die every year due to contaminated drinking water (Niti Ayog 2018). India receives 4000 billion cubic meter water per year from precipitation. An additional 500 billion cubic meter water is received by India from neighboring countries through rivers flow. A total calculated amount of replenishable groundwater is 1868 billion cubic meters per year (Singh et al. 2018; Vinke et al. 2017). India is also facing the problem related to climate change. Changing climate conditions have affected the precipitation pattern of India due to this some states receive an excess of precipitation while others receive negligible. Flood and drought are directly affecting the water resources of the country. A governing body National Water Mission, which is a branch of the National Action Plan on Climate Change, is indicating threats to Indian water resources that are discussed below:

- Lots of studies indicate that the glaciers of the Himalayan range are melting rapidly that will affect the river water flow in glacier-fed rivers.
- India is surrounded by three large water mass bodies (Arabian Sea, Indian Ocean, and the Bay of Bengal) that govern the Indian monsoon. Due to increases in sea surface temperature and uncertainty of climate, monsoon in India becomes weaker and stronger. This directly impacts on precipitation.
- Some time due to cyclonic activity, India faces torrential rain that leads to the flood hazards.
- Changes in precipitation pattern and evapotranspiration directly affect groundwater recharge (increasing urbanization with time has also adversely impacted the groundwater recharge).
- Melting of glaciers has increased water level in the sea, which results in the flooding of coastal areas with seawater and increases groundwater salinity.

CWMI is an innovative program launched by NITI Aayog for managing, monitoring, and making policy for cities which are facing water crises. CWMI is the first integrated water data set for India. Basically, it is a tool in which indexing of different cities and states takes place on the basis of different criteria:

(1) It analyzes the current situation of water resources in cities and project data for setting up a benchmark for better performance.
(2) CWMI evaluates the methods used by the cities to solve their water crisis-related issues and also categorized them according to their achievements. Cities with high achievements serve as a benchmark for other cities and government award them by giving some incentive. CWMI also encourages competition among the states for achieving incentives.
(3) This is a tool for grading the states on the basis of cities that require higher concern and higher investment. Apart from that NITI Aayog is also making an inventory of the water resource available in the country for its better management.

12.2.4 Nepal

Water resources are abundant in Nepal. Nepal is having approximately 6000 rivers with a total catchment cover area of 191,000 km^2 (Devkota and Gyawali 2015). Nepal is facing threats of climate change. Global warming results into increased in the rate of Himalayan Glaciers melting (Sherpa et al. 2019).

Glacier melting is the highest threat to Nepal. Most of the parts of Nepal lie in the mountain range and are at the originating point of rivers. Glacier's melting results in the increase the volume water and greater flow velocity in the mountain rivers. This might lead to increase in landslide and rock slides. The major source of water in Nepal is meltwater present in rivers, and if these glaciers disappeared due to global warming it will have a large impact over agriculture, tourism activities and livelihood of the population residing in the country (Budhathoki et al. 2011). Melting of glaciers may also result in the formation of artificial glacier lakes in the Himalayan region, which may burst and results in the flash flood in the human settlement areas (Lala et al. 2018; Ives et al. 2010).

12.2.5   Sri Lanka

The main sources of water in Sri Lanka are rainwater and permanent water bodies like river, lakes, reservoir, and wetland. Precipitation is affected by climate change, thereby increasing pressure on inland water resources. Apart from climate change, there are multiple factors contributing to water stress in Sri Lanka (Schulz and Kingston 2017). Sri Lanka is an island nation. Since it is surrounded by the Indian Ocean, the intrusion of saltwater into the groundwater is a major problem. Excess of groundwater is extracted near coastal areas for various activities including agricultural and domestic requirements. This has lowered the water table and converted many perennial rivers into seasonal rivers due to decrease in the groundwater contribution in the base flow. Due to the expansion of urban area, impervious surface has increased, thereby lowering the groundwater recharge. Anthropogenic activities pollute the surface water due to discharge of untreated sewage and industrial wastewater and runoff from agriculture and urban areas into the river, lakes, and pond (Udayakumara and Gunawardena 2018).

12.2.6   Afghanistan

Afghanistan is not self-sufficient for its water need, but also gets affected by the neighbor's water consumption. It is a landlocked country that shares four out of its five rivers' basins with other states. It only uses a small proportion (around one-third) of water from the four major rivers that flow into neighboring countries that originate here. In Hindu Kush Mountains, glacier height of above 2000 m, provides the bulk of water resources, and therefore, it is considered as the key importance to the country for the storage of natural water (UNFCC 2018).

Glaciers start freezing when temperature is down (winter) and melt when temperature rises (summer). When the snow melts, it contributes water flow in the rivers of Afghanistan. Amu Darya is main glacier-fed river of Afghanistan which contributes 55% of Afghanistan water resources that are originated from Hindu Kush Mountains

(Aich et al. 2017). Due to an increase in the earth's average temperature, melting of Afghanistan glaciers increases. Large glaciers of central Asia Mountains have been reported to be shrinking, and the small glaciers have been reported completely melted which affects the wetlands and the flow of rivers in the country. Figure 12.3 shows the shrinking of Sistan wetland from 1993 to 2005. Sistan wetland is fed by Helmand River, and satellite imagery study indicates that most of its area gets dry in the year 2005 (Fig. 12.3).

Due to the dryness of Sistan wetland, most of the flora and fauna got affected and the whole ecosystem of wetland collapse. Due to dryness of Sistan wetland, waterfowl's population decline rapidly. 150 species of waterfowls are reported in this area but due to drying up of Sistan wetland, most of the species not remain today (UNEP 2006). In Afghanistan, approximately 80% population depends on the agriculture sector. Climate change changes the pattern of precipitation and also causes the glaciers melting due to this irrigation system get affected and crop production will

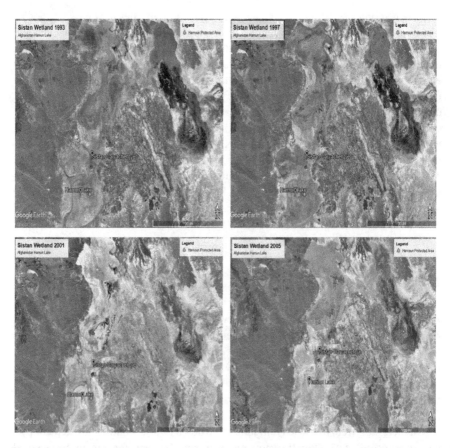

Fig. 12.3 Sistan wetland and Hamoun area dryness from 1993 to 2005. *Source* Google Earth

decline, and Afghanistan GDP will also decline (Afghanistan Online Afghanistan Online 2018).

12.3 Climate Change and Issues Related to Water Resource Management

Sustainable management of surface and groundwater is a critical challenge for South Asia's growth and livelihoods. The first key component of the challenge is the availability of reliable supply of water for different purposes (e.g., domestic supplies, agriculture, energy generation, cooling water for thermal plants, industrial production, livestock, ecosystems). Proper recycling and reuse of gray water ("waste"waters that result from different uses) are also required (Owusu and Sarkodie 2016; Souza et al. 2015). The second component is variability, including managing the changes in volume, timing, and frequency of supplies, to minimize the impact of natural hazards like floods, droughts, and storms. The South Asian country's economy and per capita water availability is given in Table 12.1 (FAO, AQUASTAT 2017). India, the South Asia's largest economy and the most populous nation, has lower per capita water availability while Pakistan, the second-largest economy, has the lowest accessibility of water to its people (Berrang-Ford et al. 2015; Mertz et al. 2009).

12.3.1 South Asia's Transboundary Water Management

Five rivers of South Asia, Ganga, Brahmaputra, Meghna, Helmand, and Indus, give a major contribution to South Asian water resources. Brief introduction of these rivers is given in Table 12.2. The decline in GDP of South Asia due to water stress condition might affect the new irrigation projects, income, human well-being, and damages to the human settlements and infrastructure (Howard and Howard 2016). As per some studies, the demand for water is going to be double in some parts of this region by 2030, leading to increase in the water stress condition the South Asian region (Zeitoun and Warner 2006). The Ganges–Brahmaputra–Meghna (GBM) and the Indus basins drain about half the South Asian region by area. Ganga is the most exploited river for irrigation in the Indian region. The glaciers are a major source of water in the rivers such as Indus and Ganga, and the seasonal changes in this area are the key driver of the river's flow (Hanasz 2017). The Helmand River shared by Afghanistan and Pakistan also originates from the high mountains. In groundwater terms, the transboundary Indo-Gangetic Basin (IGB) alluvial aquifers underlie most of Pakistan, northern India, southern Nepal, and Bangladesh and are among the most productive aquifers in the world (Rai et al. 2017; Giordano and Wolf 2003).

Table 12.1 Brief introduction of South Asian countries

	Population (million)	Rainfall/snowfall (mm/year)	Transboundary water (km³/year)	Groundwater contribution to renewable water (%)	Snowmelt contribution to renewable water (%)	Total renewable water resource (km³/year)	GDP/capita, current (US $)	Per capita water availability (m³)
Afghanistan	35.5	327	10	16	92	65.33	601	2008
Bangladesh	164.7	2666	1122	2	0	1227	1212	7622
Bhutan	0.8	2200	0	9	17	78	2532	100,645
India	1339.2	1083	635	19	10	1911	1674	1458
Nepal	29.3	1530	12	9	12	210	785	7372
Sri Lanka	21	1712	0	13	0	53	3924	2549
Pakistan	197	494	265	19	57	247	1437	1306

Source FAO AQUASTAT (2017) http://www.fao.org/nr/water/aquastat/data/query.html

Table 12.2 Brief Introduction of major Rivers of South Asia

	Total surface area of rivers (km^2)	Annual mean flow (km^3/year)	Annual mean runoff (mm/year)	Flowing countries
Ganges	1,087,000	525.32	483	India, Nepal, Bangladesh
Brahmaputra	552,000	624.78	1132	Bhutan, India, Bangladesh
Helmand	306,500	15	45	Afghanistan, Pakistan
Meghna	82,000	161.0	1963	India, Bangladesh
Indus	1,165,000	207.0	178	India, Afghanistan, Pakistan
GBM	1,721,000	1311.19	762	India, Nepal, Bhutan, Bangladesh

Source SAWI, World Bank (2018)

12.3.2 Climate Change: Management Risks and Future Costs

The climate change losses will cost 1.8% of GDP of South Asia by 2050 and projected to grow up to 8.8% in 2100 if we assume the current scenario of greenhouse gas emission (Ahmed and Suphachalasai 2014). Countries that depend mostly on agriculture will be facing huge loss due to climate change because agriculture totally based on climatic condition. Poor countries will be also affected due to lack of technology to predict natural hazards for reducing the infrastructure, life, and property loss. The climate change will have substantial impact over South Asian economy with worse effect will be on Nepal, Bangladesh followed by India and Bhutan if the average temperature rises within 2 °C (Jongman 2018; Sayers et al. 2018). A recent case study suggested that in India the productivity of crops is being decreased due to climate change which results in the increase in the poverty of the country. It was also reported that the decrease in productivity of staple crop may cause loss of US $208 billion by 2050 and if unabated the losses can grow up to US $366 billion in 2100 (Chaturvedi 2015; Jung et al. 2018). Economic impact assessment (EIA) analyzed that uncertainty of climate affects the agriculture and hydropower sector because of change in the pattern of precipitation (Schäfer et al. 2018). The EIA estimates climate change to have a direct cost to GDP of 1.5–2% (US $270–360 million/year in 2013 prices) per year rising to a GDP cost of 5% per year in extreme years (Arent et al. 2015). Bangladesh would experience these impacts through declines in agricultural production as well as through loss of land availability due to melting glaciers that cause sea-level rise (Barnosky et al. 2012).

12.3.3   The Present Strategy of South Asia's Water Resources Management

- Developing reliable supplies for meeting growing domestic, industrial, and agricultural water demand and energy use.
- Promoting sustainable use, protecting water quality, and managing watersheds, aquifers, lakes, and wetlands.
- Systematic planning, development, and management to address the systemic risks emerging of.
- Water-related disasters, i.e., droughts, floods, sedimentation.
- Promoting collaborative management of shared waters across districts, river basins and aquifers, states and provinces, and nations.
- Integrating water policies and actions with those outside the water sector (environment, land use, energy).

12.4   Role of the Different Institution for Water Resource Management

12.4.1   Role of Global Water Partnership (GWP)

GWP works with integrated water resource management (IWRM). The main function of the institute is to manage water- and land-related natural resources with an integrated approach that enhances the strength of social and economic and environment factors without compromising sustainability principle (Table 12.3). In areas of water stress, several adaptations for interventions could consist of (Mollinga 2006; Giupponi and Gain 2017).

- Limited consumption of water at the time of shortage.
- Reuse of wastewater discharged from industrial and agricultural production.
- Increased collection and storage of surface runoff.
- Recycling and reuse of wastewater after treatment.
- Desalination of salty or brackish water.
- Better recharge systems, monitoring, and use of groundwater resources.
- Rainwater harvesting to augment availability and increase groundwater recharge capacity.

In areas where water quality is affected, possible measures are:

- Improvements in drainage systems to reduce the mixing of different water types (blue, green, and gray).

Table 12.3 Some organization with their outcome in water resource management sector

Organization	Working region	Agenda	Achievements
Global Water Partnership (GWP)	Caribbean, Central America, Central Africa, Central and eastern Europe, Central Asia, China, South Asia, Southeast Asia, South America, West Africa, Eastern Africa, Southern Africa	Provide water solution with development challenges Catalyze climate resilience development Enhance transboundry co-operation	Skills women to bring improvement in the management of larger water sector-Bangladesh (2018) Community water projects focusing on improving access to water–Bhutan (2018) Completed the solid and wastewater dumped and reduced water stress in Ajmer and Kishangarh city RWH installed in government schools, flood management system prepare in Bihar-India (2018)
South Asia Water Initiative (SAWI)	Afghanistan, Bangladesh, Bhutan, India, Nepal, Pakistan, Sri Lanka	Promoting Sustainable Water Resources Management in South Asia Enhance the management of the major Himalayan river systems in South Asia for sustainable development and climate resilience	Building Trust and Confidence, Generating and Sharing Knowledge , Building Institutional and Professional Capacity, Scoping Interventions and Investments for . Indus River Basin Development Plan Ganges Basin Development Plan Brahmaputra Basin Development Plan (BSDP)

(continued)

Table 12.3 (continued)

Organization	Working region	Agenda	Achievements
International Centre for Integrated Mountain Development (ICIMOD)	Afghanistan, Bangladesh, Bhutan, China, India, Myanmar, Nepal, and Pakistan	Monitoring Hindu Kush Range and Himalayan ecosystem for healthy environment	Improved transboundry relationship between member countries and developed a model for sustenance of mountain water system Developed a program for reduce critical risk disaster in Hindu Kush Range
International Water Management Institute (IWMI)	Africa, Asia, Middle East, and North Africa	Building resilience Sustainable growth Rural–urban linkages	Climate Smart village in South Asia-2018 Strategic Basin Assessment for Brahmaputra River-2017 Groundwater use in irrigation (Sri Lanka)-2017 Climate resilience agriculture in Nepal-2017

- Upgrading or standardization of water treatment at different levels and scales.
- Better monitoring for variation in the quality of water.
- Special measures should be taken during high precipitation seasons to manage impacts on water quality.

12.4.1.1 Water Management Plan in Bangladesh

From 2012, Global Water Partnership organized numbers of workshop, college courses, trainings, conferences for encouraging sustainable water management. Data shows that thousands of participants attend these events all over the world and learned the basic principle and application of integrated water resource management (Global Water Partnership (GWP) 2018). According to Alauddin Ahmed Choudhury, who is the Deputy Team Leader of the Bangladesh Delta Plan 2100 formulation Project, "more than 200 water management expert's starts the trainings in different parts of Bangladesh to protect the rivers basin under the Bangladesh delta plan 2100." Many IWRM centers have been established and managed by local engineer of the country to facilitate the implementation of small-scale irrigation projects. The southwest region IWRM is operated and managed by Bangladesh Water Development Board (Islam et al. 2019; Zegwaarda et al. 2019).

12.4.1.2 Urban Water Planning in Rajasthan, India

Most of the population migrates toward the urban area. So it is compulsory to provide drinking water to all people in a healthy and safer way. The main role of the integrated urban water management monitoring, planning, and designing urban water system for fulfilling the Sustainable Development Goal (SDG-6). Local governing body for sustainability has been working to develop and implement integrated urban water management (IUWM) action plan for Kishangarh Town, Ajmer District of Rajasthan with the help of European Union Fund (EUF) and International Council for Local Environmental Initiatives (ICLEI). In 2017, GWP India provided financial support to update the existing plan and apply the IUWM toolkit in a pilot project focused on the collection of household waste and composting. To enhance the capacity of the urban local authority and to showcase the adaptability of the approach is the main aim of this program. After conducting a fact-finding study, the partners organized a series of training and capacity building workshops for stakeholders from local government offices, non-governmental organizations, municipal contractors, and sanitation workers. These events along with additional training courses on sanitation and waste collection gave stakeholder an innovative practice water management. If the program will achieve its objectives, then it will encourage other city to adopt these plans (Dillon et al. 2018; Pingale et al. 2016).

12.4.1.3 Kabul and Kunar Basin Development

Afghanistan and Pakistan are two neighboring countries. Kabul and Kunar basin development creates a strong system within the governments of these two countries for the establishment of law, regulation, and norms for transboundary water and basic facilities, to provide communication between the two nations and to increase mutual understanding on the development and management of the Kabul River Basin (KRB). Several agendas have been developing to build up the relationships between countries with riverbank in the Indus Basin region. For future growth and development, numbers of the seminar, workshop, and internship program will be conducted that will make an international relationship more strong (United Nations-WATER 2017; Rai et al. 2017). This program helps to consumption water of both river and creates hydropower plant for energy securities.

12.4.2 South Asia Water Initiative (SAWI)

SAWI worked in basin management of three rivers Indus, Ganga, Brahmaputra that contain seven countries detail information about this organization in Table 12.3.

12.4.2.1 Indus Basin Management

In 2013, the World Bank started a conversation with those countries which are part of Indus River Basin (IRB) countries. For the participant of conversation, Indus Forum program was launched with the objective to create credence for establishing a positive ecosystem for basin wise co-operation. Indian River Basin along with the Indus Forum makes an agenda that focuses on technical collaboration on the issues which are earlier explained by the Indus Forum. Along with these other initiatives such as Atlantic Council Indian-Pakistan Dialogue, Upper Indus Basin (UIB) Initiative, and Institutes such as International Centre for Integrated Mountain Development (ICIMOD) Himalayan Adaptation, Water and Resilience project promote the faith among the participating countries (SAWI-World Bank 2018).

12.4.2.2 Ganges

Ganges Basin Plan—For maintaining the river health of Ganga, India developed a comprehensive river basin model. Ganga basin is the most fertile basin of India, so it is necessary to make this river pollution-free and increases its ecological flow to meet the consumption of water and supports river, lakes, canal, waterway navigation (Wahid et al. 2016). Ganges Basin Plan integrated and collaborated with different government ministries that operate numerous important schemes such as

Ganga Action Plan, Namami Gange, National Groundwater Management Improvement Project (MIP), National Ganga River Basin Project (NGRBP), and the Uttar Pradesh Water Sector Restructuring Project (Srinivas et al. 2019).

Ganges Basin Development Plan—This program was launched to develop a river basin model for India and Nepal. This model requires some technical support to make basin-wise dialogue for hydrologic water resource modeling. Ganges basin made the most fertile land of India and provides agriculture and fisheries income to millions of people. So for Ganga River basin, modeling scientist and expert are invited all over the world from the different technical institutions, and they are diligently committed for the modeling in the river Ganges (Srinivas et al. 2018; SAWI-World Bank 2018).

12.4.2.3 Brahmaputra

Brahmaputra Basin Management (BBM)

Climate change is the urgent threat for all water resources, due to glaciers melting water level continuously increase in Brahmaputra River. More than 60% of Majuli Island is submerged under Brahmaputra River. BBM works on remediation and proper monitoring of climate change by creating information infrastructure of the basin with the help of climatic models and also finds a different alternative that invests on BBM project for basin development (Fischer et al. 2016). This program analyzes the scenario of dam failure during monsoon season also figures out the buffer area that will affect by the flood. The whole work is a part of NHP—a World Bank initiative program for water resources management and also Assam flood erosion and river basin management activity that is initiative of the Assam government (Borgohain 2019; SAWI-World Bank 2018).

Brahmaputra Basin Development Plan (BSDP)

This scheme is developed to expand the regional contribution by providing a platform to discuss and shared issues related to water. BSDP program encourages and enables working relationships stronger between basin bank countries to optimize the consideration of river basin management of the Brahmaputra. To achieve the objectives, teamwork on the country-specific problem is with their local bodies (Barua 2018). Brahmaputra symposium organized by BSDP suggests some reliable steps such as project team should visit the local area for the understanding of problem, and they should also participate in national and international workshops, seminars, conferences for the exchange of ideas and better understanding of the issue.

12.4.3   Role of International Water Management Institute (IWMI)

The main aim of this scientific organization is to focus on the sustainability of water resources and issues related to its crisis. Keeping this in mind, they are working on producing a large quantity of food by consuming less water so as to maintain equilibrium between mankind and natural resources. Some relevant information about this organization is given in Table 12.3.

Revitalizing irrigation—The IWMI is working on many irrigation-related plans that can be more fruitful and sustainable for the irrigation system. An innovative method of irrigation called canal irrigation has been introduced in many rural areas. The canal system is based on the division of the main canal into many secondary and tertiary canals that created a network and reach to the irrigation field. This method has increased the production of crops and thereby possesses a profitable impact on South Asian agriculture system (Giordano et al. 2018). A canal-based irrigated land has shown more yield as compared to the land that depends on rainwater.

Improving agricultural water productivity—The organization is also working on policies related to water resources that need to be modified so as to get more food without excess exploitation of natural resources. This will help to tackle the problem like poverty which has become a major problem in many developing countries. They also help to promote the necessary knowledge about the time management for planting and sowing, techniques such as drip irrigation and utilization of rainwater for irrigation and by introducing genetically modified crops to the farmers that can be utilized to achieve high production of crops (Shah et al. 2018). An advanced method that is based on GIS and computer software has also been introduced which can able to find out the current status of availability of water which will be helpful to enhance the farm remuneration by targeting the infrastructure and establishing the security of food in that region which cannot be accessed by any other methods (Qureshi 2018). According to the recent report, China with a well-managed system is producing three times more food as compared to India by minimal use of water. Furthermore, South Asia has a strong capability to increase the yield of water as evidenced by successes elsewhere (IWMI 2016).

Transboundary water transfers—On a global scale, IWMI focuses on transboundary-related issues and to develop a best-suited method for policymakers over sharing of water resources across the south Asian river basins. This practice may able to develop an amicable relationship between the countries (Zhupankhan 2018). National river linking project of India was one of the biggest contributions of IWMI on the international platform participating actively on the debate over the sharing of the surface water with the neighboring country. They also provide information to the countries sharing the common water resource that would help to achieve a proper management system on water boundaries.

Groundwater—Groundwater is an essential source of drinking water, and hence, its sustainable use is the prime objective of this organization. They are still working on developing a method that is convenient for both farmers and for governmental authorities to develop policies. To achieve this, many technologies that can recharge groundwater have been established in many rural areas. It will increase the level of groundwater and hence makes it available to utilize it for irrigation purposes. South Asia has an abundance of groundwater, yet its overexploitation and unawareness cause the depletion of the groundwater table that would probably affect the availability of water (Pavelic et al. 2018). IWMI is also working on various parts of South Asia that have plenty of groundwater by providing suitable methods and investment strategies. They are also organizing an awareness campaign to understand the water-energy relationship and sustainable use of water (Ray 2018).

Environmental flows—Natural resources directly affect the quality of life in both rural and urban areas. Hence, IWMI is committed to maintain equilibrium between the natural system and living being by the help of their computer-based environmental calculator that correlates and provides information about water demand in the region. The organization along with their local partners first time worked on environmental flow of Ganges which is well known for its cultural as well as religious importance. Also, they are working to fix the conflict between the collaborators for a peaceful exchange of the trade.

12.5 Conclusions

South Asia and their water resources' management have been studied which shows that many factors affect the utilization of water resources. These factors depend on individual countries, regions, and even continents. Increases in global temperature (climate change affects both surface and atmospheric) that melt the ice mass (cryosphere), due to which there is a seasonal variation which ultimately affects the size and duration of the monsoon that put an adverse effect on the rate of evaporation and ecological flow of the river. Groundwater resources are the major source of drinking water and for agriculture in South Asia. Nowadays, it is facing a high risk of depletion due to its overexploitation and less groundwater recharge due to increase in the rate of urban area growth. The poorly managed and uneconomical use of natural resources along with the pollution created by industries and urban centers is the prime reason for climate change that ultimately affects the annual rainfall and rise in the sea level. The uncertainty of rainfall causes catastrophic hazards that affect the lives and economic status of millions of people. This chapter has put an effort to understand the present scenario of water resources in South Asia and their management policies that have been adapted to tackle the problems related to water resources and climate change.

References

Adhikari KN (2014) Conflict and cooperation on South Asian water resources. IPRI J 14(2):45–62

Afghanistan Online (2018). www.afghan-web.com. Retrieved April 19, 2109, from www.afghan-web.com: https://www.afghan-web.com/environment/water

Ahmed AU, Appadurai N, Neelormi S (2018) Status of climate change adaptation in South Asia region. In Alam M, Lee J, Sawhney P (eds) Status of climate change adaptation in Asia and the Pacific. Springer, Switzerland (in press)

Ahmed M, Suphachalasai S (2014) Assessing the costs of climate change and adaptation in South Asia. Asian Development Bank

Aich V, Akhundzadah NA, Knuerr A, Khoshbeen AJ, Hattermann F, Paeth H, Scanlon A, Paton EN (2017) Climate change in Afghanistan deduced from reanalysis and coordinated regional climate downscaling experiment (CORDEX)—South Asia simulations. Climate 5:38

Akanda MAI (2019) Seasonal and regional limits to growth of water-intensive crop farming in Bangladesh. Sustain Water Resour Manag 2:817–830

Alam M, Huq S (2019) Measuring status of climate change adaptation: an assessment framework. In: Status of climate change adaptation in Asia and the Pacific. Springer, Cham, pp 13–26

Amrith SS (2018) Risk and the South Asian monsoon. Clim Change 151:17–28

Arent DJ, Tol RS, Faust E, Hella JP, Kumar S, Strzepek KM, Tóth FL, Yan D, Abdulla A, Kheshgi H, Xu H (2015) Key economic sectors and services. In: Climate change 2014 impacts, adaptation and vulnerability: part A: global and sectoral aspects, pp 659–708

Artuc E, Lopez-Acevedo G, Robertson R, Samaan D (2019) Exports to jobs: boosting the gains from trade in South Asia. The World Bank

Aryal JP, Sapkota TB, Khurana R, Khatri-Chhetri A (2019) Climate change mitigation options among farmers in South Asia. Environ, Dev Sustain 1–23

Barnosky AD, Hadly EA, Bascompte J, Berlow EL, Brown JH, Fortelius M, Getz WM, Harte J, Hastings A, Marquet PA, Martinez ND (2012) Approaching a state shift in Earth's biosphere. Nature 486(7401):52

Barua A (2018) Water diplomacy as an approach to regional cooperation in South Asia: a case from the Brahmaputra basin. J Hydrol 567:60–70

Berrang-Ford L, Pearce T, Ford JD (2015) Systematic review approaches for climate change adaptation research. Reg Environ Change 15:755–769

Borgohain PL (2019) Downstream impacts of the Ranganadi hydel project in Brahmaputra Basin, India: Implications for design of future projects. Environ Dev 30:114–128

Budhathoki KP, Bajracharya OR, Pokharel BK (2011) Assessment of Imja Glacier Lake outburst Flood (GLOF) risk in Dudhkoshi river basin using remote sensing techniques. Journal of Hydrology and Meteorology 7:75–90

Bush KF, Luber G, Kotha SR, Dhaliwal RS, Kapil V, Pascual M, Brown DG, Frumkin H, Dhiman RC, Hess J, Wilson ML (2011) Impacts of climate change on public health in India: future research directions. Environ Health Perspect 119(6):765–770

Chaturvedi V (2015) A working paper on 'The costs of climate change impacts for India: a preliminary analysis'. Council on Energy, Environment and Water. http://www.indiaenvironmentportal.org.in/files/file/The%20Costs%20of%20Climate%20Change%20Impacts%20for%20India.pdf

Das JK, Rizvi A, Bhatti Z, Paul V, Bahl R, Shahidullah M, Manandhar D, Stanekzai H, Amarasena S, Bhutta ZA (2015) State of neonatal health care in eight countries of the SAARC region, South Asia: how can we make a difference? Paediatr Int Child Health 35(3):174–186

Devkota LP, Gyawali DR (2015) Impacts of climate change on hydrological regime and water resources management of the Koshi River Basin, Nepal. J Hydrol: Reg Stud 4(B):502–515

Dillon P, Pavelic P, Nava AP, Weiping W (2018) Advances in multi-stage planning and implementing managed aquifer recharge for integrated water management. Sustain Water Resour Manag 145–151

Field CV (2014) Summary for policymakers. In Climate change 2014: impacts, adaptation, and vulnerability. Part A: global and sectoral aspects. Contribution of working group II to the fifth assessment report of the intergovernmental panel on climate change. Cambridge University Press, Cambridge, United Kingdom and New York: IPCC

Fischer S, Pietron J, Bring A, Thorslund J, Jarsjo J (2016) Present to future sediment transport of the Brahmaputra River: reducing uncertainty in predictions and management. Reg Environ Change 17:515–526

Giordano MA, Wolf AT (2003) Sharing waters: post-Rio international water management. Nat Resour Forum 27(2):163–171

Giordano M, Barron J, Ünver O (2018) Water scarcity and challenges for smallholder agriculture. In: Campanhola C, Pandey S (eds) Sustainable food and agriculture: an integrated approach. Academic Press, Cambridge, USA

Giupponi C, Gain AK (2017) Integrated water resources management (IWRM) for climate. Reg Environ Change 17(7):1865–1867

Global Water Partnership (GWP) (2018) GWP in action 2017 annual report. Global Water Partnership (GWP), Stockholm, SWEDEN

Haigh R, Amaratunga D, Hemachandra D (2018) A capacity analysis framework for multi-hazard early warning in coastal communities. Procedia Eng 212:1139–1146

Hanasz P (2017) Transboundary water governance and international actors in South Asia: the Ganges-Brahmaputra-Meghna Basin. Routledge

Hirji, R., Nicol, A., Davis, R. (2017). Climate Risks and Solutions: Adaptation Frameworks for Water Resources Planning, Development and Management in South Asia. South Asia 8S Climate Change Risks in Water Management. World Bank report

Howard KW, Howard KK (2016) The new "Silk Road Economic Belt" as a threat to the sustainable management of Central Asia's transboundary water resources. Environ Earth Sci 75:1–12

IPCC (2014) Assesment report. IPCC, USA

Islam, R., Jahan, S.C., Mazumder, Q.H., Miah, S., Rahaman, F. (2019). Water footprint and governance assessment for sustainable water resource management in drought-prone Barind area, NW Bangladesh. Advances in sustainable and environmental hydrology, hydrogeology, hydrochemistry and water resources, pp 371–373

Ives JD, Shrestha RB, Mool PK (2010) Formation of glacial lakes in the Hindu Kush-Himalayas and GLOF risk assessment. International Centre for Integrated Mountain Development, Kathmandu, Kathmandu

IWMI (2016) IWMI Research South Asia. CGIAR, IWMI

Jongman B (2018) Effective adaptation to rising flood risk. Nat Commun. 9 Article number: 1986

Jung J, Herbohn K, Clarkson P (2018) Carbon risk, carbon risk awareness and the cost of debt financing. J Bus Ethics 150(4):1151–1171

Kumar P, Masago Y, Mishra BK, Fukushi K (2018) Evaluating future stress due to combined effect of climate change and rapid urbanization for Pasig-Marikina River, Manila. Groundw Sustain Dev 6:227–234

Kumar M, Patel AK, Das A, Kumar P, Goswami R, Deka JP, Das N (2017) Hydrogeochemical controls on mobilization of arsenic and associated health risk in Nagaon district of the central Brahmaputra Plain, India. Environ Geochem Health 39(1):161–178

Kumar M, Herbert R Jr, Ramanathan AL, Rao MS, Deka JP, Kumar B (2013) Hydrogeochemical zonation for groundwater management in the area with diversified geological and land-use setup. Chemie der Erde-Geochemistry 73:267–274

Lacombe G, Chinnasamy P, Nicol A (2019) Review of climate change science, knowledge and impacts on water resources in South Asia. Background Paper 1. Colombo, Sri Lanka: International Water Management Institute (IWMI). 73p

Lala JM, Rounce DR, McKinney DC (2018) Modeling the glacial lake outburst flood process chain in the Nepal Himalaya: reassessing Imja Tsho's hazard. Hydrol Earth Syst Sci 22:3721–3737

Mahanta C, Mahagaonkar A, Choudhury R (2018) Climate change and hydrological perspective of Bhutan. In: Mukherjee A (ed) Groundwater of South Asia. Springer, Singapore

Muthukumara M, Bandyopadhyay S, Chonabayashi S, Markandya A, Mosier T (2018) South Asia's hotspots: impacts of temperature and precipitation changes on living standards. South Asia Development Matters. World Bank, Washington, DC. https://openknowledge.worldbank.org/handle/10986/28723

Mertz O, Halsnæs K, Oleson JE, Rasmussen K (2009) Adaptation to climate change in developing countries. Environmen Manag 43:743–752

Mollinga PP (2006) IWRM in South Asia: a concept looking for a constituency. In: Integrated water resources management in South Asia. Global theory, emerging practice and local needs; SAGE Publications: California, CA, USA, Chapter 1; pp 21–37

Mukherjee S, Patel AK, Kumar M (2020) Water scarcity and land degradation Nexus in the era of Anthropocene: some reformations to encounter the environmental challenges for advanced water management systems meeting the sustainable development. In: ---------- (ISBN 978-93-81891-41-4), Publisher Springer Nature

Mukherjee N, Rowan JS, Khanum R, Nishat A, Rahman S (2019) Climate change-induced loss and damage of freshwater resources in Bangladesh. Confronting climate change in Bangladesh. Springer, Cham, pp 23–37

Niti Ayog (2018) Composite water management index. Government of India Ministry of Drinking Water and Sanitation, New Delhi

Owusu PA, Sarkodie SA (2016) A review of renewable energy sources, sustainability issues and climate change mitigation. Cogent Eng 3(1):1–14

Pavelic P, Johnston R, McCartney M, Lacombe G, Sellamuttu SS (2018) Groundwater resources in the dry zone of Myanmar: a review of current knowledge. In: Mukherjee A (ed) Groundwater of South Asia. Springer, Berlin, Germany, pp 695–705

Pingale SM, Khare D, Jat MK, Adamowski J (2016) Trend analysis of climatic variables in an arid and semi-arid region of the Ajmer District, Rajasthan, India. J water land dev 28(1):3–18

Patel AK, Das N, Kumar M (2019) Multilayer arsenic mobilization and multimetal co-enrichment in the alluvium (Brahmaputra) plains of India: a tale of redox domination along the depth. Chemosphere 224:140–150

Qureshi AS (2018) Increasing water productivity in the agricultural sector. In: Indus River Basin edited by Khan, SI and Adams 111, TE, pp 229–244

Rahmasary AN, Robert S, Chang IS, Jing W, Park J, Bluemling B, Koop S, Leeuwen KV (2019) Overcoming the challenges of water, waste and climate change in Asian cities. Environ Manage 63(4):520–535

Rai SP, Young W, Sharma N (2017) Risk and opportunity assessment for Water Cooperation in Transboundary River Basins in South Asia. Water Resour Manag 31(7):2187–2205

Ravenscroft P, McArthur JM, Rahman MS (2018) Identifying multiple deep aquifers in the Bengal Basin: implications for resource management. Hydrol Process 32(24):3615–3632

Ray P, Alexander B, Casey M (2015) Confronting climate uncertainty in water resources planning and project design. 149: International Bank for Reconstruction and Development/The World Bank

Ray SP (2018) Major ground water development issues in South Asia: an overview. In Ray SP (ed) Ground water development—issues and sustainable solutions, pp 3–11

SAWI, World Bank (2018) The World Bank SAWI. Retrieved MAY 2019, from SAWI: http://www.worldbank.org/en/programs/sawi#3

Sayers P, Rowsell ECP, Matt Horritt M (2018) Flood vulnerability, risk, and social disadvantage: current and future patterns in the UK. Reg Environ Change 18(2):339–352

Schäfer L, Warner K, Kreft S (2018) Exploring and managing adaptation frontiers with climate risk insurance. In: Mechler R, Bouwer LM, Schinko T, Surminski S, Bayer JL (eds) Loss and damage from climate change concepts, methods and policy options, Springer, pp 317–341

Schulz L, Kingston DG (2017) GCM-related uncertainty in river flow projections at the threshold for "dangerous" climate change: the Kalu Ganga River, Sri Lanka. Hydrol Sci J 62(14):2369–2380

Seth A, Giannini A, Rojas M, Rauscher SA, Bordoni S, Singh D, Camargo SJ (2019) Monsoon responses to climate changes—connecting past, present and future. Curr Clim Change Rep 5:63–79

Shah T, Lele U (2011) Climate change, food and water security in South Asia: critical issues and cooperative strategies in an age of increased risk and uncertainty. Colombo, Sri Lanka: Global Water Partnership (GWP), IWMI

Shah T, Rajan A, Rai GP, Verma S, Durga N (2018). Solar pumps and South Asia's energy-groundwater nexus: exploring implications and reimagining its future. Environ Res Lett 13(11)

Singh A, Patel AK, Kumar M (2020) Mitigating the risk of arsenic and fluoride contamination of groundwater through a multi-model framework of statistical assessment and natural remediation techniques In: Kumar M, Snow D, Honda R (eds) Emerging issues in the water environment during Anthropocene: a South East Asian perspective (ISBN 978-93-81891-41-4), Publisher Springer Nature

Singh A, Patel AK, Deka JP, Das A, Kumar A, Manish K (2019) Prediction of arsenic vulnerable zones in groundwater environment of rapidly urbanizing setup, Guwahati, India. Geochemistry 125590. https://doi.org/10.1016/j.chemer.2019.125590

Sherpa SF, Shrestha M, Eakin H, Boone CG (2019) Cryospheric hazards and risk perceptions in the Sagarmatha (Mt. Everest) National Park and Buffer Zone, Nepal. Nat Hazards pp 1–20

Singh C, Daron J, Bazaz A, Ziervogel G, Spear D, Krishnaswamy J, Zaroug M, Kituyi E (2018) The utility of weather and climate information for adaptation decision-making: current uses and future prospects in Africa and India. Clim Dev 10:389–405

Souza KD, Kituyi E, Harvey B, Leone M, Kallur M, Ford JD (2015) Vulnerability to climate change in three hot spots in Africa and Asia: key issues for policy-relevant adaptation and resilience-building research. Reg Environ Change 15:747–753

Sovacool BK, Linnér BO, Klein RJT (2016) Climate change adaptation and the least developed countries fund (LDCF): qualitative insights from policy implementation in the Asia-Pacific. Clim Change 140:209–226

Srinivas R, Singh AP, Shankar D (2019) Understanding the threats and challenges concerning Ganges River basin for effective policy recommendations towards sustainable development. Environ, Dev Sustain 21(100)

Srinivas R, Singh AP, Dhadse K, Garg C, Deshmukh A (2018) Sustainable management of a river basin by integrating an improved fuzzy based hybridized SWOT model and geo-statistical weighted thematic overlay analysis. J Hydrol 563:92–105

Turner AG, Annamalai H (2012) Climate change and the South Asian summer. Nat Clim Change 2(8):587–595

Turner SW, Hejazi M, Kim SH, Clarke L, Edmonds J (2017) Climate impacts on hydropower and consequences for global electricity supply investment needs. Energy 141:2081–2090

Udayakumara EPN, Gunawardena UADP (2018) Cost-benefit analysis of Samanalawewa hydro-electric project in Sri Lanka: An Ex post analysis. Earth Syst Environ 2(2):401–412

UNFCC (2018) Water Resources and Adaptation programs in Afghanistan. UNFCC

UNICEF (2017) WHO JMP UNICEF 2017 progress on drinking water, sanitation and hygiene. Retrieved April 2019, from http://www.who.int: http://www.who.int/mediacentre/news/releases/2017/launch-version-report-jmp-water-sanitation-hygiene.pdf

United Nations Environment Programme (2006) History of Environmental Change in the Sistan Basin Based on Satellite Image Analysis: 1976–2005

United Nation-WATER (2017) UN WATER. Retrieved April 2019, from http://www.unwater.org/water-facts/scarcity/

Vinke K, Martin MA, Adams S, Baarsch F, Bondeau A, Coumou D, Donner RV, Menon A, Perrette M, Rehfeld K, Robinson A, Rocha M, Schaeffer M, Schwan S, Serdeczny O, Svirejeva-Hopkins A (2017) Climatic risks and impacts in South Asia: extremes of water scarcity and excess. Reg Environ Change 17(6):1569–1583

Wahid SM, Mukherji A, Shrestha A (2016) Climate change adaptation, water infrastructure development, and responsive governance in the Himalayas: the case study of Nepal's Koshi River basin. In: Increasing resilience to climate variability and change. Springer, Singapore, pp 61–80

WHO (2018) World Health Organization. Retrieved April 2019, from www.who.int: http://www. who.int/mediacentre/factsheets/fs391/en/

Zegwaarda A, Zwarteveen M, Halsemac Gv, Petersen A (2019) Sameness and difference in delta planning. Environ Sci Policy 94:237–244

Zeitoun M, Warner J (2006) Hydro-hegemony-a framework for analysis of trans-boundary water conflicts. Water Policy 8:435–460

Zevenbergen C, Khan SA, van Alphen J, Terwisscha van Scheltinga C, Veerbeek W (2018) Adaptive delta management: a comparison between the Netherlands and Bangladesh delta program. Int J River Basin Manag 16(3):299–305

Zhupankhan AK (2018) Water in Kazakhstan, a key in Central Asian water management. Hydrol Sci J 63:752–762

Chapter 13
Attenuation and Fate of Pharmaceuticals in River Environments

Seiya Hanamoto

13.1 Introduction

Pharmaceuticals have been widely detected in surface waters (Kolpin et al. 2002; Kasprzyk-Hordern et al. 2008; Nakada et al. 2008). Due to the potential risk to aquatic organisms, pharmaceuticals are recognized as contaminants of emerging concern (Cooper et al. 2008; Boxall et al. 2012; Cizmas et al. 2015). In addition, the presence of antibiotics in aquatic environments is possibly linked to antibiotic resistance (Ågerstrand et al. 2015; Kumar et al. 2019a), which is "one of the most important challenges to the health care sector in the twenty-first century" (Carvalho and Santos 2016). Therefore, in order to assess their risks and aid in their management, the environmental fate of pharmaceuticals should be modeled.

In the aquatic environment, pharmaceuticals may be attenuated by physical, chemical, and/or biological processes. Studies on the natural attenuation of pharmaceuticals during river transport suggest that rapid removal is possible for some of the pharmaceuticals (Gurr and Reinhard). To the best of our knowledge, there are no reviews on the natural attenuation of pharmaceuticals since the study conducted by Gurr and Reinhard (2006; Kumar et al. 2019b). The purpose of this review is to summarize recent field studies on the in-stream attenuation of pharmaceuticals along with the controlling factors and models for predicting in-stream attenuation.

S. Hanamoto (✉)
Environment Preservation Center, Kanazawa University, Kakumamachi, Kanazawa, Ishikawa 920-1192, Japan
e-mail: hanamoto64@staff.kanazawa-u.ac.jp

© Springer Nature Singapore Pte Ltd. 2020
M. Kumar et al. (eds.), *Resilience, Response, and Risk in Water Systems*,
Springer Transactions in Civil and Environmental Engineering,
https://doi.org/10.1007/978-981-15-4668-6_13

13.2   Natural Attenuation of Pharmaceuticals in Rivers

Field studies on in-stream attenuation of pharmaceuticals are summarized in this section. The attenuation (%) of a compound in a river stretch indicates the percentage of the compound removed from the water parcel during transport along the river stretch.

13.2.1   Santa Ana River, California (CA), USA

Lin et al. (2006) observed the natural attenuation of pharmaceuticals along a 12-km stretch of the Santa Ana River (CA, USA). The average flow rate, depth, width, and water travel time of the river stretch were 1.4 m^3/s, 0.3 m, 10–20 m, and 7.5 h, respectively. During the sampling period, flow in the river mostly consisted of tertiary treated wastewater. Water grab samples were collected from the surface 0.8, 1.6, 3.6, 5.6, and 12 km downstream of a wastewater treatment plant (WWTP), three times (i.e., morning, afternoon, and night) on September 15–16, 2004. The average attenuation and corresponding first-order decay constants in the river stretch were 94% and 0.39 h^{-1} for naproxen, 85% and 0.26 h^{-1} for gemfibrozil, and 62% and 0.13 h^{-1} for ibuprofen, respectively.

13.2.2   Trinity River, Texas (TX), USA

Fono et al. (2006) observed natural attenuation of pharmaceuticals along a 500-km stretch of the Trinity River (TX, USA). The average flow rate, depth, width, and water travel time of the river stretch were 21–23 m^3/s, 2 m, 50 m, and 13–14 d, respectively. During the sampling period, effluents from WWTPs accounted for approximately 83% of the flow of the river stretch. Water grab samples were collected from the surface at five sites on the river stretch, four times in September 2005. The average attenuation and corresponding first-order decay constants in the river stretch were 88% and 0.0067 h^{-1} for naproxen, 87% and 0.0063 h^{-1} for ibuprofen, 83% and 0.0054 h^{-1} for metoprolol, and 66% and 0.0033 h^{-1} for gemfibrozil, respectively.

13.2.3   Gründlach River, Germany

Kunkel and Radke (2012) and Li et al. (2016) observed natural attenuation of pharmaceuticals along a 12.5-km stretch of the Gründlach River in Germany. The average flow rate, depth, width, and water travel time of the river stretch were 0.029 m^3/s, 0.15 m, 3 m, and 2.1 d, respectively. Surface water samples were collected by an automatic sampler at 0.6 and 12.5 km downstream of a WWTP in July 2010 (Kunkel and Radke 2012) and June 2014 (Li et al. 2016). Sampling was carried out for a week, and samples collected every hour were combined to form composite samples. A persistent chemical, fluconazole, was used as the reference compound to account for dilution by infiltrating groundwater or minor tributaries along the stretch. The average attenuation and corresponding first-order decay constants in the river stretch were 95% and 0.059 h^{-1} for acetaminophen, 90% and 0.046 h^{-1} for furosemide, 70% and 0.024 h^{-1} for propranolol, 60% and 0.018 h^{-1} for diclofenac, 55% and 0.016 h^{-1} for metoprolol, 50% and 0.014 h^{-1} for naproxen, 20% and 0.004 h^{-1} for sulfamethoxazole, and ~0% and ~0 h^{-1} for carbamazepine, respectively.

13.2.4 Steinlach River, Germany

Guillet et al. (2019) observed natural attenuation of pharmaceuticals along a 1.3-km stretch of the Steinlach River in Germany. The average flow rate of the river stretch was 1.7 m^3/s. At the baseflow, the depth and width were 0.2 m and 7 m, respectively. During the sampling period, the flow rate and water travel time were 0.18 m^3/s and 3.5 h, respectively, and effluents from a WWTP accounted for 73–77% of the total flow in the river stretch. Water grab samples were collected from the surface at four sites on the river stretch, with fluorescent tracer experiments in August 2015. The average attenuation and corresponding first-order decay constants in the river stretch were 50% and 0.20 h^{-1} for diclofenac, 50% and 0.20 h^{-1} for metoprolol, 30% and 0.10 h^{-1} for atenolol, 25% and 0.082 h^{-1} for trimethoprim, 20% and 0.064 h^{-1} for sulfamethoxazole, 10% and 0.030 h^{-1} for primidone, 10% and 0.030 h^{-1} for sulpiride, and 5% and 0.015 h^{-1} for carbamazepine, respectively.

13.2.5 Erpe River, Germany

Jaeger et al. (2019) observed natural attenuation of pharmaceuticals along 1.6- and 3.1-km stretches of the Erpe River in Germany. During the sampling period, flow rates and water travel times of the river stretches ranged from 0.50–0.71 m^3/s and 3.1–5.4 h, respectively. Surface water samples were collected hourly by an automatic water sampler for 48 h at the beginning and end on the river stretches. A persistent chemical, boron, was used as the reference compound to account for dilution. Samplings were conducted twice in June 2016 before and after the removal of macrophytes from a riverbed of the 1.6-km stretch. The first-order decay constants in the river stretches were ~0–0.10 h^{-1} for diclofenac, 0.02–0.09 h^{-1} for metoprolol, and <0.03 h^{-1} for sulfamethoxazole and carbamazepine.

13.2.6 Three Rivers in Sweden

Li et al. (2016) observed natural attenuation of pharmaceuticals along a 7-km stretch of the Fyris River (FY), a 6-km stretch of the Rönne River (RO), and a 7-km stretch of the Viskan River (VI) in Sweden. The average flow rates and water travel times in the three river stretches were 2.7–6.0 m^3/s and 0.6–1.1 d, respectively. Surface water samples were collected by automatic water samplers 0.5 and 6–7 km downstream of WWTPs in June 2014 (FY) and August 2014 (RO and VI). Sampling was conducted for a week, and hourly samples were combined to form composite samples. A persistent chemical, fluconazole, was used as the reference compound to account for dilution by infiltrating groundwater or minor tributaries along the stretch. The average attenuation and corresponding first-order decay constants in FY, RO, and VI were 40–80% and 0.019–0.112 h^{-1} for ketoprofen, 40–60% and 0.019–0.064 h^{-1} for furosemide, 5–50% and 0.004–0.029 h^{-1} for acetaminophen, 10–35% and 0.007–0.030 h^{-1} for diclofenac, 5–30% and 0.002–0.025 h^{-1} for propranolol, 5–20% and 0.002–0.015 h^{-1} for metoprolol, ~0−5% and ~0–0.002 h^{-1} for sulfamethoxazole and carbamazepine, respectively.

13.2.7 Four Rivers in Spain

Acuña et al. (2015) observed natural attenuation of pharmaceuticals along an 18-km stretch of the Puigcerdà River, an 8-km stretch of the Gasteiz River, an 8-km stretch of the Citruénigo River, and a 12-km stretch of the Alcanyís River in Spain. The average flow rates and depth of the river stretches were 0.05–2.77 m^3/s and 0.09–2.19 m, respectively, and the average water travel time ranged from 15.8 to 28.8 h. Water grab samples were collected from the surface from four sites at increasing distances from the discharge point of the WWTP effluent in March 2012. The average first-order decay constants in the selected four river stretches were 0.43 h^{-1} for diclofenac, 0.41 h^{-1} for acetaminophen, 0.33 h^{-1} for atenolol, 0.17 h^{-1} for ketoprofen, 0.17 h^{-1} for carbamazepine, 0.12 h^{-1} for sulfamethoxazole, 0.08 h^{-1} for naproxen, 0.07 h^{-1} for trimethoprim, and 0.07 h^{-1} for furosemide.

13.2.8 Segre River, Spain

Aymerich et al. (2016) observed natural attenuation of pharmaceuticals along a 4-km stretch of the Segre River in Spain. The average depth, width, and water travel time of the river stretch were 0.22 m, 13 m, and 3.8 h, respectively. Water grab samples were collected from the surface every 4 h and combined to form 48-h composite samples from 0.5 and 4.5 km downstream of the discharge of the WWTP effluent in October 2012. The average attenuation and first-order decay constants in the river stretch were

58% and 0.21 h^{-1} for ibuprofen, 16% and 0.046 h^{-1} for sulfamethoxazole, 12% and 0.033 h^{-1} for diclofenac, and ~0% and ~0 h^{-1} for sulfapyridine and carbamazepine, respectively.

13.2.9 Glatt River, Switzerland

Golet et al. (2002) observed natural attenuation of quinolones along a 25-km stretch of the Glatt River in Switzerland. The flow rate and water travel time of the river stretch fluctuated in the ranges of 3–12 m^3/s and 15–20 h, respectively. Twenty-four-hour composite samples were collected from three sites in the river stretch and eight WWTP outlets flowing into the stretch during a 4-day period in August 2001. A mass balance approach was used to quantify attenuation during transport along the river stretch. The average attenuation in the river stretch was 66% for ciprofloxacin and 48% for norfloxacin.

13.2.10 Thames River, UK

Hanamoto et al. (2018a) observed natural attenuation of antibiotics along an 8.3-km stretch of the Thames River in the UK. During the sampling period, the average flow rate, depth, and water travel time of the river stretch were 7.6 m^3/s, 0.53 m, and 6.5 h, respectively. Surface water samples were collected every hour by an automatic water sampler as 24-h composite samples from three sites in the river stretch, and a mass balance approach was used to quantify attenuation during transport along the river stretch. Samplings were conducted once or twice in the summer between 2012 and 2015, yielding a total of seven samplings. A persistent chemical, carbamazepine, was used as the reference compound to account for dilution. The average attenuation and corresponding first-order decay constants in the river stretch were 92% and 0.37 h^{-1} for azithromycin, 48% and 0.15 h^{-1} for clarithromycin, 11% and 0.024 h^{-1} for sulfapyridine, and ~0% and ~0 h^{-1} for sulfamethoxazole, respectively.

13.2.11 Katsura River, Japan

Hanamoto et al. (2013) observed natural attenuation of pharmaceuticals along a 7.6-km stretch of the Katsura River in Japan. The average flow rate, depth, and water travel time of the river stretch were 28.7 m^3/s, 0.3–2 m, and 8.6 h, respectively. Around 30% of the water consists of treated wastewater. Surface water or effluent samples were collected every hour by automatic water samplers from a downstream site and two WWTP outlets, the major sources for the selected pharmaceuticals in the study area, and once a day by grab from the other four sites. A mass balance approach was used to quantify attenuation during transport along the river stretch. Samplings were conducted 3 times in winter in 2011–2012 and 3 times in summer in 2012. A persistent chemical, carbamazepine, was used as the reference compound to account for dilution. The average attenuation and corresponding first-order decay constants in the river stretch were 85% and 0.30 h^{-1} for azithromycin, 80% and 0.26 h^{-1} for ofloxacin, 63% and 0.17 h^{-1} for ketoprofen, 29% and 0.082 h^{-1} for furosemide, 26% and 0.050 h^{-1} for trimethoprim, 25% and 0.048 h^{-1} for roxithromycin, 19% and 0.036 h^{-1} for diclofenac, 13% and 0.035 h^{-1} for clarithromycin, 10% and 0.009 h^{-1} for sulpiride, 2% and 0.004 h^{-1} for sulfapyridine, and ~0% and ~0 h^{-1} for sulfamethoxazole and crotamiton, respectively.

13.2.12 Yodo River, Japan

Hanamoto et al. (2018b) observed natural attenuation of pharmaceuticals along a 20.0-km stretch of the Katsura and Yodo rivers in Japan. Samples were collected by grab from four river sites and two outlets of WWTPs, and by an automatic water sampler from the other two outlets of WWTPs. A mass balance approach was used to quantify attenuation during transport along the river stretch. Samplings were conducted approximately once a week between October 2009 and February 2010, yielding a total of 17 sampling events. The average attenuation in the river stretch was 91% for azithromycin, 82% for ofloxacin, 48% for ketoprofen, 29% for clarithromycin, 22% for atenolol, 18% for trimethoprim, 13% for furosemide, 11% for metoprolol, 8% for primidone, and ~0% for carbamazepine, sulpiride, roxithromycin, sulfamethoxazole, sulfapyridine, and crotamiton.

13.3 In-Stream Attenuation of Pharmaceuticals in Each Therapeutic Category

The measurements of the natural attenuation of pharmaceuticals in the Santa Ana River (Lin et al. 2006), the Trinity River (Fono et al. 2006), the Gründlach River (Kunkel and Radke 2012; Li et al. 2016), the Steinlach River (Guillet et al. 2019), the Erpe River (Jaeger et al. 2019), the Fyris River (Li et al. 2016), the Rönne River (Li et al. 2016), the Viskan River (Li et al. 2016), the Segre River (Aymerich et al. 2016), the Glatt River (Golet et al. 2002), the Thames River (Hanamoto et al. 2018a), the Katsura River (Hanamoto et al. 2013), and the Yodo River (Hanamoto et al. 2018b) (see Sect. 13.2 above for details) were classified into therapeutic classes. As for four rivers in Spain observed by Acuña et al. (2015), all pharmaceuticals including the persistent pharmaceutical, carbamazepine, were highly attenuated during the river transport. This might be due to an error made by spot sampling conducted only once without any tracer. Exceptionally high attenuation of carbamazepine in the Spanish rivers was also pointed out by Jaeger et al. (2019). Therefore, measurements by Acuña et al. (2015), are excluded from this discussion.

13.3.1 Sulfa Drugs

Attenuation of sulfamethoxazole was limited (<25%) in all reported rivers, i.e., the Gründlach River, the Steinlach River, the Erpe River, the Fyris River, the Rönne River, the Viskan River, the Segre River, the Thames River, the Katsura River, and the Yodo River. The reported attenuation of sulfapyridine was also limited (<20%) in all samples, i.e., the Segre River, the Thames River, the Katsura River, and the Yodo River. Thus, in-stream attenuation is not effective for the removal of these sulfa drugs. The factor affecting the attenuation of sulfapyridine in the Thames River and Katsura River was estimated by biodegradation, sorption, direct photolysis, and indirect photolysis experiments with a model estimation of direct and indirect photolysis (Hanamoto et al. 2013, 2018a). The attenuation of sulfapyridine in the Thames River was attributed to both direct and indirect photolysis, whereas attenuation was mostly due to direct photolysis in the Katsura River.

13.3.2 Quinolone Antibiotics

The average attenuation of ciprofloxacin and norfloxacin was 66 and 48% in the Glatt River, respectively. The average attenuation and corresponding first-order decay constant of ofloxacin were 80% and 0.26 h^{-1} in the Katsura River, respectively. Thus, all the reported quinolones were highly attenuated (>40%) during river transport, though observations were limited. Individual laboratory experiments for biodegradation, sorption, direct photolysis, and indirect photolysis revealed that direct photolysis and sorption to suspended solids (SS) and sediment could be responsible for the attenuation of ofloxacin in the Katsura River (Hanamoto et al. 2013). Given that degradation and sorption trends of ciprofloxacin and norfloxacin were similar to those of ofloxacin (Hanamoto et al. 2013), and the significant attenuation of ciprofloxacin and norfloxacin observed in the Glatt River could be attributable to direct photolysis and sorption to SS and sediment.

13.3.3 Macrolide Antibiotics

The average attenuation and corresponding first-order decay constants of azithromycin were 92% and 0.37 h^{-1} in the Thames River, and 85% and 0.30 h^{-1}

in the Katsura River, respectively. For clarithromycin, those values were 48% and 0.15 h^{-1} in the Thames River and 13% and 0.035 h^{-1} in the Katsura River, respectively; for roxithromycin, they were 25% and 0.048 h^{-1} in the Katsura River, respectively. Thus, azithromycin was highly attenuated (>80%) in both the river stretches with similar decay constants, whereas the other two macrolides were attenuated to a lesser extent (13–48%). The decay constants of clarithromycin were 4.4 times higher in the Thames River than in the Katsura River. Individual laboratory tests for biodegradation, sorption, direct photolysis, and indirect photolysis revealed that sorption to SS and sediment could be responsible for the attenuation of these macrolides in both the river stretches (Hanamoto et al. 2013, 2018a). Given the 5.5 times higher linear sorption coefficient (K_d) of clarithromycin to the Thames River sediment than to the Katsura River sediment, the probable explanation for the difference in the loss rates between the two rivers was considered to be the sediment sorption capacity (Hanamoto et al. 2018a). Similarly, given more than 10 times higher K_d value of azithromycin in the Katsura River sediment than that of clarithromycin and roxithromycin (Hanamoto et al. 2013), the much higher attenuation of azithromycin than that of the other two macrolides could be associated with their sorption affinities to the sediment.

13.3.4 Anti-inflammatory Drugs

The average attenuation and corresponding first-order decay constant of ketoprofen were 80% and 0.11 h^{-1} in the Fyris River, 63% and 0.17 h^{-1} in the Katsura River, and 40% and 0.019 h^{-1} in the Viskan River, respectively. For ibuprofen, those values were 87% and 0.0063 h^{-1} in the Trinity River, 62% and 0.13 h^{-1} in the Santa Ana River, and 58% and 0.21 h^{-1} in the Segre River, respectively. For acetaminophen, those values were 95% and 0.059 h^{-1} in the Gründlach River, 50% and 0.029 h^{-1} in the Rönne River, 35% and 0.016 h^{-1} in the Viskan River, and 5% and 0.004 h^{-1} in the Fyris River, respectively. For naproxen, those values were 94% and 0.39 h^{-1} in the Santa Ana River, 88% and 0.0067 h^{-1} in the Trinity River, and 50% and 0.014 h^{-1} in the Gründlach River, respectively. For diclofenac, those values were 60% and 0.018 h^{-1} in the Gründlach River, 50% and 0.20 h^{-1} in the Steinlach River, 35% and 0.030 h^{-1} in the Fyris River, 29% and 0.036 h^{-1} in the Katsura River, 15% and 0.007 h^{-1} in the Rönne River, 12% and 0.033 h^{-1} in the Segre River, and 10% and 0.004 h^{-1} in the Viskan River, respectively. Thus, ketoprofen, ibuprofen, and naproxen were highly attenuated (>40%) in all the reported rivers, while the reported in-stream attenuation of acetaminophen and diclofenac was highly different among the rivers (5–95% and 10–60%, respectively). The decay constants were 8.6 (ketoprofen), 16.7 (acetaminophen), 33.3 (ibuprofen), 49.6 (diclofenac), and 58 times (naproxen) different among rivers. Individual laboratory tests for biodegradation, sorption, direct photolysis, and indirect photolysis combined with model estimation of direct photolysis revealed that direct photolysis by sunlight was the major contributor to the attenuation of ketoprofen and diclofenac in the Katsura

River (Hanamoto et al. 2013). Kunkel and Radke (2012) also concluded that direct photolysis was only a relevant elimination process for diclofenac in the Gründlach River by conducting in situ photolysis experiments at several sites within the river stretch. A high contribution of photolysis to the in-stream attenuation of diclofenac was also suggested in the Steinlach River by observing its much lower attenuation in the nighttime sampling (Guillet et al. 2019). Significantly enhanced in-stream attenuation of diclofenac by macrophyte removal suggested that photolysis also played an important role in its attenuation in the Erpe River (Jaeger et al. 2019). As for naproxen, the modeling approach suggested that direct photolysis contributed to approximately 40% of the attenuation in the Santa Ana River; the factor affecting the remaining 60% of the attenuation was not identified (Lin et al. 2006). Laboratory-scale microcosms suggested that biotransformation in surface water was more important than photolysis for the attenuation of naproxen and ibuprofen in the Trinity River (Fono et al. 2006). As the concentrations of naproxen and acetaminophen in the sediments decreased relative to carbamazepine with depth, biotransformation in the sediments was considered to be an important attenuation process for those pharmaceuticals in the Gründlach River (Kunkel and Radke 2012). As the photolability of naproxen is not as high as that of ketoprofen and diclofenac (Hanamoto et al. 2013), the contribution of photolysis to the in-stream attenuation of naproxen would fluctuate according to the river conditions such as depth, turbidity, weather, latitude, and shading by plants. The importance of biodegradation for acetaminophen and ibuprofen harmonizes with their extremely high removal in WWTPs (Nakada et al. 2006; Hanamoto et al. 2018b).

13.3.5 Antiarrhythmic Drugs

The average attenuation and corresponding first-order decay constants of metoprolol were 83% and $0.0054 \ h^{-1}$ in the Trinity River, 55% and $0.016 \ h^{-1}$ in the Gründlach River, 50% and $0.20 \ h^{-1}$ in the Steinlach River, 20% and $0.015 \ h^{-1}$ in the Fyris River, 15% and $0.007 \ h^{-1}$ in the Rönne River, and 5% and $0.002 \ h^{-1}$ in the Viskan River, respectively. For propranolol, those values were 70% and $0.024 \ h^{-1}$ in the Gründlach River, 30% and $0.025 \ h^{-1}$ in the Fyris River, 30% and $0.015 \ h^{-1}$ in the Rönne River, and 5% and $0.002 \ h^{-1}$ in the Viskan River, respectively. For atenolol, those values were 30% and $0.10 \ h^{-1}$ in the Steinlach River. Thus, the reported in-stream attenuation of the frequently used beta-blockers, metoprolol, propranolol, and atenolol and varied considerably among the rivers (5–83%, 5–70%, and 30%, respectively). The differences in the decay constants were 12.7 and 101.9 times among the rivers for propranolol and metoprolol, respectively. As the concentrations of metoprolol and propranolol in the sediments decreased relative to carbamazepine with depth, biotransformation in the sediments was considered to be an important attenuation process for these pharmaceuticals in the Gründlach River (Kunkel and Radke 2012).

For the chiral chemical metoprolol, the importance of biotransformation in the sediment was also indicated by a decrease in enantiomer fractionation in the deeper sediments along the river stretch. Evidence for biotransformation of metoprolol in the surface water of the Trinity River was also provided by measurements of its enantiomeric fraction, which showed a gradual decrease as the water moved downstream (Fono et al. 2006). The attenuation of metoprolol significantly decreased by removing macrophytes in the Erpe River, and its governing process was considered to be biodegradation in the surface water (Jaeger et al. 2019).

13.3.6 Antiepileptic Drugs

The attenuation of carbamazepine was negligible (<10%) in all reported rivers, i.e., the Gründlach River, the Steinlach River, the Erpe River, the Fyris River, the Rönne River, the Viskan River, the Segre River, and the Yodo River. Reported attenuation of primidone was also limited (<10%) in the Steinlach River and the Yodo River. Thus, these two antiepileptic drugs were considered to be persistent during river transport. The persistence of carbamazepine is consistent with studies in rivers (Nakada et al. 2008) and groundwater filtration (Clara et al. 2004), WWTPs (Clara et al. 2004), and in comprehensive laboratory experiments (Yamamoto et al. 2009). Although there are few studies on the environmental fate of primidone, its persistence harmonizes with its negligible removal in WWTPs (Hanamoto et al. 2018b).

13.3.7 The Others

The average attenuation and corresponding first-order decay constants of furosemide, a diuretic drug, were 90% and 0.046 h^{-1} in the Gründlach River, 60% and 0.064 h^{-1} in the Fyris River, 40% and 0.021 h^{-1} in the Rönne River, 40% and 0.019 h^{-1} in the Viskan River, and 29% and 0.082 h^{-1} in the Katsura River, respectively. For trimethoprim, an anti-infective drug, those values were 26% and 0.050 h^{-1} in the Katsura River, and 25% and 0.082 h^{-1} in the Steinlach River, respectively. For sulpiride, an antipsychotic drug, the attenuation was limited (<10%) in the Steinlach River, the Katsura River, and the Yodo River. Thus, sulpiride was considered to be persistent during river transport, which harmonizes with its negligible removal in WWTPs (Hanamoto et al. 2018b; Singh et al. 2019). Reported in-stream attenuation of furosemide and trimethoprim varied considerably among the rivers (29–90%) and moderate (25–26%), respectively, and their decay constants were 4.3 and 1.6 times different among the rivers, respectively. The individual laboratory tests for biodegradation, sorption, direct photolysis, and indirect photolysis combined with the model estimation of direct photolysis revealed that direct photolysis by sunlight contributed mostly to the attenuation of furosemide in the Katsura River, whereas sorption to SS and sediment was a probable factor for the attenuation of trimethoprim (Hanamoto et al. 2013).

13.4 Factors Affecting the In-Stream Attenuation of Pharmaceuticals

Factors controlling the in-stream attenuation of pharmaceuticals are summarized in this section; the details are described in Sect. 13.3 above.

13.4.1 Direct Photolysis

Direct photolysis represents the degradation of a compound due to direct absorption of light by the compound. Direct photolysis by sunlight was the dominant factor for the attenuation of ketoprofen, diclofenac, furosemide, and sulfapyridine in the Katsura River, though the attenuation of sulfapyridine was limited (<20%) (Hanamoto et al. 2013, 2018a). Direct photolysis was considered to be a major contributor to the in-stream attenuation of diclofenac in the Gründlach River (Kunkel and Radke 2012) and the Steinlach River (Guillet et al. 2019). Direct photolysis partly contributed to the attenuation of ofloxacin in the Katsura River (Hanamoto et al. 2013), of naproxen in the Santa Ana River (Lin et al. 2006), and of sulfapyridine in the Thames River (Hanamoto et al. 2018a). Attenuation of the other two quinolones, ciprofloxacin and norfloxacin, in the Glatt River was also considered to be partly due to direct photolysis.

13.4.2 Indirect Photolysis

Indirect photolysis represents the degradation of a compound driven by reactive species (e.g., singlet oxygen and hydroxyl radical) produced under light irradiance to dissolved or particulate matter (e.g., humic substance and nitrate) in surface water. Indirect photolysis partly contributed to the attenuation of sulfapyridine in the Thames River, though the attenuation was limited (<20%) (Hanamoto et al. 2018a).

13.4.3 Sorption to SS/Sediment

Sorption to SS and sediment includes the mass transfer of a compound at the sediment–water interface, sorption of a compound to temporally re-suspended sediment,

and sorption of a compound to SS loaded to the river stretch followed by sedimentation of SS. Sorption to SS and sediment was considered to be the dominant factor for
the attenuation of two macrolides, azithromycin, and clarithromycin, in the Katsura
River (Hanamoto et al. 2013) and the Thames River (Hanamoto et al. 2018a), which
could also be applicable to the macrolide antibiotic, roxithromycin. Sorption to SS
and sediment played an important role in the attenuation of a quinolone antibiotic
ofloxacin in the Katsura River (Hanamoto et al. 2013). The importance of sorption to SS and sediment may also be applicable to the attenuation of quinolones,
ciprofloxacin, and norfloxacin, as observed in the Glatt River. A significant contribution by sorption to in-stream attenuation of macrolides and quinolones was also
suggested by their high concentrations in the sediment relative to those in the overly-
ing water (i.e., high pseudo-partitioning coefficients) observed in the rivers (Azuma
et al. 2017; Li et al. 2018; Tang et al. 2019; Li et al. 2019; Kumar et al. 2009).

13.4.4   Biodegradation in Surface Water

Biotransformation in surface water played an important role in the attenuation of
naproxen, ibuprofen, and metoprolol in the Trinity River (Fono et al. 2006), where
the travel time is as long as 2-weeks and the depth is around 2 m, and of metoprolol in
the macrophyte-rich zone of the Erpe River (Jaeger et al. 2019). Thus, biotransformation in surface water could be relatively important in long deep river stretches because
sunlight intensity and water-sediment contact decrease with increasing depth. Macrophytes would also enhance biotransformation. Thus, though biodegradation did not
contribute much to the reported in-stream attenuation of pharmaceuticals, it could
become a beneficial factor for reducing their risks, given the increasingly reported
toxic photoproducts of pharmaceuticals (Schmitt-Jansen et al. 2006; Nasuhoglu et al.
2011; Yuan et al. 2011; Mukherjee et al. 2020).

13.4.5   Biodegradation in Sediment

Biodegradation of a compound in the sediment followed by mass transfer of the
compound at the sediment–water interface leads to its in-stream attenuation. The
hyporheic zone, which is defined as the portion of river sediment that acts as a mixing
zone between groundwater and surface water, could dramatically enhance the in-
stream attenuation by its abundant and diverse microbial communities and frequent
water exchange at the sediment–water interface. Therefore, the hyporheic zone is
believed to act as an efficient bioreactor (Schaper et al. 2018a). It was suggested that
the hyporheic zone may play an important role in the attenuation of acetaminophen,
naproxen, metoprolol, and propranolol in the Gründlach River (Kunkel and Radke
2012). Other pharmaceuticals, such as diclofenac and sulfamethoxazole, were also
attenuated by biodegradation in the hyporheic zone in the Erpe River in Germany

(Schaper et al. 2018b) and the Sturt River in Australia (Schaper et al. 2018a; Patel et al. 2019).

13.5 Modeling the Fate of In-Stream Attenuation of Pharmaceuticals

Models for the direct photolysis rate constant, photoproduct occurrence, and sorption affinity to sediment are summarized in this section. Direct photolysis by sunlight played an important role in the in-stream attenuation of pharmaceuticals (see Sect. 13.4 for details). As the toxicities of some photoproducts were higher than those of their parent pharmaceuticals (Schmitt-Jansen et al. 2006; Nasuhoglu et al. 2011; Yuan et al. 2011), a model for predicting concentration changes of photoproducts in the aquatic environment is also included. Sorption affinity to sediment is a key factor controlling sorption to SS/sediment and biodegradation in the sediment, which also played important roles in the in-stream attenuation of pharmaceuticals as described in Sect. 13.4 above. As the hydrological conditions as well as biogeochemical factors controlling sorption to SS/sediment and biodegradation in sediment have not been fully understood, their full models are not included.

13.5.1 Direct Photolysis Rate Constants in Aquatic Environments

Considering the attenuation of sunlight in the atmosphere and water, Zepp and Cline (1977) developed the following equation for predicting the direct photolysis rate constant in the aquatic environment:

$$k_p = \varphi(1 - R) \sum_{\lambda=297.5}^{800} \frac{\left(1 - 10^{-\alpha_\lambda l}\right) L_\lambda \varepsilon_\lambda}{\alpha_\lambda D} \tag{13.1}$$

where k_p = direct photolysis rate constant of a compound in the water body (h^{-1}), φ = quantum yield of the compound (–), R = fraction of sunlight reflected at the surface of the water body (–), α_λ = decadic absorption coefficient of the water body at wavelength λ (m^{-1}), l = path length of sunlight in the water body (m), L_λ = annual average sunlight intensity at Earth's surface at wavelength λ $(10^{-3}$ einsteins $cm^{-2} h^{-1})$, ε_λ = molar absorption coefficient of the compound at wavelength λ $(M^{-1} cm^{-1})$, and D = depth of the water (m). The quantum yield, which is the ratio of the number of molecules that photoreact to the number of quanta of light absorbed, can be quantified by a sunlight exposure test with a chemical actinometer (US Environmental Protection Agency's harmonized test guideline, 1998). Hanamoto et al. (2013, 2014) verified Eq. 13.1 in river environments by combining Eq. 13.1 with sunlight intensity measurements as follows:

$$k_p = \varphi \left\{ \frac{\text{UVB}(1 - R_{\text{UVB}})}{\text{UVB}_t} \sum_{\lambda=297.5}^{315} \frac{\left(1 - 10^{-\alpha_\lambda l}\right) L_\lambda \varepsilon_\lambda}{\alpha_\lambda D} \right.$$
$$\left. + \frac{\text{UVA}(1 - R_{\text{UVA}})}{\text{UVA}_t} \sum_{\lambda=315}^{490} \frac{\left(1 - 10^{-\alpha_\lambda l}\right) L_\lambda \varepsilon_\lambda}{\alpha_\lambda D} \right\} \qquad (13.2)$$

where UVB and UVA = sunlight intensity at Earth's surface in those wavelengths (W/m^2), UVB_t and UVA_t = annual average sunlight intensity at Earth's surface in those wavelengths (W/m^2), and R_{UVB} and R_{UVA} = fraction of sunlight reflected at the surface of the water body in those wavelengths $(-)$. Diurnal and seasonal in-stream attenuation of ketoprofen, furosemide, and diclofenac, pharmaceuticals that are highly photolabile and insensitive to attenuation factors other than direct photolysis was predicted well using Eq. 13.2 (Hanamoto et al. 2013, 2014). Tixier et al. (2002) also predicted the attenuation of triclosan in a lake well by combining measured and theoretical global radiation with Eq. 13.1. Thus, the equation developed by Zepp and Cline (1977) was verified to be a useful tool for predicting direct photolysis rate constants of chemicals in aquatic environments. The limitation of the model is the difficulty in quantifying shading effects caused by river macrophytes and riverbank trees. Though shading effects were negligible in rivers and a lake studied by Hanamoto et al. (2013, 2014) and Tixier et al. (2002), they led to an 85% decrease in the photolysis rate of a chemical in the Glatt River, which was roughly estimated by comparing measured and predicted concentrations of the chemical (Kari and Giger 1995).

13.5.2 Concentration Change of Photoproducts in Aquatic Environments

Poiger et al. (2001) predicted concentration change of a photoproduct in an aquatic environment with the following equation:

$$\frac{dC_P}{dt} = x k_A C_A - k_P C_P \qquad (13.3)$$

where C = concentration (ng/L), t = time (h), x = fraction of parent compound transformed into photoproduct $(-)$, k = direct photolysis rate constant in the environment (h^{-1}), and subscripts A and P indicate parent compound and its photoproduct, respectively. The fraction of transformation (x) can be estimated from the concentration changes of the parent compound and its photoproduct in sunlight, as performed with diclofenac (Poiger et al. 2001) and ketoprofen (Hanamoto et al.

2016). Equation 13.3 was verified using 3-ethylbenzophenone, a photoproduct of ketoprofen, in the Katsura River (Hanamoto et al. 2016). As not only ketoprofen but also 3-ethylbenzophenone were insensitive to attenuation factors other than direct photolysis, Eq. 13.3 predicted the concentration of 3-ethylbenzophenone in the river water well.

13.5.3 Sorption Affinity to Sediment

Compounds that contain a basic group with a $pK_a > 7$ occur largely as cationic species in the aquatic environment. Approximately 45% of all pharmaceuticals contain a single base moiety, and >70% of these bases have a pK_a of >7 (Droge and Goss 2013a). Macrolides and quinolones, highly sorptive antibiotics that showed significant in-stream attenuation (see Sect. 13.4.3 above), also contain amino groups and occur mainly in the cationic (macrolides) and zwitterionic (quinolones) forms at neutral pH, resulting in low octanol–water distribution ratios (D_{ow}) (McFarland et al. 1997; Takacs-Novak et al. 1992). The sorption affinities of hydrophobic chemicals to soil or sediment were successfully predicted by considering partitioning into organic matter (OM) (Karickhoff et al. 1978). However, such sorption models are incapable of adequately predicting sorption affinities to soil or sediment for bases that are largely protonated at environmentally relevant pH values (Franco and Trapp 2008; Stein et al. 2008).

The sorption of organic cations could be associated with the cation exchange capacity (CEC) of sorbents. For example, sorption affinities of cationic beta-blockers for soil or sediment were highly correlated with CEC ($R^2 = 0.73$–0.84, $P < 0.01$) (Kodesova et al. 2015; Al-Khazrajy and Boxall 2016). However, such high correlations cannot be applicable to all organic cations because some showed an order-of-magnitude difference in CEC-normalized sorption affinities between mineral and organic components of soil or sediment (Droge and Goss 2013b). To overcome this discrepancy, Droge and Goss (2013a) developed a cation exchange-based sorption model incorporating individual sorption affinities for OM and clay minerals as follows:

$$K_d = K_{CEC,OM}CEC_{OM} + K_{CEC,CLAY}CEC_{CLAY} \qquad (13.4)$$

$$CEC_{OM} = 200 \times 1.7 \times OC \qquad (13.5)$$

$$CEC_{CLAY} = CEC - CEC_{OM} \qquad (13.6)$$

where K_d is the linear sorption coefficient (L/kg), K_{CEC} is K_d values normalized to CEC (so this is defined as K_d/CEC) (L/cmol$_c$), $K_{CEC,OM}$ and $K_{CEC,CLAY}$ are K_{CEC} on reference OM and clay minerals (L/cmol$_c$), respectively, CEC_{OM} and CEC_{CLAY} are CECs of the sediment contributed by OM and clay minerals (cmol$_c$/kg), respectively, and OC is organic carbon content (g/g). "1.7" in Eq. 13.5 indicates a conversion factor from OC to dry weight of OM. "200" in Eq. 13.5 indicates CEC (cmol$_c$/kg) of OM itself and derives from the CEC of Pahokee peat. As the application of 200 resulted in CEC$_{CLAY}$ of <0 for some sediments collected from Japanese rivers where peat rarely exists, Hanamoto and Ogawa (2019) used "160" instead of "200" in accordance with Charles et al. (2006). Although the application of the CEC of the same OM (i.e., "200" in Eq. 13.5) to all sediments could introduce some error, the K_d values of soils estimated by Eqs. 13.4–13.6 using sorption data on reference OM and clay minerals were mostly within a factor of 10 from the measurements for a diverse set of organic cations (Droge and Goss 2013a). The sorption of a macrolide antibiotic azithromycin and a quinolone antibiotic levofloxacin to sediments was also reasonably predictable by Eqs. 13.4–13.6 using individual sorption affinities for OM and clay minerals (Hanamoto and Ogawa 2019).

13.6 Conclusions and Areas for Further Research

According to studies summarized in this review, the following conclusions regarding in-stream attenuation and fate of pharmaceuticals could be drawn:

- Three quinolones (levofloxacin, ciprofloxacin, and norfloxacin), a macrolide antibiotic (azithromycin), and three anti-inflammatory drugs (ketoprofen, ibuprofen, and naproxen) were highly attenuated (>40%) in all the reported river stretches.
- In-stream attenuation was different in the rivers and moderate for two macrolides (clarithromycin and roxithromycin), two anti-inflammatory drugs (acetaminophen and diclofenac), three antiarrhythmic drugs (metoprolol, propranolol, and atenolol), a diuretic drug (furosemide), and an anti-infective drug (trimethoprim).
- Attenuation was limited (<25%) for two sulfa drugs (sulfamethoxazole and sulfapyridine) and negligible (<10%) for two antiepileptic drugs (carbamazepine and primidone) and an antipsychotic drug (sulpiride) in all the reported rivers.
- Direct photolysis by sunlight was considered a major contributor to the in-stream attenuation of ketoprofen, diclofenac, and furosemide, and partly to that of sulfapyridine, naproxen, and the aforementioned three quinolones, whereas indirect photolysis affected in-stream attenuation of only sulfapyridine in a river stretch.
- Biodegradation in the hyporheic zone played an important role in the in-stream attenuation of acetaminophen, ibuprofen, naproxen, and three antiarrhythmic drugs above, whereas biodegradation in surface water affected the in-stream attenuation of naproxen, ibuprofen, and metoprolol in a long, deep, or macrophyte-rich river stretch.
- Sorption to SS and sediment played an important role in the in-stream attenuation of three macrolides and three quinolones mentioned above.

- The direct photolysis rate constants of the pharmaceuticals and concentration changes of their photoproducts in rivers or lakes were predicted well by using their photochemical properties and river and meteorological data.
- Sorption affinities of pharmaceuticals to soils or sediments were reasonably predictable by the cation exchange-based sorption model using their sorption affinities for mineral and organic components.

Although in-stream attenuation of pharmaceuticals has been observed worldwide, further study is needed given the large differences in the decay constants among river stretches for some pharmaceuticals. In addition, to understand the mechanism of in-stream attenuation, laboratory tests and model-based calculations should be performed along with field measurements. Monitoring the breakdown products could also help in understanding the mechanism of attenuation. As reports of the contributions of the hyporheic zone to the in-stream attenuation of pharmaceuticals have increased recently (Schaper et al. 2018a, b, 2019), more field measurements should be focused on the hyporheic zone. Modeling approaches also should be focused on biodegradation in the hyporheic zone and mass transfer at the sediment–water interface.

References

Acuña V, Schiller D, García-Galán MJ, Rodríguez-Mozaz S, Corominas L, Petrovic M, Poch M, Barceló D, Sabater S (2015) Occurrence and in-stream attenuation of wastewater-derived pharmaceuticals in Iberian rivers. Sci Total Environ 503–504:133–141

Ågerstrand M, Berg C, Björlenius B, Breitholtz M, Brunström B, Fick J et al (2015) Improving environmental risk assessment of human pharmaceuticals. Environ Sci Technol 49:5336–5345

Al-Khazrajy OSA, Boxall ABA (2016) Impacts of compound properties and sediment characteristics on the sorption behaviour of pharmaceuticals in aquatic systems. J Hazard Mater 317:198–209

Aymerich I, Acuna V, Barcel D, García MJ, Petrovic M, Poch M, Rodriguez-Mozaz S, Rodríguez-Roda I, Sabater S, von Schiller D Corominas, Ll (2016) Attenuation of pharmaceuticals and their transformation products in a wastewater treatment plant and its receiving river ecosystem. Water Res 100:126–136

Azuma T, Arima N, Tsukada A, Hirami S, Matsuoka R, Moriwake R et al (2017) Distribution of six anticancer drugs and a variety of other pharmaceuticals, and their sorption onto sediments, in an urban Japanese river. Environ Sci Pollut Res 24:19021–19030

Boxall ABA, Rudd MA, Brooks BW, Caldwell DJ, Choi K, Hickmann S et al (2012) Pharmaceuticals and personal care products in the environment: what are the big questions? Environ Health Perspect 120:1221–1229

Carvalho IT, Santos L (2016) Antibiotics in the aquatic environments: a review of the European scenario. Environ Int 94:736e757

Charles S, Teppen BJ, Li H, Laird DA, Boyd SA (2006) Exchangeable cation hydrationproperties strongly influence soil sorption of nitroaromatic compounds. Soil Sci Soc Am J 70:1470–1479

Cizmas L, Sharma VK, Gray CM, McDonald TJ (2015) Pharmaceuticals and personal care products in waters: occurrence, toxicity, and risk. Environ Chem Lett 13:381–394

Clara M, Strenn B, Kreuzinger N (2004) Carbamazepine as a possible anthropogenic marker in the aquatic environment: Investigations on the behavior of Carbamazepine in wastewater treatment and during groundwater infiltration. Water Res 38(4):947–954

Cooper ER, Siewicki TC, Phillips K (2008) Preliminary risk assessment database and risk ranking of pharmaceuticals in the environment. Sci Total Environ 398:26–33

Droge STJ, Goss K-U (2013a) Development and evaluation of a new sorption model for organic cations in soil: contributions from organic matter and clay minerals. Environ Sci Technol 47:14233–14241

Droge STJ, Goss K-U (2013b) Sorption of organic cations to phyllosilicate clay minerals: CEC-normalization, salt dependency, and the role of electrostatic and hydrophobic effects. Environ Sci Technol 47:14224–14232

Fono LJ, Kolodziej EP, Sedlak DL (2006) Attenuation of wastewater-derived contaminants in an effluent-dominated river. Environ Sci Technol 40:7257–7262

Franco A, Trapp S (2008) Estimation of the soil-water partition coefficient normalized to organic carbon for ionizable organic chemicals. Environ Toxicol Chem 27:1995–2004

Golet EM, Alder AC, Giger W (2002) Environmental exposure and risk assessment offluoro-quinolone antibacterial agents in wastewater and river water of the Glatt Valley watershed, Switzerland. Environ Sci Technol 36(17):3645e3651

Guillet G, Knapp JL, Merel S, Cirpka OA, Grathwohl P, Zwiener C, Schwientek M (2019) Fate of wastewater contaminants in rivers: using conservative-tracer basedtransfer functions to assess reactive transport. Sci Total Environ 656:1250–1260

Gurr CJ, Reinhard M (2006) Harnessing natural attenuation of pharmaceuticals and hormones in rivers. Environ Sci Technol 40(9):2872–2876

Hanamoto S, Ogawa F (2019) Predicting the sorption of azithromycin and levofloxacin to sediments from mineral and organic components. Environ Pollut 255. https://doi.org/10.1016/j.envpol.2019. 113180

Hanamoto S, Nakada N, Yamashita N, Tanaka H (2013) Modeling the photochemical attenuation of down-the-drain chemicals during river transport by stochastic methods and field measurements of pharmaceuticals and personal care products. Environ Sci Technol 47(23):13571–13577

Hanamoto S, Kawakami T, Nakada N, Yamashita N, Tanaka H (2014) Evaluation of the photolysis of pharmaceuticals within a river by 2 year field observations and toxicity changes by sunlight. Environ Sci: Processes Impacts 16(12):2796–2803

Hanamoto S, Hasegawa E, Nakada N, Yamashita N, Tanaka H (2016) Modeling the fate of a photoproduct of ketoprofen in urban rivers receiving wastewater treatment plant effluent. Sci Total Environ 573:810–816

Hanamoto S, Nakada N, Jürgens MD, Johnson AC, Yamashita N, Tanaka H (2018a) The different fate of antibiotics in the Thames River, UK, and the Katsura River, Japan. Environ Sci Pollut Res 25:1903–1913

Hanamoto S, Nakada N, Yamashita N, Tanaka H (2018b) Source estimation of pharmaceuticals based on catchment population and in-stream attenuation in Yodo River watershed, Japan. Sci Total Environ 615:964–971

Jaeger A, Posselt M, Betterle A, Schaper J, Mechelke J, Coll C, Lewandowski J (2019) Spatial and temporal variability in attenuation of polar organic micropollutants in an urban lowland stream. Environ Sci Technol 53:2383–2395

Kari FG, Giger W (1995) Modeling the photochemical degradation of ethylenediaminetetraacetate in the River Glatt. Environ Sci Technol 29(11):2814–2827

Karickhoff SW, Brown DS, Scott TA (1978) Sorption of hydrophobic pollutants on natural sediments. Water Res 13:241–248

Kasprzyk-Hordern B, Dinsdale RM, Guwy AJ (2008) The occurrence of pharmaceuticals, personal care products, endocrine disruptors and illicit drugs in surface water in South Wales, UK. Water Res 42:3498–3518

Kodesova R, Grabic R, Kocarek M, Klement A, Golovko O, Fer M, Nikodem A, Jaksík O, Grabic R (2015) Pharmaceuticals' sorptions relative to properties of thirteen different soils. Sci Total Environ 511:435–443

Kolpin DW, Furlong ET, Meyer MT, Thurman EM, Zaugg SD, Barber LB, Buxton HT (2002) Pharmaceuticals, hormones, and other organic wastewater contaminants in U.S. streams, 1999–2000—a national reconnaissance. Environ Sci Technol 36:1202–1211

Kumar M, Furumai H, Kurisu F, Kasuga I (2009) Understanding the partitioning processes of mobile lead in soakaway sediments using sequential extraction and isotope analysis. Water Sci Technol 60(8):2085–2091

Kumar M, Ram B, Honda R, Poopipattana C, Canh VD, Chaminda T, Furumai H (2019a) Concurrence of antibiotic resistant bacteria (ARB), viruses, pharmaceuticals and personal care products (PPCPs) in ambient waters of Guwahati, India: urban vulnerability and resilience perspective. Sci Total Environ 693:133640. https://doi.org/10.1016/j.scitotenv.2019.133640

Kumar M, Chaminda T, Honda R, Furumai H (2019b) Vulnerability of urban waters to emerging contaminants in India and Sri Lanka: resilience framework and strategy. APN Sci Bull 9(1). https://doi.org/10.30852/sb.2019.799

Kunkel U, Radke M (2012) Fate of pharmaceuticals in rivers: deriving a benchmark dataset at favorable attenuation conditions. Water Res 46:5551–5565

Li Z, Sobek A, Radke M (2016) Fate of pharmaceuticals and their transformation products in four small European rivers receiving treated wastewater. Environ Sci Technol 50:5614–5621

Li S, Shi W, Liu W, Li H, Zhang W, Hu J, Ke Y, Sun W, Ni J (2018) A duodecennial national synthesis of antibiotics in China's major rivers and seas (2005–2016). Sci Total Environ 615:906–917

Li Y, Zhang L, Liu X, Ding J (2019) Ranking and prioritizing pharmaceuticals in the aquatic environment of China. Sci Total Environ 658:333–342

Lin A, Plumlee MH, Reinhard M (2006) Attenuation of pharmaceuticals and alkylphenol polyethoxylate metabolites during river transport: photochemical and biological transformation. Environ Toxicol Chem 25(6):1458–1464

McFarland JW, Berger CM, Froshauer SA, Hayashi SF, Hecker SJ, Jaynes BH, Jefson MR, Kamicker BJ, Lipinski CA, Lundy KM, Reese CP, Vu CB (1997) Quantitative structure activity relationships among macrolide antibacterial agents: In vitro and in vivo potency against Pasteurella multocida. J Med Chem 40:1340e1346

Mukherjee S, Patel AK, Kumar M (2020) Water scarcity and land degradation Nexus in the era of Anthropocene: some reformations to encounter the environmental challenges for advanced water management systems meeting the sustainable development. In: Kumar M, Snow D, Honda R (eds) Emerging Issues in the Water Environment during Anthropocene: a South East Asian perspective. ISBN 978-93-81891-41-4, Publisher Springer Nature

Nakada N, Tanishima T, Shinohara H, Kiri K, Takada H (2006) Pharmaceutical chemicals and endocrine disrupters in municipal wastewater in Tokyo and their removal during activated sludge treatment. Water Res 40(17):3297–3303

Nakada N, Kiri K, Shinohara H, Harada A, Kuroda K, Takizawa S, Takada H (2008) Evaluation of pharmaceuticals and personal care products as water-soluble molecular markers of sewage. Environ Sci Technol 42(17):6347–6353

Nasuhoglu D, Yargeau V, Berk D (2011) Photo-removal of sulfamethoxazole (SMX) by photolytic and photocatalytic processes in a batch reactor under UV-C radiation ($\lambda_{max} = 254$ nm). J Hazard Mater 186:67–75

Patel AK, Das N, Goswami R, Kumar M (2019) Arsenic mobility and potential co-leaching of fluoride from the sediments of three tributaries of the Upper Brahmaputra floodplain, Lakhimpur, Assam, India. J Geochem Exploration 203:45–58

Poiger T, Buser HR, Muller MD (2001) Photodegradation of the pharmaceutical drug diclofenac in a lake: pathway, field measurements, and mathematical modeling. Environ Toxicol Chem 20(2):256–263

Schaper JL, Posselt M, Mccallum JL, Banks E, Hoehne A, Meinikmann K, Shanafield M, Batelaan O, Lewandowski J (2018a) Hyporheic exchange controls fate of trace organic compounds in an urban stream. Environ Sci Technol 52:12285–12294

Schaper JL, Seher W, Nützmann G, Putschew A, Jekel M, Lewandowski J (2018b) The fate of polar trace organic compounds in the hyporheic zone. Water Res 140:158–166

Schaper JL, Posselt M, Bouchez C, Jaeger A, Nuetzmann G, Putschew A, Singer G, Lewandowski J (2019) Fate of trace organic compounds in the hyporheic zone: influence of retardation, the benthic biolayer, and organic carbon. Environ Sci Technol 53:4224–4234

Schmitt-Jansen M, Bartels P, Adler N, Altenburger R (2006) Phytotoxicity assessment of diclofenac and its phototransformation products. Anal Bioanal Chem 387(4):1389–1396

Singh A, Patel AK, Deka JP, Das A, Kumar A, Kumar M (2019) Prediction of arsenic vulnerable zones in groundwater environment of rapidly urbanizing setup, Guwahati, India. Geochemistry, 125590. https://doi.org/10.1016/j.chemer.2019.125590

Stein K, Ramil M, Fink G, Sander M, Ternes TA (2008) Analysis and sorption of psychoactive drugs onto sediment. Environ Sci Technol 42:6415–6423

Takacs-Novak K, Jozan M, Hermecz I, Szasz G (1992) Lipophilicity of antibacterial fluoro-quinolones. Int J Pharm 79:89e96

Tang J, Wang S, Fan J, Long S, Wang L, Tang C, Tam NF, Yang Y (2019) Predicting distribution coefficients for antibiotics in a river water-sediment using quantitative models based on their spatiotemporal variations. Sci Total Environ 2019(655):1301–1310

Tixier C, Singer HP, Canonica S, Müller SR (2002) Phototransformation of triclosan in surface waters: A relevant elimination process for this widely used biocide laboratory studies, field measurements, and modeling. Environ Sci Technol 36(16):3482–3489

U.S. Environmental Protection Agency's (1998) harmonized test guideline, Direct photolysis rate in water by sunlight, Fate, transport, and transformation test guidelines, OPPTS 835.2210; Washington, DC

Yamamoto H, Nakamura Y, Moriguchi S, Honda Y, Tamura I, Hirata Y, Hayashi A, Sekizawa J (2009) Persistence and partitioning of eight selected pharmaceuticals in the aquatic environment: laboratory photolysis, biodegradation, and sorption experiments. Water Res 43(2):351–362

Yuan F, Hu C, Hu X, Wei D, Chen Y, Qu J (2011) Photodegradation and toxicity changes of antibiotics in UV and UV/H_2O_2 process. J Hazard Mater 185(2–3):1256–1263

Zepp R, Cline D (1977) Rates of direct photolysis in aquatic environment. Environ Sci Technol 11(4):359–366

Chapter 14
Shifts and Trends in Analysis of Contaminants of Emerging Concern: Sulfonamides

M. S. Priyanka and Sanjeeb Mohapatra

14.1 Introduction

Antibiotics are commonly used to treat or control diseases and infections caused by various microorganisms, such as fungi, bacteria, and parasites. The first antibiotic, penicillin, was discovered by Alexander Fleming in the twentieth century. Chemically, antibiotics, such as penicillin and streptomycin, are a class of secondary metabolites produced by microorganisms (Ben et al. 2019; Kumar et al. 2019a). With the advancement in manufacturing techniques in the field of biomedicines, many antibiotics are produced synthetically on a large scale to treat various kinds of infections leading the way to the so-called the antibiotics era (Aminov 2010; Schlipköter and Flahault 2010). The use of antibiotics led to a rapid decline in mortality and morbidity due to deadly diseases like tuberculosis, pneumonia, syphilis, and gonorrhea. Many infections, including respiratory, gastrointestinal and urinary, are treated using a course of these specific drugs and are usually administered via both oral and intravenous routes. Apart from human use, these drugs are also commonly used in livestock and animal husbandry (Alanis 2005; Singer et al. 2003; Kumar et al. 2019b).

The molecular weight of antibiotics is less than 1000 Da, and approximately, 250 such entities were already registered for various medicinal applications (Kümmerer 2009). Based on the mode of activity against infections, antibiotics can be classified into various classes (Yim et al. 2006). According to the 2017 revision of the WHO Model List of Essential Medicines, antibiotics were grouped into three categories: Access, Watch, and Reserve (AWaRe) (WHO 2018). Antibiotics that were

M. S. Priyanka
Department of Civil Engineering, National Institute of Technology Calicut, Kozhikode, India

S. Mohapatra (✉)
Environmental Science and Engineering Department, Indian Institute of Technology Bombay, Powai, Mumbai, India
e-mail: sanjeeb.publications@gmail.com

© Springer Nature Singapore Pte Ltd. 2020
M. Kumar et al. (eds.), *Resilience, Response, and Risk in Water Systems*,
Springer Transactions in Civil and Environmental Engineering,
https://doi.org/10.1007/978-981-15-4668-6_14

not included in the list were not categorized and were reported as "Others." The first category "Access" includes 48 antibiotics of first and second choice antibiotics for the basic treatment of common infectious diseases/conditions, and they should be extensively available in the healthcare units. The "Watch" category antibiotics have the potential to develop resistance, and their use as first and second choice treatment should be restricted or limited. It includes around 110 antibiotics. Finally, the "Reserve" category includes "last resort" antibiotics whose use should be reserved for specialized settings and specific cases where alternative treatments have failed. Around 22 antibiotics fall under this category. In the year 2019, WHO has come up with a new tool to limit the use of pharmaceuticals associated with the highest risk of resistance and to increase the use of antibiotics in countries where supply and availability are low. Most of the sulfonamides (SAs), such as sulfadiazine, sulfamethizole, sulfamethoxazole, sulfametrole, and sulfamoxole in combination with trimethoprim fall under the category of "Access." However, sulfadiazine and sulfamethoxazole are not recommended these days.

Gerhard Domagk discovered antibacterial properties of the azo dye, Prontosil in the year1935 which lead to the discovery and development of sulfa drugs. Since then, numerous derivatives of SAs have been developed and used for treatmnet of various antibacterial infection. These are structural moieties of amides of sulfonic acid (Fig. 14.1). All SAs consist of aniline moiety in para-position to the sulfonyl group with variations being incorporated in the structures which are attached to the amine group, giving them a distinct five- or six-member heterocyclic ring (Jesus et al. 2016). Sulfa compounds consist of two basic and one acidic functional groups. The acidic functional group belongs to the sulfonamide group and is known to lose its proton easily ($pK_a = 6.8$) than any other groups present in the compound (Fabiańska et al. 2014). In addition to antimicrobial activity, when these groups are added to biologically relevant scaffolds, they produce versatile effects for the treatment of various chronic complex diseases like Alzheimer's disease (Mutahir et al. 2016), central nervous system (CNS) disorders, diabetes, psychosis (Zajdel et al. 2014), cancers (Gul et al. 2018), and tumors (Lu et al. 2015). SAs are also termed poly-pharmacology compounds, capable of simultaneously affecting multiple pathways or mechanisms (multi-target approach) (Winum et al. 2012).

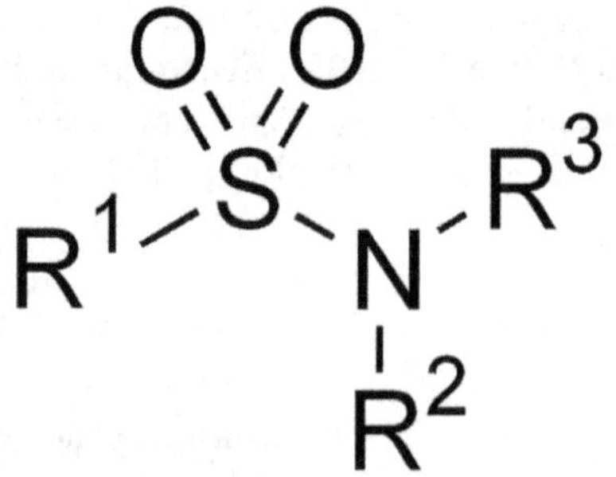

Fig. 14.1 Typical structure of sulfonamide

   Administration of SAs is successful in acting against various species, such as *Chlamydia, Clostridium, Escherichia, Neisseria, Nocardia, Salmonella, Shigella, Staphylococcus,* and *Streptococcus.* However, they are also used against other microorganisms like fungi (e.g., *Pneumocystis carinii*), protozoa (e.g., *Toxoplasma gondii*), and parasites (e.g., *Plasmodium malariae*) (Baran et al. 2011). Zhang and Messhnick (1991) have also reported that SAs exhibited the antimalarial activity by competitive inhibition of the enzyme dihydropteroate synthase (DHPS) majorly produced in bacteria. Additionally, these drugs were effective in treating the neosporosis outbreak in 1980s caused by *Neospora caninum* (protozoa) in cattle, goats, sheep, dogs, and other domestic animals (Dubey and Lindsay 1996). In humans, they are usually prescribed to treat bronchitis, eye and ear infections, bacterial meningitis, *Pneumocystis carinii* pneumonia, urinary tract infections, travelers' diarrhea, and many other bacterial infections (García-Galán et al. 2008). Prescribed quantities of SAs are used as growth promoters to enhance performance and production in poultry sector, swine husbandry, aquaculture, agricultural activities, and other livestock farming sectors. Few of the antibiotics belonging to the SAs family along with their molecular formula, molecular weight, solubility, dissociation constant (pK_a,), and octanol–water partition coefficient (Log K_{ow}) values are listed in Table 14.1.

14.2   Available Techniques to Monitor Sulfonamides

Extensive production and consumption have increased the occurrence of SAs in various environmental matrices including surface water, wastewater, sediments, and soil (Menon et al. 2020). However, in natural environmental settings, these antibiotics are commonly present in the order of ng/L to μg/L. Thus, monitoring and analysis of these compounds are challenging and require tedious sample preparation and sophisticated analytical instrumentation. As a result of which environmental samples go through a rigorous sample collection, transportation, preparation, and storage before final quantification. The high priority step is sample preparation that minimizes interference arising due to background matrix components. It further helps in attending a low limit of detection (LOD) and quantification (LOQ). This section mainly focuses on various research conducted worldwide on sample preparation and analysis of SAs through state-of-the-art chromatographic instruments. A table summarizing sample preparation for a set of SAs in different environmental matrices is presented in Table 14.2. Sample preparation is mainly done through solid-phase extraction (SPE), magnetic solid-phase extraction (MSPE), supported liquid extraction (SLE), and pressurized liquid extraction (PLE) techniques.

   Pharmaceuticals belonging to SAs class may have structural similarities, but their solubility, hydrophobicity, and polarities vary to a greater extent (Table 14.1). Except for sulfanitran, the majority of SAs experience low hydrophobicity (Log K_{ow} < 1.5) which makes the isolation and pre-concentration step exigent. The selection of pre-concentration conditions is very crucial as the polarity of SAs gets affected at a low pH range. As a result of which, SAs reported to experience a significant

Table 14.1 List of sulfonamides along with their molecular formula, molecular weight, solubility, pK_a, Log K_{ow} values

Pharmaceuticals	Abbreviation	Molecular formula	Molecular weight	Solubility (in water, mg/L)	pK_a	Log K_{ow}
Sulfamethoxazole	SMX	$C_{10}H_{11}N_3O_3S$	253.2	610	1.39, 5.8	0.89
Sulfamethizole	SMI	$C_9H_{10}N_4O_2S_2$	270.3	105	2.1, 5.3	0.54
Sulfapyridine	SP	$C_{11}H_{11}N_3O_2S$	249.2	0.235	2.58, 8.43	0.35
Sulfathiazole	STZ	$C_9H_9N_3O_2S_2$	255.3	373	2.0, 7.2	0.05
Sulfaquinoxaline	SQX	$C_{14}H_{12}N_4O_2S$	300.3	7.5	2.3, 5.1	1.68
Sulfachlorpyridazine	SCP	$C_{10}H_9ClN_4O_2S$	284.7	7000	2, 6.6	0.31
Sulfamerazine	SMR	$C_{11}H_{12}N_4O_2S$	264.3	202	1.64, 7.34	0.14
Sulfadimethoxine	SDM	$C_{12}H_{14}N_4O_4S$	310.3	343	3, 6.21	1.63
Sulfamonomethoxine	SMM	$C_{11}H_{12}N_4O_3S$	280.3	10,000	2.0, 6.0	0.7
Sulfameter/Sulfamethoxydiazine	SME	$C_{11}H_{12}N_4O_3S$	280.3	313	1.98, 7.06	0.41
Sulfamethazine	SMZ	$C_{12}H_{14}N_4O_2S$	278.3	1500	1.69, 7.89	0.89
Sulfadiazine	SDZ	$C_{10}H_{10}N_4O_2S$	250.2	77	1.64, 6.81	-0.09
Sulfisomidine	SIM	$C_{12}H_{14}N_4O_2S$	278.3	229	5, 6.12	-0.33
Sulfisoxazole	SIX	$C_{11}H_{13}N_3O_3S$	267.3	300	2.17, 5.8	1.01
Sulfanitran	SNT	$C_{14}H_{13}N_3O_5S$	335.3	34.3	4.4, 7.44	–
Sulfamethoxypyridazine	SMP	$C_{11}H_{12}N_4O_3S$	280.3	147	2.02, 6.84	0.32
Sulfacetamide	SAC	$C_8H_{10}N_2O_3S$	214.2	12,500	2.14, 4.3	-0.96
Sulfanilamide	SNL	$C_6H_8N_2O_2S$	172.2	7500	10.43, 11.63	-0.62
Sulfadoxine	SDX	$C_{12}H_{14}N_4O_4S$	310.3	296	2.55, 6.12	0.7
Sulfacarbamide	SCB	$C_7H_9N_3O_3S$	215.2	–	–	–
Sulfathiourea	STU	$C_7H_9N_3O_2S_2$	231.3	159	2.27, 6.31	–

(continued)

Table 14.1 (continued)

Pharmaceuticals	Abbreviation	Molecular formula	Molecular weight	Solubility (in water, mg/L)	pK_a	Log K_{ow}
Sulfaguanidine	SGD	$C_7H_{10}N_4O_2S$	214.2	805	7.72, 10.53	−1.22
Sulfafurazole	SFZ	$C_{11}H_{13}N_3O_3S$	267.3	300	1.5, 5	1.01
Sulfamoxole	SXL	$C_{11}H_{13}N_3O_3S$	267.3	1680	1.94, 6.81	–

Table 14.2 Sample preparation techniques in water, wastewater, soil and sludge reported in various literature

SAs	Samples	Extraction technique		Eluent	% Recovery	References
		Method	Sorbent			
SMX SMZ SMR	River water	SPE	Oasis HLB	ACN : MeOH :: 50:50	50–120	Qian et al. (2016)
STZ SMZ SDZ	River water	SPE	Oasis HLB	10 mL MeOH	88–98 49–75 77–104	Zhou et al. (2012a, b)
SMX	Surface water	SPE	Oasis HLB	10 mL MeOH	100–104	Soran et al. (2017)
SDZ SME SMX SMR STZ SMI SCP SDM SMZ	Surface water	SPE	Oasis HLB (Conditioning: 6 mL MeOH followed by 10 mL of 0.1% EDTA)	10 mL ethyl acetate/MeOH (9/1, v/v) and 10 mL MeOH + 2% ammonium hydroxide;	86–100 72–95 76–104 74–102 81–116 79–87 88–100 83–86 82–85	Tso et al. (2011)
SAC SP SDZ SMX STZ SMR SIX SMZ	Wastewater	Magnetic-SPE	Core–shell Fe_3O_4 @MoS_2 composites	3 mL MeOH with 1% ammonium hydroxide	84–90 81–96 80–99 84–90 88–105 84–109 84–89 85–96	Zhao et al. (2020)
SMX SMZ SMR	Wastewater	SPE	SAX + HLB 6 mL MeOH and 6 mL 4.38 mM H_3PO_4	10 mL of 95% MeOH, 5% 4.38 mM H_3PO_4	37–65	Renew and Huang (2004)

(continued)

Table 14.2 (continued)

SAs	Samples	Extraction technique		Sorbent	Eluent	% Recovery	References
		Method					
STZ SMZ SDZ	Wastewater	SPE		Oasis HLB 10 mL of MeOH and water	10 mL MeOH	101–104 85–102 71–94	Zhou et al. (2012a, b)
SDZ SME SMM SQX	Soil	Dynamic microwave-assisted extraction (DMAE) coupled online with SPE		Anhydrous sodium sulfate	ACN and microwave heating (power of 320 W)	83 89 91 94	Chen et al. (2009)
SDZ SMZ SMX SMM SME SDM SQX	Soil	MSPE		Alumina-coated magnetite nanoparticles (Fe_3O_4/Al_2O_3 NPs)	15 mL ACN Ultrasonic frequency (40 kHz)	71–93 42–60	Sun et al. (2010)
SDZ, STZ SMZ, SMX SDM SQX, SCP	Soil	SLE Water : ACN:: 50: 50		Oasis HLB	Water : MeOH :: 80:20	70–100	Tetzner et al. (2016)
22 SAs	Soil and Sludge	PLE		HLB	MeOH : Water :: 90:10, ACN : Water :: 25:75	60–130	García-galán et al. (2013)

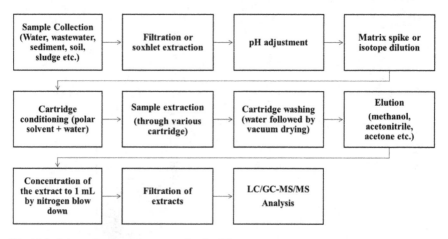

Fig. 14.2 Schematic of sample preparation for SAs

difference in their recovery values ranging from 50 to 120% (Qian et al. 2016). In such a scenario, compound-specific sample preparation techniques were developed involving a variety of solvents and cartridges. A schematic of sample preparation techniques for SAs is shown in Fig. 14.2.

14.2.1 Sample Preparation of Water and Wastewater Samples

The most commonly used sample preparation technique for the concentration of antibiotics is SPE. Before the sample passes through the sorbents/cartridges, water samples are pre-treated as per the acid–base dissociation of the target analytes to obtain a better recovery. Immediately after sample collection, samples are usually acidified to pH 2 by adding a suitable acidifying reagent to avoid degradation of SAs by the native microbial population (Qian et al. 2016). However, acidification poses a risk of acid-catalyzed hydrolysis and also increases the adsorption of target molecules on to organic matter. Apart from acidification, chemicals like sodium azide (Zhou et al. 2012a, b) and formaldehyde (Matějíček and Kubáň 2008) are used as a preservative. An antibiotic like SMX has shown no degradation even after 10 days of storage (Nebot et al. 2007). Whereas in tap water, use of sodium thiosulfate restricts oxidation of analytes with residual chlorine (Rosa Boleda et al. 2011). Similarly, a variety of sorbents are available commercially for SPE. Most commonly used sorbents are Oasis HLB (Shaaban and Górecki 2012; Kim et al. 2013), Strata-X (Iglesias et al. 2012), or Bond Elut-ENV (Vosough and Mashhadiabbas Esfahani 2013), Hysphere C18 EC, and PRLP cartridges. Cartridges like subsidiary anion exchange cartridge (SAX) are used for the removal of negatively charged humic and fulvic acids (Pan et al. 2011). In a study conducted by García-Galán et al. (2010), Hysphere C18 EC has given a better efficiency in extracting SAs (García-Galán et al.

2010). Recent developments were made in sorbents for the isolation of SAs from the water samples such as carbon nanotubes, composites based on Fe_3O_4 and either carbon nanotubes (Herrera-Herrera et al. 2013a) or graphene magnetic iron oxide nanoparticles coated with octadecyl trimethyl ammonium bromide (Sun et al. 2009). Sun et al. (2014) have developed a portable graphene-based micropipette device for flow micro-SPE of SAs (Sun et al. 2014). Zhoa et al. (2020) have synthesized a molybdenum disulfide (MoS_2)-based core–shell magnetic nanocomposite (Fe_3O_4 @MoS_2) for the extraction of SAs and obtained > 80% recoveries (Zhao et al. 2020). Extracting SAs from natural water can be performed by many other methods like single-drop liquid-phase microextraction (Guo et al. 2012), dispersive liquid–liquid microextraction (Herrera-Herrera et al. 2013b), and membrane microextraction (Tong et al. 2013). Commonly used eluents are polar solvents like methanol (MeOH), acetonitrile (ACN), acetone (ACE), diatomite, and hexane. Methanol, acetone, and combination of methanol–acetone mixture (1:1v/v) are considered as the best eluents for SAs. Additionally, recovery of SAs depends upon the calibration methods. Isotope dilution method (ISDM)-based approach reported to give better recovery for sulfamethoxazole when sulfamethoxazole-d_8 was spiked to wastewater samples. Studies comparing the effect of ISDM and without ISDM on recovery of SAs were studied elsewhere (Mohapatra et al. 2018).

14.2.2 Sample Preparation of Soil Samples

The presence of SAs in soils can affect the microorganisms and soil functions by causing resistant genes or organisms. Relative low sorption coefficients of SAs, as given in Table 14.1, indicate poor interactions with soil binding sites. In recent times, extraction of SAs from soil has increased tremendously due to their pseudo-persistence in the environment. Techniques like ultrasonic extraction (Ho et al. 2012; Zhou et al. 2012a, b), microwave-assisted extraction (Raich-Montiu et al. 2010), pressurized liquid extraction (PLE) (Raich-Montiu et al. 2010; Pamreddy et al. 2013) and matrix solid-phase dispersion (MSPD), microwave-assisted micellar extraction (MAME) are commonly employed for the extraction of SAs from the soil.

Solvents like acetonitrile (Salvia et al. 2012), methanol (Karci and Balcioğlu 2009), nonionic surfactant Triton X-114 (Chen et al. 2010), and mixtures of these solvents with different buffers (Ho et al. 2012; Karci and Balcioğlu 2009) were also employed for the extraction of SAs from soil and sediment samples. SPE is operated in offline or online mode, and offline SPE requires time-consuming steps of extensive sample handling which ultimately increases the method uncertainty. In contrast, online SPE involves a repetitive injection of samples resulting in increased detection frequency. Offline SPE is commonly performed for the purification and pre-concentration of analytes of interest. For better recovery, it is suggested to keep the organic content of the extract to less than 5% (Tetzner et al. 2016).

14.2.3   Sample Analysis Through Chromatographic Technique

The shift from the gas chromatography (GC) to liquid chromatography (LC) coupled with high-resolution mass spectrometry (HRMS) has increased the reliability with which such SAs can be detected with better precision and accuracy in various samples like soil, surface water, and wastewater. Such methods are competent in terms of less consumption of solvents and reduced cost of analysis. Usually, the cost is attributed to the extraction, purification, and analysis of target compounds. Table 14.3 summarizes the use of various chromatographic techniques, LOD, and LOQ values for a set of SAs detected in water, wastewater, soil and sludge matrices. Among the listed techniques, both Tso et al. (2011) and Zhao et al. (2020) developed extensive analysis protocols to analyze a set of SAs in surface water and wastewater samples, respectively. Similarly, García-galán et al. (2013) developed a robust method for the analysis of 21 SAs in sludge and soil samples. They could attend LOD and LOQ ranging from 0.01 to 17.4, and 0.05 to 14 ng/L, respectively. García-Galán et al. (2009) also made several attempts to develop sensitive methods to analyze SAs in different water matrices. The most commonly used techniques such as LC-MS and GC-MS have also opened up routes for the legislation departments to monitor, understand, and frame regulatory standards for the disposal of such compounds into the environment.

14.3   Consumption and Occurrences of SAs in the Asian Region

Growing economic opulence, rising incomes, increasing population, and diseases have contributed to the increased consumption of antibiotics across the world. Globally, antibiotics consumption has increased by 65% from the year 2000–2015. The consumption is estimated at 4.4–6.4 defined daily doses (DDD) per 1000 inhabitants per day. In 2000, global consumption increased due to the higher consumption in high-income countries (HICs) like the USA, France, New Zealand, Spain, and Hong Kong. Between 2000 and 2010, global consumption of antibiotics increased by 76%, and three-quarters of this increase were accounted for by Brazil (68%), Russia (19%), India (66%), China (37%), and South Africa (219%) (BRICS). In BRICS countries, 23% of the increase in the retail sales volume was attributable to India and up to 57% of the increase in the hospital sector was attributable to China. In 2015, the increase in global consumption was mainly driven by the increased consumption in low–middle-income countries (LMICs) like India, China, Pakistan, Turkey, Tunisia, Algeria, and Romania. Countries like India, China, and Pakistan stood as the highest consumers under LMICs in 2015. Between 2000 and 2015, antibiotic consumption increased from 3.2 to 6.5 billion DDDs (103%) in India, from 2.3 to 4.2 billion DDDs (79%) in China, and from 0.8 to 1.3 billion DDDs (65%) in Pakistan (Klein et al. 2018). China was the highest antibiotic consumer in livestock followed by the

Table 14.3 Analytical techniques for analysis of SAs from water, wastewater, soil and sludge samples reported in various literature

SAs	Sample	Analytical technique			LOD (ng/L)	LOQ (ng/L)	References
		Technique	Mobile phases	m/z			
SMX SMZ SMR	River water	LC-MS/MS Kinetex C18 column	For −ve ESI 5 mmol/L ammonium acetate in water and MeOH For +ve ESI water (0.1% formic acid) and MeOH	–	–	1.07 1.86–2.15 –	Qian et al. (2016)
STZ SMZ SDZ	River water	LC-MS/MS Zorbax Bonus-RP column	0.1% formic acid in ultrapure water, and ACN with 0.1% formic acid	256 279 251	0.4 0.5 0.4	–	Zhou et al. (2012a, b)
SMX	Surface water	HPLC-MS Grace Alltima RP18 column	90:10 ACN–ultrapure water and 0.1% aqueous formic acid	254	1.1	2.2	Soran et al. (2017)
SDZ SME SMX SMR STZ SMI SCP SDM SMZ	Surface water	HPLC-MS Beta basic C18	Water/MeOH (96/4, v/v) with 5 mM ammonium hydroxide and water/MeOH/ACN(10/10/80, v/v/v) with 5 mM ammonium hydroxide	–	0.4 0.3 0.3 0.3 0.3 0.3 0.3 0.3 0.3	1.2 1.0 1.0 1.0 1.0 1.0 1.0 1.0 1.0	Tso et al. (2011)

(continued)

Table 14.3 (continued)

SAs	Sample	Analytical technique		m/z	LOD (ng/L)	LOQ (ng/L)	References
		Technique	Mobile phases				
SAC	Wastewater	HPLC-MS/MS C18 Column	0.1% formic acid water and ACN	–	1.15	3.50	Zhao et al. (2020)
SP					0.35	0.95	
SDZ					0.32	0.92	
SMX					0.35	0.95	
STZ					0.30	0.90	
SMR					0.30	0.85	
SIX					0.25	0.80	
SMZ					0.20	0.70	
SMR	Wastewater	HPLC-MS Zorbax SB-C18 column	1 mM ammonia acetate, 0.007% (v/v) glacial acetic acid and 10% ACN	265	–	–	Renew and Huang (2004)
SMZ				279			
SMX				254			
STZ	Wastewater	LC-MS/MS Zorbax Bonus-RP column	0.1% formic acid in ultrapure water, and ACN with 0.1% formic acid	256	1		Zhou et al. (2012a, b)
SMZ				279	1		
SDZ				251	1		
SDZ	Soil	LC-MS/MS C18 Column	0.2% acetic acid and ACN	251.1	2.3–4.8	7.7–16	Chen et al. (2009)
SME				281.1	2.3–3.2	7.7–10.7	
SMM				281.1	1.4–1.9	4.6–6.5	
SQX				301.1	2.2–2.6	7.2–8.8	

(continued)

Table 14.3 (continued)

SAs	Sample	Analytical technique		Mobile phases	m/z	LOD (ng/L)	LOQ (ng/L)	References
		Technique						
SDZ	Soil	LC–QT/MS/MS ZORBAX SB-C18 column		0.5% acetic acid aqueous solution and ACN	251.1	1.30	–	Sun et al. (2010)
SMZ					283.2	3.72		
SMX					254.1	6.74		
SMM					281.1	0.37		
SME					281.1	0.73		
SDM					311.1	2.30		
SQX					301.1	0.94		
SDZ	Soil	UHPLC–MS/MS Acquity UPLC BEH C18 column		Water and MeOH with formic acid (0.1% v/v)	250.9	–	0.5	Tetzner et al. (2016)
STZ					255.9			
SMZ					279.0			
SMX					253.9			
SDM					311.0			
SQX					301.0			
SCP					284.7			
22 SAs	Soil and Sludge	LC-MS/MS C18 LC-column		Water acidified with 10 mM HCOOH, and ACN with 10 mM HCOOH	–	0.01–4.2 0.03–17.4	0.05–14 0.50–58	García-galán et al. (2013)

m/z: molecular ion

USA, Brazil, Germany, and India. BRICS utilization of antibiotics in livestock is anticipated to be doubled by 2030 as their population may increase by 13% (Done and Halden 2015; Gelband et al. 2015).

SAs are one of the largest groups among the critically important antibiotic family. Since 1950s huge quantities of SAs were employed across the globe, due to its specific antimicrobial activity, to treat, and prevent diseases with low or micro-level concentrations. Along with inhibiting or influencing pathogens in humans and animals, they also act as non-therapeutic agents (animal growth promoters) in cattle, poultry, and swine livestock (Sarmah et al. 2006). These chemotherapeutic agents once administered, based on their efficacy may remain in the host and move out in their native form or transformed form (metabolites). Only 15% of the SAs are absorbed and metabolized in the body. Approximately, 30–90% of the drugs are excreted as un-metabolized (Carvalho and Santos 2016) of which 70–80% is excreted in urine and 20–30% in fecal matter. These excreted antibiotics can be recalcitrant, non-biodegradable, and persistent in the environment. Annually, over 20,000 tons of sulfonamides enter into the environment and accumulate in natural water bodies. Consequently, sulfonamide-resistant bacteria were developed and reported (Díaz-Cruz and Barceló 2005; Klauson et al. 2019). SAs residues were detected in different environmental compartments across the globe. A study conducted by Shimizu et al. (2013) reported higher concentrations of SAs in the tropical regions of Asia (Vietnam, Philippines, India, Indonesia, and Malaysia) compared to USA, Europe, Canada, and Japan. Table 14.4 summarizes the occurrence of SAs in a set of Asian countries. The most frequently detected predominant member of SAs in wastewater was SMX with an average highest reported concentration of 1720 ng/L in Vietnam (Hanoi, Ho Chi Minh, Can Tho) followed by 802 ng/L in Philippines (Manila). SAs were also detected in India (Kolkata-538 ng/L), Indonesia (Jakarta-282 ng/L), and Malaysia (Kuala Lumpur-76 ng/L) (Shimizu et al. 2013). Another study has also reported a concentration of 16 ng/L for SMX in the environment (Tran et al. 2019). The concentration of SAs in Sri Lankan hospital wastewater effluents ranged from 1000 to 18,000 ng/L. The concentration of SMX in the Kshipra River of India was found to be 4660 ng/L. The lowest concentrations were reported in Malaysia and Singapore, ranging from 0.6 to 2.6 ng/L.

14.4 Conclusions

Since 1935, SAs have been used extensively as antibiotics to treat bacterial infections, Alzheimer's disease, central nervous system (CNS) disorders, psychosis, cancers, and tumors. These are the oldest and largest group to be used in clinical practices. This class of antibiotics is also widely used as a growth promoter to enhance performance and production in the poultry sector, swine husbandry, aquaculture, agricultural activities, and other livestock farming sectors. The shift from the GC/MS to LC/MS has increased the reliability with which such SAs were detected with better precision

Table 14.4 Occurrences of antibiotics in various matrices of Asian countries

Country	Source	Compound	Concentration (ng/L)/(µg/kg)	References
China	River estuary	SMX	1.1	Qian et al. (2016)
		SMZ	1.9	
	WWTP	SMX	5597	Zhang and Li (2011)
	H 1 WW	SDZ	119.2	Wang et al. (2018)
	H-2 WW	SMX	434.3	
	H-3 WW	SMZ	11.6	
		SDZ	124.9	
		SMX	587.9	
		SMZ	BLD	
		SDZ	125.8	
		SMX	587.9	
		SMZ	BLD	
	Soil	SMX	0.03–0.9	Hu et al. (2010)
		SDO	1.2–9.1	
		SCP	0.2–2.5	
	Soil	SDZ	97.2	Chen et al. (2014)
		SMX	90.0	
India (Madhya Pradesh)	River water Sediments	SMX	40–4660	Diwan et al. (2018)
			0–8230	
Korea	River water	SMX	57	Lee et al. (2008)
		SMZ	111	
Malaysia	HWW	SAC	0.9	Lye et al. (2019)
		SDZ	18.4	
		STZ	6.7	
		SP	2.35	
		SMX	1.4–91	
		SDM	60.2	

(continued)

Table 14.4 (continued)

Country	Source	Compound	Concentration (ng/L)/(μg/kg)	References
Singapore	Surface water soil	SMR SMZ SMX SMZ	5.9 6.4 2.6 1.0	Yi et al. (2019)
Sri Lanka	HWW	SDZ SMX	1000–3000 10,000–18,000	Liyanage and Manage (2016)
Taiwan	STPs Regional discharge	SAs	2–226 10–282	Lin et al. (2008)
Thailand	HWW	SMX	1499	Sinthuchai et al. (2016)
Turkey	WWTP 1 WWTP 2 Industrial Influent	SMX SMX SMX	6130 5350 19,700	Ata and Yıldız Töre (2019)
Vietnam	Lakes Urban Canals	SMX SMZ	108–3508 <MQL–209 310–15,591 <MQL–128	Tran et al. (2019)

WWTP Wastewater treatment plant; *HWW* Hospital wastewater

and accuracy in various environmental matrices ranging from ng/L to µg/L. However, optimization of sample preparation technique, including choice of solvent, and cartridges, extraction protocol, pH, concentration of organic matter and a combination of eluents plays a critical role in deciding the absolute recovery SAs. Some of these shortcomings can be overcome by adopting the online SPE technique. Use of HR LC-MS/MS over LC-MS has also increased the accurate quantification of such compounds in complex environmental matrices. The concentration of SAs in water, wastewater, soil, and sludge varies from 1.1–1559 to 0.9–19,700 ng/L, and 0.03–97, 0–8230 µg/kg, respectively. At this environmental concentrations, SAs are reported to produce various adverse impacts on the environment including the rise of superbugs. Hence, the development of assay-based monitoring technologies such as immunoassays and biosensors followed by advanced treatment technologies such as membrane bioreactors and advanced oxidation processes must be adopted.

References

Alanis AJ (2005) Resistance to antibiotics: are we in the post-antibiotic era? Arch Med Res 36:697–705. https://doi.org/10.1016/j.arcmed.2005.06.009

Aminov RI (2010) A brief history of the antibiotic era: lessons learned and challenges for the future. Front Microbiol 1:1–7. https://doi.org/10.3389/fmicb.2010.00134

Ata R, Yıldız Töre G (2019) Characterization and removal of antibiotic residues by NFC-doped photocatalytic oxidation from domestic and industrial secondary treated wastewaters in Meric-Ergene Basin and reuse assessment for irrigation. J Environ Manage 233:673–680. https://doi.org/10.1016/j.jenvman.2018.11.095

Baran W, Adamek E, Ziemiańska J, Sobczak A (2011) Effects of the presence of sulfonamides in the environment and their influence on human health. J Hazard Mater 196:1–15. https://doi.org/10.1016/j.jhazmat.2011.08.082

Carvalho IT, Santos L (2016) Antibiotics in the aquatic environments: a review of the European scenario Isabel. Environ Int 94:736–757. https://doi.org/10.1016/j.scitotenv.2018.01.271

Chen L, Zeng Q, Wang H, Su R, Xu Y, Zhang X, Yu A, Zhang H, Ding L (2009) On-line coupling of dynamic microwave-assisted extraction to solid-phase extraction for the determination of sulfonamide antibiotics in soil. Anal Chim Acta 648:200–206. https://doi.org/10.1016/j.aca.2009.07.010

Chen L, Zhao Q, Xu Y, Sun L, Zeng Q, Xu H, Wang H, Zhang X, Yu A, Zhang H, Ding L (2010) A green method using micellar system for determination of sulfonamides in soil. Talanta 82:1186–1192. https://doi.org/10.1016/j.talanta.2010.06.031

Chen C, Li J, Chen P, Ding R, Zhang P, Li X (2014) Occurrence of antibiotics and antibiotic resistances in soils from wastewater irrigation areas in Beijing and Tianjin. China Environ Pollut 193:94–101. https://doi.org/10.1016/j.envpol.2014.06.005

Díaz-Cruz MS, Barceló D (2005) LC-MS2 trace analysis of antimicrobials in water, sediment and soil. TrAC - Trends Anal Chem 24:645–657. https://doi.org/10.1016/j.trac.2005.05.005

Diwan V, Hanna N, Purohit M, Chandran S, Riggi E, Parashar V, Tamhankar AJ, Stålsby Lundborg C (2018) Seasonal variations in water-quality, antibiotic residues, resistant bacteria and antibiotic resistance genes of Escherichia coli isolates from water and sediments of the Kshipra River in Central India. Int J Environ Res Public Health 15:1–16. https://doi.org/10.3390/ijerph15061281

Done HY, Halden RU (2015) Reconnaissance of 47 antibiotics and associated microbial risks in seafood sold in the United States. J Hazard Mater 282:10–17

M. S. Priyanka and S. Mohapatra

Dubey JP, Lindsay DS (1996) A review of Neospora caninum and neosporosis. Vet Parasitol 67:1–59. https://doi.org/10.1016/S0304-4017(96)01035-7

Fabiańska A, Białk-bieli A, Stepnowski P, Stolte S, Siedlecka EM (2014) Electrochemical degradation of sulfonamides at BDD electrode: kinetics, reaction pathway and eco-toxicity evaluation. J Hazard Mater 280:579–587. https://doi.org/10.1016/j.jhazmat.2014.08.050

García-Galán MJ, Silvia Díaz-Cruz M, Barceló D (2008) Identification and determination of metabolites and degradation products of sulfonamide antibiotics. TrAC—Trends Anal Chem 27:1008–1022. https://doi.org/10.1016/j.trac.2008.10.001

García-Galán MJ, Silvia Díaz-Cruz M, Barceló D (2009) Combining chemical analysis and eco-toxicity to determine environmental exposure and to assess risk from sulfonamides. Trends Anal Chem 28:804–819. https://doi.org/10.1016/j.trac.2009.04.006

García-Galán MJ, Díaz-Cruz MS, Barceló D (2010) Determination of 19 sulfonamides in environmental water samples by automated on-line solid-phase extraction-liquid chromatography-tandem mass spectrometry (SPE-LC-MS/MS). Talanta 81:355–366. https://doi.org/10.1016/j.talanta.2009.12.009

García-galán MJ, Díaz-cruz S, Barceló D (2013) Multiresidue trace analysis of sulfonamide antibiotics and their metabolites in soils and sewage sludge by pressurized liquid extraction followed by liquid chromatography—electrospray-quadrupole linear ion trap mass spectrometry. J Chromatogr A 1275:32–40. https://doi.org/10.1016/j.chroma.2012.12.004

Gelband H, Miller-Petrie M, Pant S, Gandra S, Levinson J, Barter D, White A, Laxminarayan R (2015) The state of the world's antibiotics 2015

Gul HI, Yamali C, Sakagami H, Angeli A, Leitans J, Kazaks A, Tars K, Ozgun DO, Supuran CT (2018) New anticancer drug candidates sulfonamides as selective hCA IX or hCA XII inhibitors. Bioorg Chem 77:411–419. https://doi.org/10.1016/j.bioorg.2018.01.021

Guo X, Yin D, Peng J, Hu X (2012) Ionic liquid-based single-drop liquid-phase microextraction combined with high-performance liquid chromatography for the determination of sulfonamides in environmental water. J Sep Sci 35:452–458. https://doi.org/10.1002/jssc.201100777

Herrera-Herrera AV, Hernández-Borges J, Afonso MM, Palenzuela JA, Rodríguez-Delgado MÁ (2013a) Comparison between magnetic and non magnetic multi-walled carbon nanotubes-dispersive solid-phase extraction combined with ultra-high performance liquid chromatography for the determination of sulfonamide antibiotics in water samples. Talanta 116:695–703. https://doi.org/10.1016/j.talanta.2013.07.060

Herrera-Herrera AV, Hernández-Borges J, Borges-Miquel TM, Rodríguez-Delgado M (2013b) Dispersive liquid-liquid microextraction combined with ultra-high performance liquid chromatography for the simultaneous determination of 25 sulfonamide and quinolone antibiotics in water samples. J Pharm Biomed Anal 75:130–137. https://doi.org/10.1016/j.jpba.2012.11.026

Ho YB, Zakaria MP, Latif PA, Saari N (2012) Simultaneous determination of veterinary antibiotics and hormone in broiler manure, soil and manure compost by liquid chromatography-tandem mass spectrometry. J Chromatogr A 1262:160–168. https://doi.org/10.1016/j.chroma.2012.09.024

Ben Y, Fu C, Hu, M, Liu L, Wong MH, Zheng C (2019) Human health risk assessment of antibiotic resistance associated with antibiotic residues in the environment: a review. Environ Res 169:483–493. https://doi.org/10.1016/j.envres.2018.11.040

Hu X, Zhou Q, Luo Y (2010) Occurrence and source analysis of typical veterinary antibiotics in manure, soil, vegetables and groundwater from organic vegetable bases, northern China. Environ Pollut 158:2992–2998. https://doi.org/10.1016/j.envpol.2010.05.023

Iglesias A, Nebot C, Miranda JM, Vázquez BI, Cepeda A (2012) Detection and quantitative analysis of 21 veterinary drugs in river water using high-pressure liquid chromatography coupled to tandem mass spectrometry. Environ Sci Pollut Res 19:3235–3249. https://doi.org/10.1007/s11356-012-0830-3

Jesus V De, Cardoso VV, Benoliel MJ, Almeida CMM (2016) Chlorination and oxidation of sulfonamides by free chlorine: identification and behaviour of reaction products by UPLC-MS/ MS. J Environ Manage 166:466–477

Karci A, Balcioğlu IA (2009) Investigation of the tetracycline, sulfonamide, and fluoroquinolone antimicrobial compounds in animal manure and agricultural soils in Turkey. Sci Total Environ 407:4652–4664. https://doi.org/10.1016/j.scitotenv.2009.04.047

Kim H, Hong Y, Park JE, Sharma VK, Cho S Il (2013) Sulfonamides and tetracyclines in livestock wastewater. Chemosphere 91:888–894. https://doi.org/10.1016/j.chemosphere.2013.02.027

Klauson D, Romero N, Krichevskaya M, Kattel E, Dulova N, Dedova T, Trapido M (2019) Advanced oxidation processes for sulfonamide antibiotic sulfamethizole degradation: process applicability study at ppm level and scale-down to ppb level. J Environ Chem Eng 7:103287

Klein EY, Van Boeckel TP, Martinez EM, Pant S, Gandra S, Levin SA, Goossens H, Laxminarayan R (2018) Global increase and geographic convergence in antibiotic consumption between 2000 and 2015. Proc Natl Acad Sci 115:E3463–E3470. https://doi.org/10.1073/pnas.1717295115

Kumar M, Ram B, Honda R, Poopipattana C, Canh VD, Chaminda T, Furumai H (2019a) Concurrence of antibiotic resistant bacteria (ARB), viruses, pharmaceuticals and personal care products (PPCPs) in Ambient Waters of Guwahati, India: Urban vulnerability and resilience perspective. Sci Total Environ 693:133640. https://doi.org/10.1016/j.scitotenv.2019.133640

Kumar M, Chaminda T, Honda R, Furumai H (2019b) Vulnerability of urban waters to emerging contaminants in India and Sri Lanka: resilience framework and strategy. APN Sci Bull 9(1). https://doi.org/10.30852/sb.2019.799

Kümmerer K (2009) Antibiotics in the aquatic environment—a review—part I. Chemosphere 75:417–434. https://doi.org/10.1016/j.chemosphere.2008.11.086

Lee YJ, Lee SE, Lee DS, Kim YH (2008) Risk assessment of human antibiotics in Korean aquatic environment. Environ Toxicol Pharmacol 26:216–221. https://doi.org/10.1016/j.etap.2008.03.014

Lin AY-C, Yu TH, Lin CF (2008) Pharmaceutical contamination in residential, industrial, and agricultural waste streams: risk to aqueous environments in Taiwan. Chemosphere 74:131–141. https://doi.org/10.1016/j.chemosphere.2008.08.027

Liyanage GY, Manage PM (2016) Occurrence, fate and ecological risk of antibiotics in hospital effluent water and sediments in Sri lanka. Int J Agric Environ Res 02:909–935

Lu T, Laughton CA, Wang S, Bradshaw TD (2015) In vitro antitumor mechanism of (E) - N - (2-methoxy-5- (((2, 4, 6- trimethoxystyryl) sulfonyl) methyl) pyridin-3-yl) methanesulfonamide s. Mol Pharmacol 18–30

Lye YL, Bong CW, Lee CW, Zhang RJ, Zhang G, Suzuki S, Chai LC (2019) Anthropogenic impacts on sulfonamide residues and sulfonamide resistant bacteria and genes in Larut and Sangga Besar River. Perak Sci Total Environ 688:1335–1347. https://doi.org/10.1016/j.scitotenv.2019.06.304

Matějíček D, Kubáň V (2008) Enhancing sensitivity of liquid chromatographic/ion-trap tandem mass spectrometric determination of estrogens by on-line pre-column derivatization. J Chromatogr A 1192:248–253. https://doi.org/10.1016/j.chroma.2008.03.061

Menon NG, Mohapatra S, Padhye LP, Sarma SV, Mukherji S (2020) Review on occurrence and toxicity of pharmaceutical contamination in Southeast Asia. In: Emerging issues in the water environment during anthropocene: a south east Asian perspective. Springer, pp 63–91

Mohapatra S, Padhye LP, Mukherji S (2018) Challenges in detection of antibiotics in wastewater matrix. In: Environmental contaminants: measurement, modelling and control, pp 3–20

Mutahir S, Khan IU, Khan MA, Riaz S, Hussain S, Jończyk J, Bajda M, Ullah N, Ashraf M, Yar M (2016) Novel biphenyl bis-sulfonamides as acetyl and butyrylcholinesterase inhibitors: synthesis, biological evaluation and molecular modeling studies. Bioorg Chem 64:13–20. https://doi.org/10.1016/j.bioorg.2015.11.002

Nebot C, Gibb SW, Boyd KG (2007) Quantification of human pharmaceuticals in water samples by high performance liquid chromatography-tandem mass spectrometry. Anal Chim Acta 598:87–94. https://doi.org/10.1016/j.aca.2007.07.029

Pamreddy A, Hidalgo M, Havel J, Salvadó V (2013) Determination of antibiotics (tetracyclines and sulfonamides) in biosolids by pressurized liquid extraction and liquid chromatography–tandem mass spectrometry. J Chromatogr A 1298:68–75

Pan X, Qiang Z, Ben W, Chen M (2011) Simultaneous determination of three classes of antibiotics in the suspended solids of swine wastewater by ultrasonic extraction, solid-phase extraction and liquid chromatography-mass spectrometry. J Environ Sci 23:1729–1737. https://doi.org/10.1016/S1001-0742(10)60590-6

Qian S, Yan L, MingYue L, Ashfaq M, Min L, HongJie W, Anyi H, ChangPing Y (2016) PPCPs in Jiulong river estuary (China): spatiotemporal distributions, fate, and their use as chemical markers of wastewater. Chemosphere 150:596–604

Raich-Montiu J, Beltrán JL, Prat MD, Granados M (2010) Studies on the extraction of sulfonamides from agricultural soils. Anal Bioanal Chem 397:807–814. https://doi.org/10.1007/s00216-010-3580-4

Renew JE, Huang CH (2004) Simultaneous determination of fluoroquinolone, sulfonamide, and trimethoprim antibiotics in wastewater using tandem solid phase extraction and liquid chromatography-electrospray mass spectrometry. J Chromatogr A 1042:113–121. https://doi.org/10.1016/j.chroma.2004.05.056

Rosa Boleda M, Huerta-Fontela M, Ventura F, Galceran MT (2011) Evaluation of the presence of drugs of abuse in tap waters. Chemosphere 84:1601–1607. https://doi.org/10.1016/j.chemosphere.2011.05.033

Salvia MV, Vulliet E, Wiest L, Baudot R, Cren-Olivé C (2012) Development of a multi-residue method using acetonitrile-based extraction followed by liquid chromatography-tandem mass spectrometry for the analysis of steroids and veterinary and human drugs at trace levels in soil. J Chromatogr A 1245:122–133. https://doi.org/10.1016/j.chroma.2012.05.034

Sarmah AK, Meyer MT, Boxall ABA (2006) A global perspective on the use, sales, exposure pathways, occurrence, fate and effects of veterinary antibiotics (VAs) in the environment. Chemosphere 65:725–759. https://doi.org/10.1016/j.chemosphere.2006.03.026

Schlipköter U, Flahault A (2010) Communicable diseases: achievements and challenges for public health. Public Health Rev 32:90–119

Shaaban H, Górecki T (2012) Optimization and validation of a fast ultrahigh-pressure liquid chromatographic method for simultaneous determination of selected sulphonamides in water samples using a fully porous sub-2 μm column at elevated temperature. J Sep Sci 35:216–224. https://doi.org/10.1002/jssc.201100754

Shimizu A, Takada H, Koike T, Takeshita A, Saha M, Rinawati Nakada N, Murata A, Suzuki T, Suzuki S, Chiem NH, Tuyen BC, Viet PH, Siringan MA, Kwan C, Zakaria MP, Reungsang A (2013) Ubiquitous occurrence of sulfonamides in tropical Asian waters. Sci Total Environ 453:108–115

Singer RS, Finch R, Wegener HC, Bywater R, Walters J, Lipsitch M (2003) Antibiotic resistance-the interplay between antibiotic use in animals and human beings. Lancet Infect Dis 3:47–51

Sinthuchai D, Boontanon SK, Boontanon N, Polprasert C (2016) Evaluation of removal efficiency of human antibiotics in wastewater treatment plants in Bangkok. Thailand Water Sci Technol 73:182–191. https://doi.org/10.2166/wst.2015.484

Soran ML, Lung I, Opriş O, Floare-Avram V, Coman C (2017) Determination of antibiotics in surface water by solid-phase extraction and high-performance liquid chromatography with diode array and mass spectrometry detection. Anal Lett 50:1209–1218. https://doi.org/10.1080/00032719.2016.1209516

Sun L, Chen L, Sun X, Du X, Yue Y, He D, Xu H, Zeng Q, Wang H, Ding L (2009) Analysis of sulfonamides in environmental water samples based on magnetic mixed hemimicelles solid-phase extraction coupled with HPLC-UV detection. Chemosphere 77:1306–1312. https://doi.org/10.1016/j.chemosphere.2009.09.049

Sun L, Sun X, Du X, Yue Y, Chen L, Xu H, Zeng Q, Wang H, Ding L (2010) Determination of sulfonamides in soil samples based on alumina-coated magnetite nanoparticles as adsorbents. Anal Chim Acta 665:185–192. https://doi.org/10.1016/j.aca.2010.03.044

Sun N, Han Y, Yan H, Song Y (2014) A self-assembly pipette tip graphene solid-phase extraction coupled with liquid chromatography for the determination of three sulfonamides in environmental water. Anal Chim Acta 810:25–31. https://doi.org/10.1016/j.aca.2013.12.013

Tetzner NF, Maniero MG, Rodrigues-Silva C, Rath S (2016) On-line solid phase extraction-ultra high performance liquid chromatography-tandem mass spectrometry as a powerful technique for the determination of sulfonamide residues in soils. J Chromatogr A 1452:89–97. https://doi.org/10.1016/j.chroma.2016.05.034

Tong F, Zhang Y, Chen F, Li Y, Ma G, Chen Y, Liu K, Dong J, Ye J, Chu Q (2013) Hollow-fiber liquid-phase microextraction combined with capillary electrophoresis for trace analysis of sulfonamide compounds. J Chromatogr B 942–943:134–140. https://doi.org/10.1016/j.jchromb.2013.10.038

Tran NH, Hoang L, Nghiem LD, Nguyen NMH, Ngo HH, Guo W, Trinh QT, Mai NH, Chen H, Nguyen DD, Ta TT, Gin KYH (2019) Occurrence and risk assessment of multiple classes of antibiotics in urban canals and lakes in Hanoi. Vietnam Sci Total Environ 692:157–174. https://doi.org/10.1016/j.scitotenv.2019.07.092

Tso J, Dutta S, Inamdar S, Aga DS (2011) Simultaneous analysis of free and conjugated estrogens, sulfonamides, and tetracyclines in runoff water and soils using solid-phase extraction and liquid chromatography-tandem mass spectrometry. J Agric Food Chem 59:2213–2222. https://doi.org/10.1021/jf104355x

Vosough M, Mashhadiabbas Esfahani H (2013) Fast HPLC-DAD quantification procedure for selected sulfonamids, metronidazole and chloramphenicol in wastewaters using second-order calibration based on MCR-ALS. Talanta 113:68–75. https://doi.org/10.1016/j.talanta.2013.03.049

Wang Q, Wang P, Yang Q (2018) Science of the total environment occurrence and diversity of antibiotic resistance in untreated hospital wastewater. Sci Total Environ 621:990–999. https://doi.org/10.1016/j.scitotenv.2017.10.128

WHO (2018) WHO report on surveillance of antibiotic consumption

Winum JY, Maresca A, Carta F, Scozzafava A, Supuran CT (2012) Polypharmacology of sulfon-amides: Pazopanib, a multitargeted receptor tyrosine kinase inhibitor in clinical use, potently inhibits several mammalian carbonic anhydrases. Chem Commun 48:8177–8179. https://doi.org/10.1039/c2cc33415a

Yi X, Lin C, Ong EJL, Wang M, Zhou Z (2019) Occurrence and distribution of trace levels of antibiotics in surface waters and soils driven by non-point source pollution and anthropogenic pressure. Chemoisphere 216:213–223

Yim G, Wang HH, Davies J (2006) The truth about antibiotics. Int J Med Microbiol 296:163–170. https://doi.org/10.1016/j.ijmm.2006.01.039

Zajdel P, Marciniec B, Pawlowski W (2014) Analogs of Aripiprazole : novel antipsychotic agents ? Futur Med Chem 57–75

Zhang T, Li B (2011) Occurrence, transformation, and fate of antibiotics in municipal wastew-ater treatment plants. Crit Rev Environ Sci Technol 41:951–998. https://doi.org/10.1080/10643380903392692

Zhang Y, Messhnick SR (1991) Inhibition of *Plasmodium falciparum* dihydropteroate synthetase and growth in vitro by sulfa drugs. Antimicrob Agents Chemother 35:267–271. https://doi.org/10.1128/AAC.35.2.267

Zhao Y, Wu R, Yu H, Li J, Liu L, Wang S, Chen X, Chan TD (2020) Magnetic solid-phase extraction of sulfonamide antibiotics in water and animal-derived food samples using core-shell magnetite and molybdenum disulfide nanocomposite adsorbent. J Chromatogr A 1610:460–543. https://doi.org/10.1016/j.chroma.2019.460543

Zhou JL, Maskaoui K, Lufadeju A (2012a) Optimization of antibiotic analysis in water by solid-phase extraction and high performance liquid chromatography-mass spectrometry/mass spectrometry. Anal Chim Acta 731:32–39. https://doi.org/10.1016/j.aca.2012.04.021

Zhou LJ, Ying GG, Liu S, Zhao JL, Chen F, Zhang RQ, Peng FQ, Zhang QQ (2012b) Simultaneous determination of human and veterinary antibiotics in various environmental matrices by rapid res-olution liquid chromatography-electrospray ionization tandem mass spectrometry. J Chromatogr A 1244:123–138. https://doi.org/10.1016/j.chroma.2012.04.076

Chapter 15
Fate of Micropollutants in Engineered and Natural Environment

Tejaswini Eregowda and Sanjeeb Mohapatra

15.1 Introduction

The consumption of synthetic compounds, such as nutrition supplements, natural and synthetic hormones, artificial sweeteners, pharmaceuticals and personal care products (PPCPs), detergents, industrial chemicals, and products, pesticides, biocides, disinfectants, cleaning agents, solvents and other chemicals, collectively termed as micropollutants, has seen an exponential increase at the turn of this century. The term micropollutants encompass both legacy contaminants (established toxic effects and control measures) and the emerging contaminants (currently unregulated and potential threat to the environment and human health and safety) (Yang et al. 2014; Singh et al. 2019). The USA has the largest pharmaceutical market in the world with a value of $333, 694 million followed by Japan ($94, 025 million) and China ($86, 774 million). Between 2003 and 2011, the production of pharmaceuticals in China has increased to 2 million tones which were around 20% of the global average (Liu and Wong 2013). The percent contribution to the total pharmaceuticals production in China from 2003 to 2011 is shown in Fig. 15.1a. However, in Europe, Germany's pharmaceutical market is valued highest at around $45,828 million, followed by France at $37,156 million. The average per capita pharmaceuticals consumption per day in Western Europe is around 300 mg of which 60 pharmaceuticals majorly contribute 90% of the mass. In Europe, around 3000 different types of pharmaceuticals are commonly sold at various pharmaceuticals outlets (Margot et al. 2015). Similarly, the USA, Japan, and China are the top three countries with the highest consumption of personal care products (PCPs). With an average increase in the production rate

T. Eregowda
IHE Delft Institute of Water Education, Delft, The Netherlands

S. Mohapatra (✉)
Environmental Science and Engineering Department, Indian Institute of Technology Bombay, Mumbai, India
e-mail: sanjeeb.publications@gmail.com

© Springer Nature Singapore Pte Ltd. 2020
M. Kumar et al. (eds.), *Resilience, Response, and Risk in Water Systems*,
Springer Transactions in Civil and Environmental Engineering,
https://doi.org/10.1007/978-981-15-4668-6_15

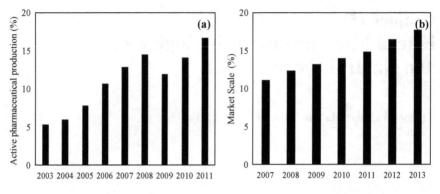

Fig. 15.1 Percentage increase **a** active pharmaceutical ingredient production (2003–2011) and **b** market scale of PCPs (2007–2013) in China (Reprinted with permisssion from Liu and Wong, copyright Elsevier, 2013)

of 8% between 2011 and 2013, China's average PCPs market value reached $21.3 billion, accounting for about 10% of the world (Liu and Wong 2013) (Fig. 15.1b).

The main source of micropollutants in water bodies is wastewater (domestic and industrial). The design and operation of existing wastewater/effluent treatment facilities are inadequate to handle micropollutants. Furthermore, awareness and instructions for monitoring and removal are not established (Luo et al. 2014; Kumar et al. 2019b). As a result, of improper treatment, most of the micropollutants inevitably wind up in aquatic bodies and sediments, strongly impact the environment. The fact that the micropollutants range in a concentration of parts per trillion to billion (ng/L to μg/L) hinders their detection, analysis, and treatment (Mohapatra et al. 2018). Furthermore, with an advance in the chemical technology sector, there is a continuous inclusion of new synthetic compounds to the family of micropollutants. Since 1998, the US EPA has been releasing contaminant candidate lists (CCL) belonging to various categories of environmental concern. The number of chemical contaminants made into the CCL list in the year 1998, 2005, 2009, and 2016 was 50, 42, 116, and 97, respectively. The European CAS database has registered >144 million (nearly 300 million tonnes) of synthetic chemicals that potentially find their way into aquatic bodies (Schwarzenbach et al. 2006; Kumar et al. 2019a). The occurrence of micropollutants in the water bodies is an increasing concern not only because of their adverse impact but for the unforeseen chronic toxic effect on the ecosystem.

15.2 Origin and Classification of the Micropollutants

Micropollutants can be classified in several ways based on their origin, use, potential effects, or environmental fate. Some major groups are: (a) pharmaceutical and veterinary products, (b) disinfectants and biocides, (c) illicit drugs, (d) personal care chemicals and other lifestyle products, (e) industrial chemicals, (f) food additives,

(g) water disinfection by-products, (h) nanomaterials, (i) waterborne pathogens, (j) biological toxins, and (k) heavy metals. A list of micropollutants belonging to various classes and their sources are shown in Table 15.1. EU FP7 project provides a comprehensive list of 242 chemicals as micropollutants, among which 70% are pharmaceuticals and personal care products (PPCP) and 30% are industrial agents including perfluoro compounds, pesticides, herbicides, and food additives (Schueth 2014). PPCPs contributors in the wastewater are household (70%), livestock farming (20%), hospital effluent (5%), and non-particular runoff (5%) (Das et al. 2016). Similarly, US EPA published a list of 97 organic pollutants relevant to drinking water quality and their common application under Contaminant Candidate List-4 (CCL-4) (Table 15.2). Such contaminants fall under the category of chemicals used in commerce, pesticides, biological toxins, disinfection by-products (DBPs), and PPCPs.

Table 15.1 Classification of Micropollutants (Adapted with permission from Luo et al., copyright Elsevier, 2014)

Category	Subclasses	Sources
Pharmaceuticals	Antibiotics, β-blockers, NSAIDs, antiepileptic, psychoactive, anesthetic, antihistamine, anticoagulant	Household discharge (through excretion) and medicine discards Hospital effluent Illegal discharges from pharmaceuticals industries Animal farming, aquaculture facility
Hormones	Lipid-soluble Amino acid-derived hormones Peptide hormones Glycoprotein	Household discharge (through excretion) Animal farming
Personal care products	Fragrances, sweeteners, disinfectants, UV filters, and insect repellents	Domestic wastewater (cleaning, washing, bathing, spraying)
Pesticides	Organochlorine insecticides, organophosphorus insecticides, herbicides and fungicides, algaecide	Agricultural runoff Discharges from gardens, lawns
Industrial chemicals	Solvents, intermediates, petrochemicals	Industries
Industrial products	Additives, flame retardants, lubricants, plasticizers	Industries
Surfactants	Anionic Non-ionic	Domestic wastewater Laundry facilities
Heavy metals	As, Cr, Cd, Pb, etc.	Food containers, baby blankets, towels, children's toys, medical bandages, gym socks, t-shirts
Nanomaterials	Fullerenes, nanotubes, nZnO, nCeO$_2$, nAu, nFe(0)	Paint and pharmaceutical industries

Table 15.2 List of chemical contaminants and their application listed under the contaminant candidate list-4 (CCL-4) (Adapted from https://www.epa.gov/ccl/chemical-contaminants-ccl-4)

Substance name	Application
1,1-Dichloroethane	Industrial solvent
1,1,1,2-Tetrachloroethane	Industrial solvent
1,2,3-Trichloropropane	Industrial solvent
1,3-Butadiene	Production of rubber and plastics
1,4-Dioxane	Solvent for cellulose formulations, resins, oils, waxes, other organic substances, wood pulping, textile processing, degreasing, in lacquers, paints, varnishes, and stains; paint and varnish removers
17 alpha-estradiol	Estrogenic hormone
1-Butanol	As a solvent for several commercial products such as perfumes
2-Methoxyethanol	Synthetic cosmetics, perfumes, fragrances, hair preparations, and skin lotions
2-Propen-1-ol	Production of other chemicals
3-Hydroxycarbofuran	Insecticide
4,4'-Methylenedianiline	Production of polyurethane foams, glues, rubber and spandex fiber
Acephate	Insecticide
Acetaldehyde	Disinfection by-product from ozonation and also used in the production of other chemicals
Acetamide	Solvent and plasticizer
Acetochlor	Herbicide
Acetochlor ethanesulfonic acid (ESA)	Environmental degradation of acetochlor
Acetochlor oxanilic acid (OA)	Environmental degradation of acetochlor
Acrolein	Herbicide, rodenticide and industrial chemical
Alachlor ethanesulfonic acid (ESA)	Environmental degradation of the pesticide alachlor
Alachlor oxanilic acid (OA)	Environmental degradation of acetochlor
alpha-Hexachlorocyclohexane	It is a component of benzene hexachloride (BHC) and was formerly used as an insecticide
Aniline	Industrial solvent for the synthesis of explosives, rubber products, and in isocyanates
Bensulide	Herbicide
Benzyl chloride	It is used in the production of other substances, such as plastics, dyes, lubricants, gasoline, and pharmaceuticals

(continued)

Table 15.2 (continued)

Substance name	Application
Butylated hydroxyanisole	Food additive (antioxidant)
Captan	Fungicide
Chlorate	Used as defoliants or desiccants and may occur in drinking water because of the use of disinfectants such as chlorine dioxide and hypochlorites
Chloromethane (Methyl chloride)	Foaming agent
Clethodim	Herbicide
Cobalt	Used as cobaltous chloride in medicines and as a germicide
Cumene hydroperoxide	It is used as a catalyst
Cyanotoxins	Toxins naturally produced and released by cyanobacteria ("blue-green algae"). The group of cyanotoxins includes, but is not limited to anatoxin-a, cylindrospermopsin, microcystins, and saxitoxin
Dicrotophos	Insecticide
Dimethipin	Herbicide and plant growth regulator
Diuron	Herbicide
Equilenin	Estrogenic hormone used in hormone replacement therapy
Equilin	Estrogenic hormone used in hormone replacement therapy
Erythromycin	Antibiotic
Estradiol (17-beta estradiol)	An isomer of estradiol found in some pharmaceuticals
Estriol	Estrogenic hormone used in veterinary pharmaceuticals
Estrone	Precursor of estradiol used in veterinary and human pharmaceuticals
Ethinyl estradiol (17-alpha ethynyl estradiol)	Estrogenic hormone used in veterinary and human oral contraceptives
Ethoprop	Insecticide
Ethylene glycol	Antifreeze, pesticide
Ethylene oxide	Fungicidal and insecticidal fumigant
Ethylene thiourea	Vulcanizing polychloroprene (neoprene) and polyacrylate rubbers. It is a metabolite of some fungicides
Formaldehyde	Ozonation disinfection by-product can occur naturally and has been used as a fungicide

(continued)

Table 15.2 (continued)

Substance name	Application
Germanium	Used in transistors and diodes, and electroplating. Sold as a dietary supplement in some cases
HCFC-22	Refrigerant, as a low-temperature solvent, and in fluorocarbon resins, especially in tetrafluoroethylene polymers
Halon 1011 (bromochloromethane)	Fire-extinguishing fluid to suppress explosions, a solvent in the manufacturing of some pesticides. May also occur as a disinfection by-product in drinking water
Hexane	Component of gasoline and used as a solvent
Hydrazine	Production of rocket propellants, and plastics
Manganese	Steel production to improve hardness, stiffness and strength
Mestranol	Precursor to ethinylestradiol used in veterinary and human pharmaceuticals
Methamidophos	Insecticide
Methanol	Industrial solvent, a gasoline additive and as an antifreeze ingredient
Methyl bromide (bromomethane)	Fumigant and fungicide
Methyl tert-butyl ether (MTBE)	Octane booster in gasoline, in the manufacturing of isobutene and as an extraction solvent
Metolachlor	Herbicide
Metolachlor ethanesulfonic acid (ESA)	Environmental degradate of metolachlor
Metolachlor oxanilic acid (OA)	Environmental degradate of metolachlor
Molybdenum	As a steel alloy. It is an essential dietary nutrient found in foods and nutritional supplements
Nitrobenzene	Production of aniline and also as a solvent in the manufacturing of paints, shoe polishes, floor polishes, metal polishes, explosives, dyes, pesticides, and drugs (such as acetaminophen)
Nitroglycerin	Production of explosives, and in rocket propellants and pharmaceutical for the treatment of angina
N-Methyl-2-pyrrolidone	Solvent in the chemical industry, and is used in the formulation of pharmaceuticals for oral and dermal delivery

(continued)

Table 15.2 (continued)

Substance name	Application
N-nitrosodiethylamine (NDEA)	Additive in gasoline and lubricants, as an antioxidant and as a stabilizer in plastics. It is formed in cured foods and during high-temperature cooking of meats and fish, and maybe a disinfection by-product
N-nitrosodimethylamine (NDMA)	Production of rocket fuels, antioxidants and softeners for copolymers. It is formed in cured foods and during high-temperature cooking
N-nitroso-di-n-propylamine (NDPA)	Formed in cured foods and during high-temperature cooking of meats and fish and maybe a disinfection by-product. It is a contaminant in dinitrofluralin herbicides
N-Nitrosodiphenylamine	Vulcanization of rubber and as an inhibitor of polymerization in the production of polystyrene
N-nitrosopyrrolidine (NPYR)	Rubber production. It is also formed in cured foods and during high-temperature cooking of meats and fish and maybe a disinfection by-product
Nonylphenol2	Manufacture of nonylphenol ethoxylates, commonly found in laundry detergents, cleaners, degreasers, paints and coatings
Norethindrone (19-Norethisterone)	Synthetic hormone used in oral contraceptives and for hormone replacement therapy
n-Propylbenzene	A constituent of asphalt and naphtha and used in the manufacture of methyl styrene and as a solvent for printing and dyeing of textiles
o-Toluidine	Production of dyes, rubber, pharmaceuticals, and pesticides
Oxirane, methyl	Industrial chemical (pesticide)
Oxydemeton-methyl	Insecticide
Oxyfluorfen	Herbicide
Perfluorooctanesulfonic acid (PFOS)	To make carpets, leathers, textiles, fabrics for furniture, paper packaging, and other materials that are resistant to water, grease, or stains. It is also used in firefighting foams at airfields
Perfluorooctanoic acid (PFOA)	To make carpets, leathers, textiles, fabrics for furniture, paper packaging, and other materials that are resistant to water, grease, or stains. It is also used in firefighting foams at airfields
Permethrin	Insecticide

(continued)

Table 15.2 (continued)

Substance name	Application
Profenofos	Insecticide and an acaricide
Quinoline	Component of coal tars and as a pharmaceutical (antimalarial)
RDX (Hexahydro-1,3,5-trinitro-1,3,5-triazine)	Explosive
sec-Butylbenzene	As a solvent for coatings in organic synthesis, as a plasticizer and in surfactants
Tebuconazole	Fungicide
Tebufenozide	Insecticide
Tellurium	As a sodium tellurite in bacteriology and medicine
Thiodicarb	Insecticide
Thiophanate-methyl	Fungicide
Toluene di-isocyanate	Manufacturing of plastics
Tribufos	Insecticide and cotton defoliant
Triethylamine	Production of chemicals, as a stabilizer in herbicides and pesticides, in consumer products, in photographic chemicals, and carpet cleaners
Triphenyltin hydroxide (TPTH)	Pesticide
Urethane	Paint and coating ingredient (polyurethanes)
Vanadium	As a catalyst
Vinclozolin	Fungicide
Ziram	Fungicide

15.3 The Fate of Micropollutants in Water/Wastewater

The input of the micropollutants to the wastewater depends on several factors such as the use of the consumable products, seasonal variation, disposal methods, rainfall pattern, and lifestyle choice of the population. Climatic conditions could also affect the overall concentration of micropollutants to a greater extent (Kolpin et al. 2004). Use of pesticides is usually regional and seasonal due to the prevalence of pests in different agro-climatic conditions. Rainfall is another important factor that affects the flow pattern of wastewater when a combined sewer system is employed. The concentration of most PPCPs in the raw wastewater was doubled when the flow was halved during dry weather, suggesting that rainwater diluted the concentrations of the compounds in the sewage (Kasprzyk-Hordern et al. 2009). Other weather conditions, such as temperature and intensity of sunlight, also can affect the fate of micropollutants at STPs.

For PPCPs, the concentrations in wastewater correlated well with their production amounts and usage/consumption patterns. For orally ingested products containing potential contaminants (pharmaceuticals), rate of excretion plays a crtical role in determining the introduction of pharmaceuticals into raw wastewater as they are metabolized in the human body and are subsequently excreted. The rate of excretion may be less than 5% for acetylsalicylic acid, carbamazepine, gemfibrozil, ibuprofen, 5–40% for diclofenac, metoprolol, primidone, sulfamethoxazole, 40–70% for bezafibrate, norfloxacin, trimethoprim, and more than 70% for various antibiotics such as amoxicillin, ciprofloxacin, and tetracycline (Luo et al. 2014). Pharmaceuticals with low excretion rates are not necessarily present at low levels in raw wastewater, possibly because low excretion rates are offset by the massive use of these compounds (Luo et al. 2014). Moreover, incidence of common diseases can induce a higher consumption of specific pharmaceuticals in certain periods.

Decrease of the micropollutants during wet conditions and an increase in the occurrence level and concentration during dry weather conditions have been consistently reported. Pharmaceuticals in summer water samples showed lower occurrence than in winter, possibly due to increased biodegradation in warmer temperature and elevated dilution during wet summer (Luo et al. 2014). However, rainfall did not always reduce the concentration levels of micropollutants. For instance, few studies revealed that the chemicals (e.g., bisphenol A and biocides) used in building material (pavement materials, facades, and roof paintings) leached in rain and accumulated to remarkable levels in roof runoff and subsequently ended up in surface water (Kolpin et al. 2004).

Existing STPs usually employ a process consisting of primary, secondary, and occasionally tertiary treatment for the optimal removal of suspended solids, dissolved organics, and nutrients from the wastewater. In primary treatment, colloidal and suspended particles are removed using coagulants such as alum, ferric chloride, and polymers and polymeric coagulant aids, during which, the organic compounds attached with dissolved humic substances are removed. Secondary treatment aims at aerobic removal of dissolved organics using a consortium of microorganisms in suspension. Sludge from both primary and secondary clarifiers is digested anaerobically before disposal. Additionally, a tertiary treatment such as activated carbon adsorption, ozonation, or filtration is used as a final polishing step to improve the water quality before reuse, recycle, or discharging to the environment.

Micropollutants in an STP usually get adsorbed on suspended particles, humic substances, primary and secondary sludge and are removed by processes such as coagulation and sedimentation, biodegradation, and adsorption. Adsorption of micropollutants on suspended solids (during both primary and secondary treatment) is important in determining the fate of micropollutants in wastewater. Adsorption occurs due to the hydrophobic interaction between the aliphatic and aromatic groups of the fat and lipid fractions in primary sludge and the lipophilic cell membrane of the microorganisms in secondary sludge. Furthermore, electrostatic interactions also occur between the positively charged micropollutants and the negatively charged microorganisms in secondary sludge. Several acidic pharmaceuticals are negatively charged at neutral pH, and their sorption on sludge is negligible (Das et al. 2016). The sorption of pharmaceuticals and antibiotics widely varies due to the non-polar core and polar moiety and is difficult to estimate (Kinney et al. 2006). PPCPs show a varied property compared to conventional persistent organic pollutants whose source has been banned or limited, making it more difficult to estimate, monitor, and treat. Adsorption of micropollutants is also affected by pH of the system (e.g., sulfonamide antibiotics). Besides, the desorption from the sludge seemed to be reversible, i.e., sorbed micropollutants can be introduced into the environment if no further treatment is employed to remove them from the biomass (Yang et al. 2011).

Although studies show that the removal of micropollutants, especially PCCPs by the conventional method, is limited (Santos et al. 2007), the process of biodegradation and adsorption is mainly responsible for the removal of PPCPs in an STP. Biodegradation of PPCPs by active sludge is affected by factors, such as initial substrate concentration, temperature, and biodegradation process type (Liu and Wong 2013). A study comparing the biodegradation of 12 PPCPs (in 2 STPs in Beijing, China) by three processes, viz. conventional activated sludge process (ASP), biological nutrient removal and membrane reactor (MBR) revealed that MBR showed the maximum removal (Sui et al. 2011). As a result of ineffective and unregulated treatment, the micropollutants in the effluent from STP are released to the water bodies.

15.4   Environmental Effects of the Micropollutants

The occurrence and persistence of a wide range of micropollutants in different environmental medium strongly raised concern about their potential threat to the environment and human health. Besides, few studies pointed out severe adverse effects on wildlife population, such as the residues of veterinary diclofenac causing a significant decline of vulture population (Oaks et al. 2004), the risk of acute toxic effects of the PPCPs is believed to be unlikely based on the toxicological data and environmental concentration levels. However, the chronic effects of micropollutants cannot be disregarded. The continuous input of micropollutants in the environment results in their accumulation and irreversible harm to the entire ecosystem (Brausch and Rand 2011). Chronic exposure of antibiotics to the environment can accelerate persistence or emergence of antibiotic resistance genes (ARGs) that encode resistance to a broad range of antibiotic species (Zhang et al. 2009), resulting in a "superbug" (a multidrug-resistant bacteria such as methicillin-resistant *Staphylococcus aureus*, MRSA).

Hormones possess 10,000–100,000 times higher estrogenic potency compared with exogenous endocrine-disrupting chemicals (Khanal et al. 2006) which are the most significant estrogenic compounds commonly present in sewage effluents (Jobling et al. 2006). Several aquatic species, including but not limited to, crucian carp, trout, minnow, and turtle, have been reported to be sexually inhibited or reversed in the presence of estrogens hormones. Endocrine disruptors can induce a wide range of reproductive and developmental issues including reduced fertility, the feminization of males, induction of vitellogenesis in male (plasma vitellogenin induction), and development of intersex individuals (Liu et al. 2012). Along with antibiotics and endocrine disruption by hormones, gemfibrozil (blood lipid regulator), triclosan, and triclocarban (antimicrobial agents) are reported to inhibit the growth of algae. Caffeine (stimulant drug) results in endocrine disruption of goldfish (Liu et al. 2012), and propranolol (β-blocker) can reduce the viable eggs of Japanese medaka. Carbamazepine (antiepileptic drug) and HHCB (polycyclic musk) may result in oxidation stress to rainbow trout and goldfish. Diclofenac (a non-steroidal anti-inflammatory drug) may cause renal lesions and gill alterations to rainbow trout (Liu and Wong 2013). Carbamazepine and diclofenac are reported to have antiestrogenic effects based on in vitro studies conducted with estrogen receptor-positive human breast cancer cell line, MCF-7 (Mishra et al. 2018). Similarly, species sensitivity distribution (SSD) conducted by Menon et al. (2020) predicted reproductive failure and/or vitellogenin induction in aquatic species at ethinylestradiol (EE2) concentration of the order of ng/L. For sulfamethoxazole, carbamazepine, and atenolol, such effects were only observed in the concentration range of μg/L. Several groups of micropollutants such as heavy metals, inorganic compounds, and PPCPs (UV filters, disinfectants, and synthetic musks) are likely to bioaccumulate (Brausch and Rand 2011). Additionally, a synergistic effect of these toxic chemicals is of substantial concern. Compared to the effects measured individually, tests with combinations of various pharmaceuticals (carbamazepine, diclofenac, and ibuprofen) had stronger effects on the target aquatic organism, i.e., *Daphnia magna* than anticipated (Cleuvers 2004). To make the matter worse, the concern about the bioaccumulation and biomagnification of the micropollutants in the aquatic food web undermines the above-mentioned issues.

15.5 Treatment

Conventional STPs are although effective for the treatment of wastewater in terms of suspended particles, dissolved organics, and nutrients, they are inefficient for micropollutants. Given their diverse properties, a specific treatment will not suffice the removal of various micropollutants. Biodegradation and adsorption are very efficient for the removal of a few groups of PPCPs such as fluoroquinolones, hormones, caffeine, antimicrobial agents, and preservatives. However, groups such as macrolides and sulfonamides, penicillin, fluconazole, and carbamazepine are recalcitrant to the conventional treatment processes and are subsequently discharged to the water bodies. Based on their percentage removal in a conventional STP, in fact, pharmaceuticals can be broadly categorized into three classes. PPCPs such as carbamazepine, metoprolol, and diclofenac are poorly removed (<40%), trimethoprim, ketoprofen, clofibric acid, sulfamethoxazole, atenolol, and nonylphenol are moderately removed (40–70%) while estrone, bisphenol A, triclosan, naproxen, gemfibrozil, caffeine, and ibuprofen are removed at a greater extent (>70%) (Luo et al. 2014). Overall, for the treatment of micropollutants, especially PPCPs, advanced technologies need to be introduced (Liu and Wong 2013).

15.5.1 Activated Carbon Adsorption

Adsorption by activated carbon, a commonly employed method for controlling taste and odor in drinking water, has potential for the treatment of secondary effluent and also shown effective in removing micropollutants in comparison with the coagulation–flocculation process. The adsorption is affected by the properties of both adsorbates (K_{OW}, pKa, molecular size, aromaticity versus aliphaticity, and presence of specific functional groups) and adsorbent (surface area, pore size and texture, surface chemistry, and mineral matter content) (Eregowda et al. 2019).

Both powdered activated carbon (PAC) and granular activated carbon (GAC) have been widely used in adsorption processes. PAC is an effective adsorbent for treating persistent/non-biodegradable organic compounds. The main application of PAC is its addition in activated sludge tank or post-treatment configurations in the full-scale STP, which showed to reduce the micropollutants by more than 80% (Luo et al. 2014). An advantage of employing PAC is that it can provide fresh carbon continuously or can be used seasonally or occasionally when the risk of trace organics is present at a high level. Considerable removals of steroidal estrogens from sewage effluent were observed during the GAC tertiary treatment. By comparison, the reduction of pharmaceutical concentrations varied. Higher removal (84–99%) was observed for mebeverine, indomethacin, and diclofenac, while the removal was low for compounds such as carbamazepine and propranolol (17–23%) (Grover et al. 2011).

15.5.2 Attached Growth Treatment

There are several factors which make attached growth-based treatment technology distinct from conventional activated sludge-based processes. Some of these factors include better oxygen transfer, high nitrification rate, higher biomass concentrations, effective removal of organic matter at higher organic loading rate and shorter HRT, development of microorganisms at relatively low specific growth rates (e.g., methanogens), and better performance under variable or intermittent loadings rates. More importantly, attached growth-based treatment is a compelling biological technique for micropollutant removal (Luo et al. 2014). Several laboratory/pilot-scale attempts were made to evaluate the potential of the attached growth system for removal of micropollutants including endocrine-disrupting hormones, 17α-ethinylestradiol. In an aerated submerged fixed-bed bioreactor (ASFBBR), 96% of this compound was removed at an initial loading concentration of 11 μg/L. However, a reduced removal of 81 and 74% was obtained when the loading was increased to 40 and 143 μg/L, respectively. However, retro-fitting such treatment systems by employing ammonia-oxidizing bacteria (AOB) has excellent potential for the removal of micropollutants (Forrez et al. 2009). In such a system, micropollutant removal was solely governed by co-metabolism and the removal can be enhanced through the addition of packing/moving carriers which will facilitate the growth of slow-growing specific microorganisms. However, additional research is necessary to make attached growth processes an excellent treatment alternative for effective removal of micropollutants.

15.5.3 Membrane Process

Removal of micropollutants during the membrane process is governed by several mechanisms including size exclusion, adsorption on the membrane, and charge repulsion. However, such mechanisms are controlled by a set of factors such as type of membrane process, characteristics of the membrane, and operating conditions, characteristics of micropollutant, and membrane fouling (Schäfer et al. 2011). Even though microfiltration (MF) and ultrafiltration (UF) can efficiently remove water/wastewater turbidity, micropollutants are reported to be poorly removed due to their lower molecular size compared to the pore size of the membranes used in UF and MF. Partial removal of micropollutants in such systems attributed to adsorption of contaminants on to the membrane polymers and hydrogen bridging, and Van der Waals force of attraction with the organic matter present in water/wastewater. However, nanofiltration (NF) and reverse osmosis (RO) are reported to remove micropollutants to a greater extent due to their inherent "tighter" structures. In contrast, some ionized micropollutants can pass through such membranes as reported elsewhere (Forrez et al. 2009; Steinle-Darling et al. 2010).

15.5.4 Membrane Bioreactor

Membrane bioreactor (MBR)-based STPs are gaining more acceptability compared to conventional STPs due to their high effluent quality, excellent microbial separation ability, absolute control of SRTs and HRTs, high biomass content and less sludge bulking problem, and low-rate sludge production. Such MBRs can effectively remove a wide spectrum of micropollutants including compounds that are partially removed during activated sludge processes (ASP) (Radjenović et al. 2009; Spring et al. 2007). When the removal of 57 micropollutants in an MBR was compared with ASP, higher removal in MBR was attributed to higher retention of sludge to which many micropollutants adhered, interception of micropollutants by membrane surface and longer SRT in MBRs that promoted microbial degradation of these compounds (Goswami et al. 2018). Higher biodegradation rate in the MBR led to significant removal of bezafibrate, ketoprofen, and atenolol while triclocarban, ciprofloxacin, levofloxacin, and tetracycline were primarily removed by adsorption on to MBR sludge.

15.5.5   Advanced Oxidation Processes (AOPs)

AOPs majorly involve the generation of hydroxyl radical (OH$^\bullet$) and other non-selective oxidants having high reaction potential of the order 2.8 V either in the presence or absence of light sources followed by degradation of various contaminants present in the water or wastewater matrices (Silva et al. 2017). Broadly, it is categorized into homogenous or heterogeneous depending upon the reacting phases. In case of homogenous systems such as O_3, UV/O_3, UV/O_3/H_2O_2, UV/H_2O_2, Fe^{2+}/H_2O_2, and UV/Fe^{2+}/H_2O_2, both catalyst and substrate or only substrate exist in a single phase while in the heterogeneous system catalyst and substrate exist in two different phases (Muruganandham et al. 2014). In the later system such as TiO_2/O_2/UV and TiO_2/H_2O_2/UV processes, catalyst mostly exists in solid form. Based on various routes of radical generation, AOPs are again classified as chemical, electrochemical, sonochemical, and photochemical processes (Ortiz et al. 2019). Several authors have explored the application of the above-mentioned techniques for the removal of micropollutants in water and wastewater as discussed in this section.

Among the chemical methods, ozonation is used as a tertiary treatment process in various water treatment plants across the globe where the degradation of PPCPs takes place via direct oxidation (at low pH) and/or through OH$^\bullet$ generated (at high pH) due to decomposition of ozone (Ikehata and El-Din 2004). More than 90% removal of the parent compound was reported for some of the persistent micropollutants including carbamazepine, gemfibrozil, and DEET in a water treatment plant (Sui et al. 2011). During the O_3/H_2O_2 process, HO$^\bullet$ is formed via a chemical reaction between ozone and H_2O_2 (Eq. 15.1) and degradation of the contaminant is achieved by the combined actions of ozone and HO$^\bullet$, whereas during UV/O_3/H_2O_2 process, HO$^\bullet$ is generated by the photolysis of ozone in the aqueous medium (Eq. 15.2). In such a system, 100% removal for PPCPs including bisphenol A, β-blocker (metoprolol), antibiotics (ciprofloxacin and sulfamethoxazole) was seen at ozone, and H_2O_2 concentration of 5 and 1000 mg/L, respectively (Richard et al. 2014). Similarly, the process of HO$^\bullet$ formation during Fenton (Fe^{2+}/H_2O_2) and photo-Fenton processes (UV/Fe^{2+}/H_2O_2) is shown in Eqs. 15.1, and 15.4, respectively (Muruganandham et al. 2014; Oturan and Aaron 2014). Atrazine was reported to be removed more than 90% at an experimental Fe^{3+}, Fe^{2+}, and H_2O_2 concentration of 0.06–0.5, 0.28, and 0.14–5 mM, respectively, when the reactor studies were conducted for a period of 7 min at a fixed pH of 3 (De Laat et al. 1999). An extensive review article on the application of ozonation for the treatment of PPCPs is also available (Gomes et al. 2017). Although ozonation is widely used as a tertiary treatment process at water treatment plants and STPs, accumulation of degradation by-products due to low mineralization cannot be ignored. Some of these by-products may cause acute and chronic toxicity. In such a scenario, application of H_2O_2 or UV/O_3/H_2O_2 not only enhances the removal of PPCPs but also improves the mineralization.

$$H_2O_2 + 2O_3 \rightarrow 2OH^\circ + 3O_2 \qquad (15.1)$$

$$O_3 + hv + H_2O \rightarrow 2OH^\circ + O_2 \qquad\qquad (15.2)$$

$$Fe^{2+} + H_2O_2 \rightarrow OH^\circ + Fe^{3+} + OH^- \qquad\qquad (15.3)$$

$$Fe^{3+} + H_2O_2 + UV \rightarrow OH^\circ + Fe^{2+} + H^+ \qquad\qquad (15.4)$$

During heterogeneous photocatalysis, TiO_2 is irradiated with a light source such that the incident photon energy is equivalent to or greater than the energy corresponding to the bandgap of the nanoparticle. As a result of which the excited electron promoted to the conduction band, generating both negative electron (e^-) and a positive hole (h^+) in the valence band. As the hole carries high oxidative power, it not only oxidizes the absorbed water and produces $OH^\bullet$ but also oxidizes hydroxide ions, $OH^{\bullet-}$, or any available substrate (Cincinelli et al. 2015). Both bench and pilot-scale studies were conducted to explore the potential of this technology to treat a variety of micropollutants including diclofenac and naproxen (Fernández-Ibáñez et al. 2003). An attempt was made for 100% elimination of diclofenac in a system of $O_3/TiO_2/UVA$ at an initial O_3 and TiO_2 concentration of 10 and 1.5 g/L, respectively, irradiated at 313 nm for a period of 30 min (Aguinaco et al. 2012). Water and wastewater parameters, including the concentration of DOM, dissolved oxygen, and pH, cations, anions, and radical scavengers hinder the large-scale application of this technology. Additional research may be conducted with TiO_2 doped with zinc (Zn) or silver (Ag) to overcome this limitation to make the photocatalytic process more effective with complete mineralization using sunlight or visible light.

15.5.6 River Bed Filtration (RBF)

During RBF systems, water is forced to infiltrate into the subsurface towards a series of abstraction wells located near-surface water bodies such as river or lake. Temperature, nature, and concentration of organic matter, residence time, river flow, and redox conditions decide the fate of micropollutants during RBF. Depending upon the physicochemical properties of micropollutants, both biodegradation and sorption play a major role in such systems. The initial infiltration phase of RBF is usually the oxic zone that favors the degradation of readily biodegradable micropollutants such as phenazone, dimethoate, diuron, and metoprolol (Bertelkamp et al. 2016). Next to the oxic zone, i.e., in the nitrate-reducing and iron/manganese reducing zones, higher removal was reported for more than 70 micropollutants including atrazine, primidone, sulfamethoxazole, and iopamidol (Storck et al. 2012). Based on laboratory-scale column study, it was found that the presence of humic acids enhanced depletion of caffeine and 17-β estradiol regardless of the temperature and the level of oxygen. However, the attenuation was restricted to only 10 and 30% for carbamazepine and gemfibrozil, respectively (D'Alessio et al. 2015). When experiments were conducted for an extended period of 45 days under oxic, suboxic (partial nitrate removal) and anoxic (complete nitrate removal) conditions, higher removal was observed for most of the compounds in the oxic zones. Even though strong correlation could not be obtained between physicochemical properties (hydrophobicity, charge and molecular weight) or functional groups of pharmaceuticals with the varying redox conditions, consistently, the removal was low for persistent PPCPs such as atrazine, carbamazepine, hydrochlorothiazide, and simazine. Dimethoate, chloridazon, lincomycin, sulfamethoxazole, and phenazone have shown adaptive behavior, and such behavior has to be considered while modeling RBF (Bertelkamp et al. 2016).

Hamann et al. (2016) simulated the fate of 247 micropollutants during long-term/long-distance RBF at a temporal scale for several years by incorporating both linear adsorption and first-order rate equations. The modeled equation can be expressed as:

$$R\frac{\partial C}{\partial t} = D\frac{\partial^2 C}{\partial x^2} - \vartheta\frac{\partial C}{\partial x} - \lambda C$$

where R—retardation coefficient; C—aqueous concentration of a contaminant; x—spatial dimension in flow direction; t—time; ϑ—pore velocity; D—longitudinal dispersion coefficient; and λ—first-order degradation rate constant.

Retardation due to liner adsorption can be further expressed as:

$$R = 1 + \rho_b\frac{K_d}{\theta}$$

where K_d—distribution coefficient; ρ_b—bulk density; and θ—total porosity.

Similarly, Henzler et al. (2014) investigated the fate of several micropollutants during RBF combining two-dimensional numerical flow and reactive transport model. Out of twelve organic micropollutants, primidone, ethylenediaminetetraacetic acid (EDTA), and 1-acetyl-1-methyl2-dimethyl-oxamoyl-2-phenylhydrazide (AMDOPH) showed neither biodegradation nor sorption. Slight degradation was seen for 1.5 naphthalenedisulfonic acid (1.5 NDSA) and adsorbable organic halogens (AOX); carbamazepine degraded with a half-life time of about 66 days. The model was well fitted for clindamycin, phenazone, diclofenac, and sulfamethoxazole attributed to their characteristics temperature and redox conditions-dependent biodegradation.

15.6 Conclusions

Urbanization and change in lifestyle have increased the consumption and occurrence of various micropollutants in different environmental compartments. Even at a concentration ranging from few ng/L to μg/L, micropollutants produce both chronic and toxic effects in the living organisms. Several factors including, seasonal consumption pattern, hydrogeological conditions, intensity of solar radiation, rainfall pattern, type of sewage system, and zonal planning decide load of micropollutants in the influent stream of STPs. As conventional STPs are unable to remove such micropollutants and their toxic by-products, more research should be conducted on heterogeneous advanced oxidation-based tertiary treatment technology. In general, a combination of MBR and NF/RO, GAC/MF and NF, and TiO_2 and UF can remove micropollutants at an efficiency ranging from 50 to 100%. However, economically viable and easily scalable synthesis routes with high yield for TiO_2 should be the focus. Immobilization techniques aiding easy separation of the nanoparticles from the system also need special interest. In natural environmental settings, micropollutants are present at a very low concentration and RBF has shown potential in attenuating readily biodegradable micropollutants. Both site-specific factors and properties of individual contaminants strongly influence the overall dynamics during RBF.

References

Aguinaco A, Beltrán FJ, García-Araya JF, Oropesa A (2012) Photocatalytic ozonation to remove the pharmaceutical diclofenac from water: influence of variables. Chem Eng J 189:275–282

Bertelkamp C, Verliefde ARD, Schoutteten K, Vanhaecke L, Vanden Bussche J, Singhal N, van der Hoek JP (2016) The effect of redox conditions and adaptation time on organic micropollutant removal during river bank filtration: A laboratory-scale column study. Sci Total Environ 544:309–318

Brausch JM, Rand GM (2011) A review of personal care products in the aquatic environment: environmental concentrations and toxicity. Chemosphere 82:1518–1532

Cincinelli A, Martellini T, Coppini E, Fibbi D, Katsoyiannis A (2015) Nanotechnologies for removal of pharmaceuticals and personal care products from water and wastewater. A Review. J Nanosci Nanotechnol 15(5):3333–3347

Cleuvers M (2004) Mixture toxicity of the anti-inflammatory drugs diclofenac, ibuprofen, naproxen, and acetylsalicylic acid. Ecotoxicol Environ Saf 59:309–315

D'Alessio M, Yoneyama B, Ray C (2015) Fate of selected pharmaceutically active compounds during simulated riverbank filtration. Sci Total Environ 505:615–622

Das S, Ray NM, Wan J, Khan A, Chakraborty T, Ray MB (2016) Micropollutants in wastewater: fate and removal processes. Physico-Chemical Wastewater Treat Resour Recover 75–117

De Laat J, Gallard H, Ancelin S, Legube B (1999) Comparative study of the oxidation of atrazine and acetone by H2O2/UV, Fe (III)/UV, Fe (III)/H2O2/UV and Fe (II) or Fe (III)/H2O2. Chemosphere 39(15):2693–2706

Eregowda T, Rene ER, Rintala J, Lens PNL, Rene ER (2019) Volatile fatty acid adsorption on anion exchange resins: kinetics and selective recovery of acetic acid. Sep Sci Technol 1–13

Fernández-Ibáñez P, Blanco J, Malato S, De Las Nieves FJ (2003) Application of the colloidal stability of TiO2 particles for recovery and reuse in solar photocatalysis. Water Res 37(13):3180–3188

Forrez I, Carballa M, Boon N, Verstraete W (2009) Biological removal of 17α-ethinylestradiol (EE2) in an aerated nitrifying fixed bed reactor during ammonium starvation. J Chem Technol Biotechnol 84:119–125

Gomes J, Costa R, Quinta-Ferreira RM, Martins RC (2017) Application of ozonation for pharmaceuticals and personal care products removal from water. Sci Total Environ 586:265–283

Goswami L, Kumar RV, Borah SN, Manikandan NA, Pakshirajan K, Pugazhenthi G (2018) Membrane bioreactor and integrated membrane bioreactor systems for micropollutant removal from wastewater: A Review J Water Process Eng 26:314–328

Grover DP, Zhou JL, Frickers PE, Readman JW (2011) Improved removal of estrogenic and pharmaceutical compounds in sewage effluent by full scale granular activated carbon: Impact on receiving river water. J Hazard Mater 185:1005–1011

Hamann E, Stuyfzand PJ, Greskowiak J, Timmer H, Massmann G (2016) The fate of organic micropollutants during long-term/long-distance river bank filtration. Sci Total Environ 545–546:629–640

Henzler AF, Greskowiak J, Massmann G (2014) Modelling the fate of organic micropollutants during river bank. J Contam Hydrol 156:78–92

Ikehata K, El-Din MG (2004) Degradation of recalcitrant surfactants in wastewater by ozonation and advanced oxidation processes: A review. Ozone Sci Eng 26:327–343

Jobling S, Williams R, Johnson A, Taylor A, Gross-Sorokin M, Nolan M, Tyler CR, Van Aerle R, Santos E, Brighty G (2006) Predicted exposures to steroid estrogens in U.K. Rivers correlate with widespread sexual disruption in wild fish populations. Environ Health Perspect 114:32–39

Kasprzyk-Hordern B, Dinsdale RM, Guwy AJ (2009) The removal of pharmaceuticals, personal care products, endocrine disruptors and illicit drugs during wastewater treatment and its impact on the quality of receiving waters. Water Res 43:363–380

Kinney CA, Furlong ET, Zaugg SD, Burkhardt MR, Werner SL, Cahill JD, Jorgensen GR (2006) Survey of organic wastewater contaminants in biosolids destined for land application. Environ Sci Technol 40(23):7207–7215

Kolpin DW, Skopec M, Meyer MT, Furlong ET, Zaugg SD (2004) Urban contribution of pharmaceuticals and other organic wastewater contaminants to streams during differing flow conditions. Sci Total Environ 328:119–130

Kumar M, Ram B, Honda R, Poopipattana C, Canh VD, Chaminda T, Furumai H (2019a) Concurrence of antibiotic resistant bacteria (ARB), viruses, pharmaceuticals and personal care products (PPCPs) in ambient waters of Guwahati, India: Urban vulnerability and resilience perspective. Sci Total Environ 693:133640. https://doi.org/10.1016/j.scitotenv.2019.133640

Kumar M, Chaminda T, Honda R, Furumai H (2019b) Vulnerability of urban waters to emerging contaminants in India and Sri Lanka: Resilience framework and strategy. APN Science Bulletin 9(1). https://doi.org/10.30852/sb.2019.799

Liu JL, Wong MH (2013) Pharmaceuticals and personal care products (PPCPs): a review on environmental contamination in China. Environ Int 59:208–224

Liu J, Wang R, Huang B, Lin C, Zhou J, Pan X (2012) Biological effects and bioaccumulation of steroidal and phenolic endocrine disrupting chemicals in high-back crucian carp exposed to wastewater treatment plant effluents. Environ Pollut 162:325–331

Luo Y, Guo W, Ngo HH, Nghiem LD, Hai FI, Zhang J, Liang S, Wang XC (2014) A review on the occurrence of micropollutants in the aquatic environment and their fate and removal during wastewater treatment. Sci Total Environ 473:619–641

Margot J, Rossi L, Barry DA, Holliger C (2015) A review of the fate of micropollutants in wastewater treatment plants. Water 2:457–487

Menon NG, Mohapatra S, Padhye LP (2020) Review on occurrence and toxicity of pharmaceutical contamination in Southeast Asia: emerging issues in the water environment during Anthropocene. Springer, pp 63–91

Mishra MN, Menon NG, Sarma S, Tatiparti V, Mukherji S (2018) Assessment of endocrine disruption potential of selected pharmaceuticals using an in vitro assay. J Indian Chem Soc 95:269–277

Mohapatra S, Padhye LP, Mukherji S (2018) Challenges in detection of antibiotics in wastewater matrix: environmental contaminants. Springer, Singapore, pp 3–20

Muruganandham M, Suri RPS, Jafari S, Sillanpää M, Lee GJ, Wu JJ, Swaminathan M (2014) Recent developments in homogeneous advanced oxidation processes for water and wastewater treatment. Int. J. Photoenergy

Oaks JL, Gilbert M, Virani MZ, Watson RT, Meteyer CU, Rideout BA, Shivaprasad HL, Ahmed S, Jamshed M, Chaudhry I, Arshad M, Mahmood S, Ali A, Khan AA, Lindsay Oaks J et al (2004) Diclofenac residues as the cause of vulture population decline in Pakistan. Nature 427:630–633

Ortiz I, Rivero MJ, Margallo M (2019). Advanced oxidative and catalytic processes: sustainable water and wastewater processing. Elsevier, pp 161–201

Oturan MA, Aaron JJ (2014) Advanced oxidation processes in water/wastewater treatment: principles and applications. A Review Crit Rev Environ Sci Technol 44:2577–2641

Radjenović J, Petrović M, Barceló D (2009) Fate and distribution of pharmaceuticals in wastewater and sewage sludge of the conventional activated sludge (CAS) and advanced membrane bioreactor (MBR) treatment. Water Res 43:831–841

Richard J, Boergers A, vom Eyser C, Bester K, Tuerk J (2014) Toxicity of the micropollutants Bisphenol A, Ciprofloxacin, Metoprolol and Sulfamethoxazole in water samples before and after the oxidative treatment. Int J Hyg Environ Health. https://doi.org/10.1016/j.ijheh.2013.09.007

Santos JL, Aparicio I, Alonso E (2007) Occurrence and risk assessment of pharmaceutically active compounds in wastewater treatment plants. A case study: seville city (Spain). Environ Int 33:596–601

Schäfer AI, Akanyeti I, Semião AJ (2011) Micropollutant sorption to membrane polymers: a review of mechanisms for estrogens. Adv Colloid Interface Sci 164(1–2):100–117

Schueth C (2014) MARSOL: demonstrating managed aquifer recharge as a solution to water scarcity and drought

Schwarzenbach RP, Escher BI, Fenner K, Hofstetter TB, Johnson CA, Von Gunten U, Wehrli B (2006) The challenge of micropollutants in aquatic systems. Science 313:1072–1077

Silva LLS, Moreira CG, Curzio BA, da Fonseca FV (2017) Micropollutant removal from water by membrane and advanced oxidation processes—a review. J Water Resour Prot 09:411–431

Singh A, Patel AK, Deka JP, Das A, Kumar A, Manish Kumar (2019) Prediction of Arsenic Vulnerable Zones in Groundwater Environment of Rapidly Urbanizing Setup, Guwahati, India. Geochemistry, 125590. https://doi.org/10.1016/j.chemer.2019.125590

Spring AJ, Bagley DM, Andrews RC, Lemanik S, Yang P (2007) Removal of endocrine disrupting compounds using a membrane bioreactor and disinfection. J Environ Eng Sci 6:131–137

Steinle-Darling E, Litwiller E, Reinhard M (2010) Effects of sorption on the rejection of trace organic contaminants during nanofiltration. Environ Sci Technol 44:2592–2598

Storck FR, Schmidt CK, Lange F, Henson JW, Hahn K (2012) Factors controlling micropollutant removal during riverbank filtration. J Am Water Works Assoc 104:643–652

Sui Q, Huang J, Deng S, Chen W, Yu G (2011) Seasonal variation in the occurrence and removal of pharmaceuticals and personal care products in different biological wastewater treatment processes. Environ Sci Technol 45(8):3341–3348

Yang SF, Lin CF, Yu-Chen Lin A, Andy Hong PK (2011) Sorption and biodegradation of sulfon-amide antibiotics by activated sludge: experimental assessment using batch data obtained under aerobic conditions. Water Res 45:3389–3397

Yang W, Zhou H, Cicek N (2014) Treatment of organic micropollutants in water and wastewater by UV-based processes: a literature review. Crit Rev Environ Sci Technol 44:1443–1476

Zhang XX, Zhang T, Fang HHP (2009) Antibiotic resistance genes in water environment. Appl Microbiol Biotechnol 82:397–414

Chapter 16
Impact of Solid Municipal Waste Landfills on Groundwater Resources: Need for Integrated Solid Waste Management Aligned with the Conservation of Groundwater

Medhavi Srivastava and Manish Kumar

16.1 Introduction

When the well is dry, we'll know the worth of water—Benjamin Franklin

It was times like the present that were dreaded by wise men in the past, and times like this when the wells are finally dry or poisoned, demand for urgent focus. A majority of modern problems revolve around water, not only the water that is there for human use but also the water that is the fuel to earth system as a whole. Perturbations in the water cycle by human actions have been recorded since the beginning of civilization. These perturbations are overgrown now, caused by changes in infiltration; runoff and evaporation patterns due to land use and land cover change; increases in streamflow due to channelization and changes in groundwater levels due to overexploitation of aquifers. The present-day problems of climate change and urbanization added on with the complexity of the hydrological cycle make groundwater more polluted by the day.

The accessibility of water is not just a function of quantity, but also of quality, space and time. Out of all the water that is present on our planet, only 2.5% of water is fresh and even out of that just 1% is easily accessible (National Geographic, Freshwater Crisis, Patel et al. 2019b; Das et al. 2015). Actually, only 0.007% of all the water is available for 7 billion people of the world, which is obviously on a sharing basis in our environment. People might not understand this quite well and these may merely seem as figures, but the truth is, we are running out of time. Groundwater, one of the most important resources of a nation, is a source of about 35% of water which

M. Srivastava · M. Kumar (✉)
Discipline of Earth Sciences, Indian Institute of Technology Gandhinagar, Gandhinagar, Gujarat, India
e-mail: manish.kumar@iitgn.ac.in

© Springer Nature Singapore Pte Ltd. 2020
M. Kumar et al. (eds.), *Resilience, Response, and Risk in Water Systems*,
Springer Transactions in Civil and Environmental Engineering,
https://doi.org/10.1007/978-981-15-4668-6_16

is used by us. Other than overexploitation of this groundwater, which has caused aquifers to run dry all over the world, the little of it that is left is being contaminated by modern human practices (Saikia et al. 2017; Kumar et al. 2013; Kim et al. 2011; Patel et al. 2019b).

The issue of water resource among such a large population is not the only problem; increasing population generates immense amounts of waste and with it comes the concern of waste disposal. The rates of waste generation are rising all around the world. According to a data of The World Bank, the worlds' cities generated 2.01 billion tonnes of solid waste, which amounts to the unbelievable footprint of 0.74 kg per person per day (Singh et al. 2020) This figure is only about to increase manifold in the coming years! What is even more disturbing is the absence of sanitary waste disposal methods and poor waste management. According to EPA's Waste Management Strategy, the most preferred waste management is source reduction and reuse, followed by waste recycling and energy recovery. Waste disposal is the least preferred waste management strategy, but unfortunately it is also the most commonly used method all over the world. Even in waste disposal methods, without treatment disposal is predominant in developing countries. Landfilling in poorly monitored and managed open landfills is a major threat to the water resources (Mukherjee et al. 2020; Singh et al. 2019; Das et al. 2015) (Fig. 16.1).

There is a lack of proper waste disposal methods which has left most cities with vast open landfill as the only way of waste disposal, supporting tonnes of solid waste. Landfills in densely populated cities which are pressurising resources over the top should have groundwater management strategies like monitoring on regular basis, designing of landfill in such a way as to reduce leachate migration to groundwater and collection and treatment of leachate. Moreover, groundwater in and around the landfill sites should be subjected to strict scrutiny and shall not be used for drinking purposes unless it meets specific standards. Indiscriminate dumping of wastes should be stopped without proper solid waste management practices. The leachate formed from the decomposition of these open landfill waste materials is highly polluting and finds its way to the underground water supply which is backed up by many reports in scientific journals around the world. The leachate percolates through these piles of a city's concentrated garbage and reaches the groundwater on which the city eventually

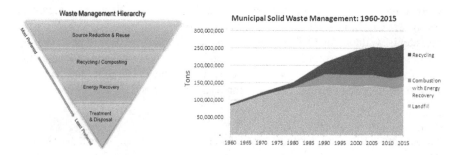

Fig. 16.1 Facts and figures on municipal solid waste management (from http://www.epa.gov)

relies. The leachate becomes a weighty source of pollution for the already dwindling groundwater (Kumar et al. 2019a, b).

The importance of an efficient solid waste management is strengthened by the knowledge and proof of threat posed to the very little amounts of groundwater left to us by leachate. Risks of groundwater contamination and the impact of landfill sites upon groundwater can be judged by monitoring the concentration of potential contaminants monitoring points, and such studies have come into limelight recently, all of which show an average increase in contaminants like heavy metals, total dissolved solids, nitrates, sulphates, etc. The current situation calls for the dire need of shaping the idea of integrated sustainable solid waste management. It includes the management of solid waste in a way as to sustain the existing water resources without its further contamination. The management of water resources is to be aligned with waste management in order solve these problems. The first step of this is the understanding about the intricate relations between the both, because all the components of our earth system are dynamically interactive. If there is a change in one component, the effect of this change propagates to the other components. This means that the environment has to be dealt with as a cohesive whole (or as an integrated system) which often is not managed as a continuum in practice. Each component of the continuum is considered separately, often with little accounting for its interaction with other components. Due to this, there are huge gaps in the understanding of sustainable solutions especially in developing countries such as India. As the pressure on our resources continues to increase by the day, all the aspects of the system should be considered in an integrated manner which could help in improvise better environmental planning and policy making for sustainable development.

16.2   Overview of the Current Situation

16.2.1   Facts and Figures

Groundwater comprises 97% of the worlds' accessible freshwater, and the pressure on it is growing at an alarming rate. The unplanned, unmanaged and accelerated increase in groundwater contamination has been neglected by governments especially in developing countries around the globe. At a time like this, when climate change continues to affect surface water sources, the quantity and quality of groundwater resource are becoming extremely valuable. The development of groundwater conservation strategies is utmost at present and we cannot risk the limited water resource merely due to improper solid waste management.

Approximately, 2.01 billion metric tonnes of municipal solid waste (MSW) is produced annually worldwide according to The World Bank report, 2018. According to a report of EPA, the total generation of municipal solid waste in the USA in 2015 was 262.4 million tonnes which was a result of 4.48 lb per person per day increase over last year (National Overview, EPA 2015). According to the Press Information

Table 16.1 MSW generated by metro cities in India

City-wise municipal solid waste generation in metro cities/state capitals in India (1999–2000, 2004–2005, 2010–2011 and 2015–2016)

City	Population (Census-2011)	Waste generation (TPD)			
		1999–2000	2004–2005	2010–2011	2015–2016
Mumbai	12,442,373	5355	5320	6500	11,000
Delhi	11,034,555	400	5922	6800	8700
Bengaluru	8,443,675	200	1669	3700	3700
Chennai	7,088,000	3124	3036	4500	5000
Hyderabad	6,731,790	1566	2187	4200	4000
Ahmedabad	5,577,940	1683	1302	2300	2500
Kolkata	4,496,694	3692	2653	3670	4000
Pune	3,124,458	700	1175	1300	1600
Lucknow	2,817,105	1010	475	1200	1200

Source: http://www.mospi.gov.in/

Bureau, India generates 62 million tonnes of waste every year, with an average annual growth rate of 4%. The amount of MSW generated by major metro cities in India is given in Table 16.1.

The World Bank also gives an estimate for overall waste generation which shows to increase to 3.40 billion metric tonnes by 2050. The report also shows that almost 40% of the waste generated is not managed properly and dumped in open landfills or improperly burned (The World Bank Report on Global Solid Waste management, 2018). The total quantity of waste generated in developing countries is expected to rise even more than three times due to lack of proper management.

16.2.2 Challenges with Solid Waste Management

Solid waste management is plausibly the most important municipal service of a city, and though the service level, environmental impacts and costs vary sensationally, it serves as a prerequisite for other municipal actions. More than one-third of waste is recycled or composted. A global review of solid waste management by The World Bank states that municipal solid waste (MSW) management is the most important service a city provides. The waste generation of a country greatly depends on its economy along with other factors like population. In developing countries, MSW is the largest single budget item for cities and one of the largest employers. Waste collection rates vary widely from place to place. For instance, in developed or high-income countries, there is a universal, proper and efficient waste collection, but developing and low-income countries tend to collect up to 48% of waste in cities and almost half of it outside urban areas making even the waste collection very inefficient.

Fig. 16.2 Waste management across the globe

Globally, about 37% of waste is disposed in landfills, 33% is openly dumped, 19% undergoes recycling and composting and only 11% is treated through modern incineration. Overall, most of the waste ends up in some type of landfill which is the most harmful method of waste disposal. Out of the global average, the developing countries can have the percentage of open dumping as high as up to 90% (Hoornweg and Bhada-Tata 2012) (Fig. 16.2).

MSW composition also varies greatly across the globe. The waste generated by low- and middle-income countries mostly comprises of organics originated from food and greens, whereas in high-income countries the fraction of organics goes down to about one-third because of larger amounts of packaging and non-organic wastes. The quantity of recyclables also varies; as countries rise in development, the quantity of recyclables in waste stream increases. It is a clear sign indicating towards very primitive solution of waste management by manufacturing more recyclable products.

If a city is unable to manage its waste effectively, it would not be able to manage more complex services such as health, education or transportation and of course, sustainable and good water supply.

In the case of India, the basic problem is excessive urbanization and the resultant slumming of cities. Municipal bodies cannot make the large investments required for proper waste management because of poor finances (India Today 1994) (Fig. 16.3).

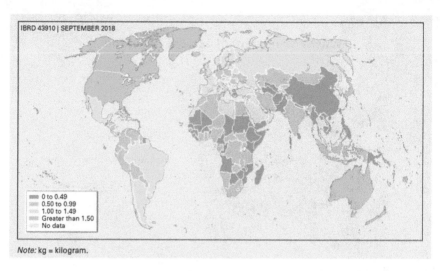

Note: kg = kilogram.

Fig. 16.3 Waste generation per kg per capita

16.2.3 Groundwater Contamination

Groundwater plays a vital role for urban and agricultural water supply. It occurs in permeable geologic formation known as aquifers. They are formations having structure that can store and transmit water at rates fast enough to supply reasonable amounts to wells (Afolayan et al. 2012). The impact of landfill on groundwater quality is dependent on many factors such as type and nature of waste, rainfall pattern, local geology, hydraulic gradient and so on (Kumar et al. 2008). Groundwater contamination is usually uncommon compared to contamination of other water bodies because contaminants get removed while infiltration through soil strata while replenishing of groundwater systems by rainfall. Groundwater contamination observed all over the globe these days is majorly anthropogenic. Mostly, in countries, where groundwater is intensively used for irrigation and industrial purposes, land use change and overexploitation are causing its contamination (Banerjee et al. 2012).

Municipal, industrial and agricultural, all activities affect groundwater quality. Sources of groundwater contamination include accidental spills, surface waste ponds, pipelines and injection wells, saltwater intrusion, landfills and waste disposal sites, improperly constructed wells, land application of pesticides, acid mine drainage and underground storage tanks. Out of all these, landfill practices, disposal of solid waste by filling depressions on land with waste, have been identified as the most potent threat to groundwater resources (Taylor and Allen 2006; Fatta et al. 1999). Waste dumped in landfills is subject to underflow and/or infiltration from precipitation resulting in the formation of leachate which seeps into the aquifers and contaminates groundwater. Major groundwater pollutants due to leachate include heavy metals such as cadmium, lead and arsenic. Increased total dissolved solids, ammoniacal nitrogen and organic matter in groundwater are also result of leachate migration.

Pesticides and fertilizers can accumulate and migrate to the water table. Contamination of aquifers due to rapid urbanization coupled with poor sanitization is the most omnipresent pollution problem in countries like India. Additionally, unlike other water resources, when groundwater becomes contaminated, it is very difficult and if at all possible, then very difficult to clean up.

16.3 Municipal Solid Waste Landfills in the Developing World

The heterogeneous waste produced in urban areas is collectively called as municipal solid waste (MSW). MSW typically consists of organics, metals, glass, paper, plastics, wood and composite products (Kumar et al. 2015). The nature of MSW varies from region to region and depends on many factors. The characteristics and quantity of MSW generated are function of economy, region, population, lifestyle and abundance and type of natural resources there. The composition of MSW also varies from country to country. The World Bank Overview on solid waste Management shows that there is an increase in organic waste as we go from developed to developing countries.

MSW is disposed of in thousands of municipal landfills throughout the world. Landfills are engineered areas where waste is deposited, compacted and covered, but the problem with them is that they remain widely open especially in developing countries. Out of all municipal landfills, vast open landfills are most common in developing countries like India. Out of all global waste, 40% of waste is disposed of in MSW Landfills. In developing countries like India, this percentage hikes up to almost 90% (Kaza et al. 2018) (Fig. 16.4).

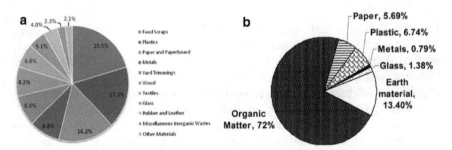

Fig. 16.4 MSW composition **a** of the USA (from: https://www.epa.gov/landfills/municipal-solid-waste-landfills) **b** of Indian landfill (Singh et al. 2008)

16.3.1 Fate of Solid Waste in Urban Municipalities

Municipal solid waste (MSW) is collected by sanitation services from three types of MSW generators—households, commercial and institutional. Out of these, the major portion of MSW comes from households, accounting for 55–65% of total MSW, followed by commercial sector which includes solid waste from offices, stores, restaurants, etc. (Singh et al. 2008). The industrial sector is not included in MSW because it mostly manages its own waste by recycling, reuse or disposal in separate industrial landfills. However, these practices are not so common in developing countries yet, and industrial waste also contributes to MSW.

After collection, MSW goes to landfills that accept "household waste". But in countries like India, municipal landfills may include various kinds of waste. In developing countries, many rag-pickers come to the picture at this time who crowd the open dumps and collect recyclable waste to sell it. They even roam around in residential areas to collect recyclable waste directly. All MSW management systems have collection and transportation of MSW as an important component. Collections of MSW vary from regions by service of local government and collectors, by extent of collection and by level of service provided to households. Collection vehicles (usually tractor trailer trucks) are assigned for the transport of waste to landfills at centrally located transfer stations. In developed countries, much sorted and pre-treated wastes enter landfills and the site is subjected to leachate treatment regularly, later to be covered up. But in countries like India, once the waste is in the landfills, it becomes a part of a huge open pile of waste. The waste disposal in landfills is unscientific and chaotic in India (Gupta et al. 1998).

16.3.2 Types of Municipal Solid Waste Landfills

Landfills are engineered containments which are made in depressions in earth, designed to minimize the impact of MSW on the environment and health. In MSW landfills, there is predominantly a liner system. The liner isolates the landfill contents from surrounding and protects the environment this way; they also control the migration of leachate from waste to groundwater to a limit. Landfill liners protect the soil and groundwater from contamination by waste and leachate. Different kinds of liners are installed in different types of MSW landfills which are defined by the characteristics of waste to be dumped in it. Most of the landfills contain a leachate collection system as well as a treatment unit along with liner system.

Landfill liner design focuses on creating a barrier between waste and soil and draining the leachate in collection systems. In landfills without collection systems, specialized liners may be present which reduce the contamination by leachate in some way. The potential threat posed by type of MSW is the main criteria to determine the type of liner required for each landfill because different wastes generate different kinds of leachate. Liners may be characterized into single, double or composite types.

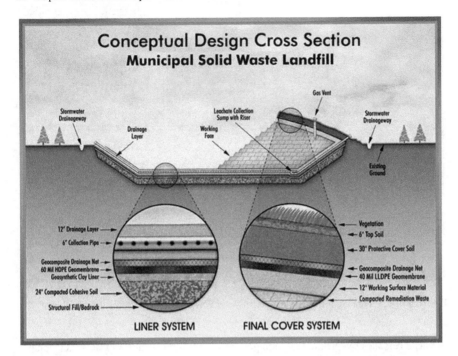

Fig. 16.5 A conventional landfill structure (from: http://www.semcolandfill.com/faq.html)

Single liners are used in designs which hold construction and demolition waste or wastes which are mostly composed of earth materials (Bouazza and Van Impe 1998).

The purposes of single liner are just to create a barrier between waste and environment. Composite liners consist of a geomembrane with a compacted-clay liner. They are made for containment of MSW with large organic content, and they limit leachate migration. Double liners consist of a combination of single liners or composite liners which functions to collect and control leachate. The collection part of landfill is usually composed of sand and gravel or geonet. There are leachate collection pipes to hold storage tanks. There are leak detection systems in some advanced double liner landfills. Lining materials include clay, geomembrane, geotextiles, geosynthetic clay and geonet, all of which are specially engineered for lining (Hughes et al. 2008) (Fig. 16.5).

16.3.3 Landfill Lifespan

The lifespan of landfills depends on their liners which act as barriers to leachate migration. It is widely acknowledged that liners deteriorate over time and eventually fail to prevent leachate migration (Jagloo 2002). So all landfills have a lifespan based on life of its liner. The process of degradation (physical, biological and chemical)

Fig. 16.6 The Pirana landfill site (a municipal Solid Waste Landfill) in Ahmedabad, Gujarat

of waste releases by-products which get dissolved in rainwater and flow as leachate. This chemical-laden leachate interacts with the liner of landfills and eventually deteriorates it over time. It then reaches to groundwater. Though this movement is very slow and cannot even be identified for a long time, leachate causes severe groundwater contamination and reacts in unanticipated ways to affect the ecosystem as a whole. In many developing countries, this fact is overall ignored and no attention is paid to the lifespan of landfills (Abarca-Guerrero et al. 2015). In India, landfill sites even in large metro cities are as old 30–40 years, with the very basic structure of the landfills with short lifespans and no leachate collection or treatment systems, just massive mounds of garbage (Fig. 16.6).

16.3.4   Existing Landfills and the Focus Towards the Problem

According to an article on World Atlas, six out of seven largest landfill sites in the world are in Asia, though the largest landfill in the world is *Apex Regional* in Las Vegas, the USA. The landfill designs and management vary considerably across the globe. The largest landfill in Nevada is so well developed and fully planned that it can handle even tonnes of more solid waste daily; it even has a gas collection system which collects methane gas released from the landfills which are then used by power plants to generate energy! EPA report states that the amount of methane generated from this site accounts for almost 17% of methane emission of the USA. So the utilization of this and its conversion in energy is a great achievement and boon for environment protection as methane is a greenhouse gas. On the contrary, we also have the *Ghazipur* landfills in Delhi, India, which come in this list but are most poorly managed of all. The Delhi landfills have reached their limit a long time ago, but still tonnes of waste is dumped on it without any proper waste management. The methane gas emissions from this landfill site could have generated 25 MW of power if tapped effectively for such purpose according to a report of International Energy Agency Reports. Classical unlined or poorly lined landfills pose a serious threat to environment and groundwater resources via leachate and landfill gas directly. Though there are many landfills constructed and managed very prudently like the landfills in developed nations which use various lining materials to limit and even restrict the leachate movement, these constructions, engineering and researches vary widely across the globe due to many factors discussed earlier.

16.4 Leachate and Groundwater

16.4.1 Characteristics of Landfill Leachate

When precipitation infiltrates MSW in open landfills, leachate laden with by-products of chemical, biological and physical degradation of waste is generated. Leachate is mainly characterized by high dissolved solids and organics, inorganic macro-compounds such as chlorides, iron and aluminium, heavy metals and large organic xenobiotics like halogenated compounds. Other chemicals like pesticides and solvents might also be present. Leachate from MSW landfill sites often include major elements like calcium, magnesium, potassium, nitrogen and ammonia, heavy metals like iron, copper, manganese, chromium, nickel, lead and organic compounds (Freeze and Cherry 1979). The leachate changes the physicochemical characteristics of groundwater eventually. The concentration and proportion of these wastes in leachate depend on the characteristics of MSW from which it is generated (Singh et al. 2008).

16.4.2 Groundwater Quality Standards in India and Impact of Leachate on Groundwater Quality

Groundwater quality is checked by physical, chemical and biological properties of water which include temperature, colour, taste, turbidity, odour, pH, alkalinity, acidity, total hardness, coliform count, etc. As the groundwater is getting contaminated by the day, the deviations in its regular properties are observed. There are specific standards set by government for the limit of specific concentrations in water. In India, these standards are set by Bureau of Indian Standards (BIS) which are further kept in check by Central Pollution Control Board (CPCB) or State Pollution Control Board

Table 16.2 Extent of groundwater pollution in India

Extent of groundwater pollution in India		
Pollutant	Permissible limit	Exceeded in number of states of India
Fluoride	1.5 ppm	14 Indian states, namely Andhra Pradesh, Bihar, Gujarat, Haryana, Karnataka, Kerala, Madhya Pradesh, Maharashtra, Orissa, Punjab, Rajasthan, Tamil Nadu, Uttar Pradesh and West Bengal affecting a total of 69 districts, according to some estimates Some other estimates find that 65% of India's villages are exposed to fluoride risk
Iron	0.3 ppm	23 districts from four states, namely Bihar, Rajasthan, Tripura and West Bengal and coastal Orissa and parts of Agartala valley in Tripura
Arsenic	50 ppb	Alluvial plains of Ganges covering six districts of West Bengal [a]Presence of heavy metals in groundwater is found in 40 districts from 13 states, viz., Andhra Pradesh, Assam, Bihar, Haryana, Himachal Pradesh, Karnataka, Madhya Pradesh, Orissa, Punjab, Rajasthan, Tamil Nadu, Uttar Pradesh, West Bengal and five blocks of Delhi
Nitrate	45 ppm	11 states, covering 95 districts and two blocks of Delhi

Source: Kumar and Shah (2004)

(SPCB). The evolving measures to control groundwater pollution are to check and monitor groundwater to make sure the parameters do not exceed the permissible limits. This involves careful water quality monitoring (WQM) done by WQM stations which are not enough in the first place unfortunately. Our country runs short of proper monitoring which creates huge gaps even in primary analysis. WQM involves expensive and sophisticated equipment which is not very advanced in our country. Studies indicate that the concentrations of contaminants in groundwater are higher than the standard limits and are backed up by experiments all across the nation, especially in and around landfill sites (Table 16.2).

16.4.3 Factors Affecting Leachate Migration

There are many factors which affect leachate migration till the groundwater and then its transport further. The geology of area, season of the year, natural discharge, etc., all has a serious impact on leachate migration. It is very important to understand all of these intricate relations in order to understand the leachate migration and groundwater contamination. In a landfill system, even more functions are added to the leachate movement like the type of liner material, the characteristics of waste, amount of waste, amount of precipitation, structure and design of landfill and many others. When the leachate flows through the site to the groundwater, some of the contaminants maybe adsorbed onto soil, and when it finally mixes with groundwater, it forms a plume which spreads by flowing in the direction of flow of groundwater. The extent of contamination of the groundwater also depends on a number of factors like the transport rate, infiltration rate, depository conditions, flow direction, nature of media, etc. (Vasanthi et al. 2008).

16.5 Importance of Efficient Solid Waste Management

16.5.1 Risks and Threats Posed by Contaminated Groundwater

Landfills pose a serious threat to the environment and human health through leachate, landfill gas and open waste. In landfills, MSW or other wastes serve as input and leachate serves as an output. The compound attributes of groundwater are controlled by several chemical reactions and natural responses during the movement of water through many zones (Kumar et al. 2015). Through leachate plume and groundwater contamination, many disease outbreaks are likely, even the past eradicated diseases because of degradation of buried wastes that initially caused those outbreaks (Nagendran et al. 2006). Degradation of MSW also creates landfill gases which are a source of air pollution. Methane is predominantly released from such sites, and as it is a GHG, it causes global problems on large scale. Open dumps also serve as breeding grounds for pests, flies and mosquitoes causing further health issues indirectly.

Contamination of groundwater with heavy metals is of major concern because it makes groundwater highly toxic and may render it totally unusable. Not only in human health, but also the groundwater contamination has severe impacts on plant life and vegetation. It can cause shedding and other diseases in vegetation as groundwater sustains many other elements of our ecosystem too!

Apart from poor drinking water quality and loss of water supply, contaminated groundwater can cause irreversible losses. For example, estuaries that are affected by high nitrogen concentrates from groundwater have shown loss of habitat for some shellfish species. Groundwater contamination affects the economy of a country, as it is very expensive and close to impossible to clean contaminated groundwater, high costs for clean up along with high costs of alternate water supply.

16.5.2 Integrated Sustainable Solid Waste Management

The preferred method for solid waste management should not be landfilling in the first place, but in the current scenario, where landfills are prominently used for municipal solid waste management, the ideal landfill should be confined to a small area with proper lining to control leachate migration, leachate collection and treatment systems, groundwater monitoring wells and stations and even the gas collection systems

Fig. 16.7 Integrated waste management

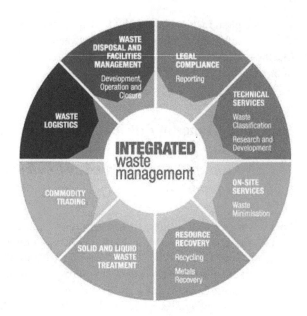

for landfill gas. Over this, the idea of landfill lifespan should be carefully considered and the landfill should be covered in due time. Multitudinous management strategies, new designs, ideas and technologies have come into limelight to reduce such serious impacts of solid waste landfills on groundwater resources. Integrated waste management basically refers to implementation and designing novel waste management systems for analysing the existing ones. It denotes that all the aspects of waste management system should be considered while analysis as they all are actually interrelated and any change in one affects the practices or activities in another. There is a dire need for a better solution to the impact of landfills on water resources in densely populated cities around the globe and for integrated solid waste management aligned with the conservation of groundwater (Fig. 16.7).

16.5.3 Groundwater Resource Management

The problem with is that because of its slow movement, contamination often remains undetected for long periods of time. But even in the known contamination sources, like MSW landfills, groundwater contamination widely remains unchecked especially in developing and low-income countries.

In India, the legal framework for groundwater protection is not simple. Distribution of groundwater is highly uneven in our country, and this clears that no single management strategy can be adopted for the whole country as geomorphology, climate, hydrology, water utilization patterns should be kept into account (Jha and Sinha 2009). Neglecting the fact of low finance in these areas, enforcement of laws is also not easy. The Water Act of 1974 and EPA of 1986 deal with most pollution issues related to groundwater in India with the support of many more policies, ordinances and boards like CPCB and SPCB. But there is a wide gap between knowledge and implication. Policies must be developed to improve finances and quality of ground-water. The solution can be up-scaling and spread of community-based groundwater management. This approach along with integrated waste management would go hand in hand in the development of integrated resource management.

16.6 Conclusion

Through this chapter, we have known the threat from landfill leachate to groundwater pollution is real and big. Aquifers are being depleted, but before they run dry, they will become unusable because of pollutants. As unfeasible it may seem, these problems are not completely without a solution. Backed up by the proof provided by the case studies, a proper groundwater resource management along with integrated solid waste management is not very difficult to achieve. There is a need for integration of data on water quality with data on water supplies and also with waste management for various economic and environmental objectives. If the barriers between government and water using sectors are removed, and awareness is increased, approaches towards resource protection will mutually be made by all groundwater users.

References

Abarca-Guerrero L, Maas G, Hogland W (2015) Solid waste management challenges for cities in developing countries. Revista Tecnología en Marcha 28(2):141–168

Afolayan OS, Ogundele FO, Odewumi SG (2012) Hydrological implication of solid waste disposal on groundwater quality in urbanized area of Lagos state, Nigeria. Int J Appl 2(5)

Banerjee DM, Mukherjee A, Acharyya SK, Chatterjee D, Mahanta C, Saha D, Dubey CS (2012) Contemporary groundwater pollution studies in India. Proc Indian Nat Sci Acad 78(3):333–342

Bouazza A, Van Impe WF (1998) Liner design for waste disposal sites. Environ Geol 35(1):41–54

Das N, Patel AK, Deka G, Das A, Sarma KP, Kumar M (2015) Geochemical controls and future perspective of arsenic mobilization for sustainable groundwater management: a study from Northeast India. J Groundwater Sust Dev 1(1–2):92–104

EPA (2015) National overview: facts and figures on materials, waste and recycling. Retrieved from https://www.epa.gov/facts-and-figures-about-materials-waste-and-recycling/national-overview-facts-and-figures-materials

Fatta-Kassinos D, Papadopoulos A, Loizidou M (1999) A study on the landfill leachate and its impact on the groundwater quality of the greater area

Freeze RA, Cherry JA (1979) Groundwater: Englewood Cliffs. New Jersey

Gupta S, Mohan K, Prasad R, Gupta S, Kansal A (1998) Solid waste management in India: options and opportunities. Resour Conserv Recycl 24(2):137–154

Hoornweg D, Bhada-Tata P (2012) What a waste: a global review of solid waste management, vol 15. World Bank, Washington, DC, p 116

Hughes KL, Christy AD, Heimlich JE (2008) Landfill types and liner systems. Ohio State University Extension Fact Sheet CDFS-138-05, 4

Jagloo K (2002) Groundwater risk analysis in the vicinity of a landfill: a case study in Mauritius. Royal Inst. of Technology

Jha BM, Sinha SK (2009) Towards better management of ground water resources in India. Q J 24(4):1–20

Kaza S, Yao L, Bhada-Tata P, Van Woerden F (2018) What a waste 2.0: a global snapshot of solid waste management to 2050. World Bank Publications

Kim J, Kuwahara Y, Kumar M (2011) A DEM-based evaluation of potential flood risk to enhance decision support system for safe evacuation. Nat Hazards 59:1561–1572

Kumar MD, Shah T (2004) Groundwater pollution and contamination in India. Emerging challenges, Hindu survey of environment. Kasturi and Sons

Kumar M, Herbert Jr R, Ramanathan AL, Rao MS, Deka JP, Kumar B (2013) Hydrogeochemical zonation for groundwater management in the area with diversified geological and land-use setup. Chemie der Erde- Geochemistry 73:267–274

Kumar M, Ramanathan AL, Chidambram S, Goswami R (2015) Criterion, indices, and classification of water quality and water reuse options. In: Eslamian S (ed) Urban water reuse handbook (UWRH), 1st ed, Chap 6. CRC Press, pp 163–176. ISBN: 13-978-1-4822-2915-8

Kumar M, Ram B, Honda R, Poopipattana C, Canh VD, Chaminda T, Furumai H (2019a) Concurrence of antibiotic resistant bacteria (ARB), viruses, pharmaceuticals and personal care products (PPCPs) in ambient waters of Guwahati, India: urban vulnerability and resilience perspective. Sci Total Environ 693:133640. https://doi.org/10.1016/j.scitotenv.2019.133640

Kumar M, Chaminda T, Honda R, Furumai H (2019b) Vulnerability of urban waters to emerging contaminants in India and Sri Lanka: resilience framework and strategy. APN Sci Bull 9(1). https://doi.org/10.30852/sb.2019.799

Mukherjee S, Patel AK, Kumar M (2020) Water scarcity and land degradation nexus in the era of Anthropocene: some reformations to encounter the environmental challenges for advanced water management systems meeting the Sustainable development. In: Kumar M, Snow D, Honda R (eds) Emerging issues in the water environment during Anthropocene: a South East Asian perspective. Springer Nature. ISBN 978-93-81891-41-4

Nagendran R, Selvam A, Joseph K, Chiemchaisri C (2006) Phytoremediation and rehabilitation of municipal solid waste landfills and dumpsites: a brief review. Waste Manage 26(12):1357–1369

Patel AK, Das N, Kumar M (2019b) Multilayer arsenic mobilization and multimetal co-enrichment in the alluvium (Brahmaputra) plains of India: a tale of redox domination along the depth. Chemosphere 224:140–150

Saikia R, Goswami R, Bordoloi N, Pant KK, Kumar Manish, Kataki R (2017) Removal of arsenic and fluoride from aqueous solution by biomass based activated biochar: optimization through response surface methodology. J Environ Chem Eng 5:5528–5539

Shibaji R (2014) NHRC seeks reports from environment ministry, Delhi govt on waste management. India Today, Living Media India Limited, 2 May 2014, http://www.indiatoday.in/india/north/story/nhrc-seeks-reports-from-environment-ministry-delhi-govt-on-waste-management-191250-2014-05-02.

Singh UK, Kumar M, Chauhan R, Jha PK, Ramanathan AL, Subramanian V (2008) Assessment of the impact of landfill on groundwater quality: a case study of the Pirana site in western India. Environ Monit Assess 141(1–3):309–321

Singh A, Patel AK, Deka JP, Das A, Kumar A, Kumar M (2019) Prediction of arsenic vulnerable zones in groundwater environment of rapidly urbanizing setup, Guwahati, India. Geochemistry 125590. https://doi.org/10.1016/j.chemer.2019.125590

Singh A, Patel AK, Kumar M (2020) Mitigating the risk of arsenic and fluoride contamination of groundwater through a multi-Model framework of statistical assessment and natural remediation techniques. In: Kumar M, Snow D, Honda R (eds) Emerging issues in the water environment during Anthropocene: a South East Asian perspective. Springer Nature. ISBN 978-93-81891-41-4

Taylor R, Allen A (2006) Waste disposal and landfill: information needs. IWA

Vasanthi P, Kaliappan S, Srinivasaraghavan R (2008) Impact of poor solid waste management on ground water. Environ Monit Assess 143(1–3):227–238

Part III
Response in Operational Water Management

Chapter 17
Sustainable Water Management in the Kelani River Basin

T. R. S. B. Bokalamulla, G. G. T. Chaminda, Y. Otaki, M. Otaki, and Manish Kumar

17.1 Introduction

Kelani River is the most sensitive river in Sri Lanka, and the Kelani River serves as primary source of water covering 80% of the portable water supply. The coverage includes millions of people, more than ten thousands of industrial and business establishments located around Colombo and suburb areas. The river upstream has minimum exposure to the domestic and industrial waste discharges; however, agriculture pollutions are present. The Kelani River is 144 km long and drains an area of 2292 km^2 (Herath and Amarasekara 2004). The most active part of the river lies between Avissawella (59 km chainage) to downstream Ambathale (14 km chainage). The poor local authority service delivery, weak environment management and governance (Arewgoda 1986; Ileperuma 2000; CEA 2014; Patel et al. 2019; Das et al. 2017) coupled with inadequate awareness and education have lead pollution in Kelani River (Mallawatantri et al. 2016). The sources of pollutant entered into the Kelani River include agricultural discharges, forest disturbances, grazing land fecal bacteria, industrial discharges, discharges from septic systems, sewage treatment plant effluent and urban runoff (CEA 2016; Kumar et al. 2019a; Das et al. 2017).

T. R. S. B. Bokalamulla (✉) · G. G. T. Chaminda
Department of Civil and Environmental Engineering, Faculty of Engineering, University of Ruhuna, Galle, Sri Lanka
e-mail: bokalamulla@gmail.com

Y. Otaki
Faculty of Sociology, Hitotsubashi University Graduate School of Sociology, Tokyo, Japan

M. Otaki
Department of Human Environmental Sciences, Ochanomizu University, Tokyo, Japan

M. Kumar
Discipline of Earth Sciences, Indian Institute of Technology Gandhinagar, Gandhinagar, Gujarat 382355, India

© Springer Nature Singapore Pte Ltd. 2020
M. Kumar et al. (eds.), *Resilience, Response, and Risk in Water Systems*,
Springer Transactions in Civil and Environmental Engineering,
https://doi.org/10.1007/978-981-15-4668-6_17

There are eight number of water purification plants extracting water from the Kelani River in the section between Avissawella to Kelaniya for producing drinking water to the community. At present, those plants extract around one million cubic meters of raw water per day. It has been forecasted that the future demand can increase up to 1.85 million cubic meters per day (Rajapaksha et al. 2016) with ongoing development activities, population growth and amplified urbanization in the Colombo and suburban areas. The flow of the river varies between 800 and 1500 m³/s during the monsoon and 20–25 m³/s in the dry season (Abeysinghe and Samarakoon 2017; Kumar et al. 2019b; Das et al. 2015). The shortages in river flow have already been experienced during the dry season, and conditions are heading toward the critical level with alteration in the global climate pattern and the increasing demand for freshwater. The study focused on two types of analysis to review the present condition in downstream of the Kelani River and future water quality characteristics. Also, water consumption details of the consumers and their perceptions over the future challenges in the water supply to the area were investigated. The results obtained from the analysis are used to predict the extreme conditions of the river, possibility of introducing alternative water source to reduce the raw water demand from the Kelani River. (Kim et al. 2011)

Water quality investigations: There are diffident types of water quality assessment tools used in water quality assesments in rivers, WASP3 (Himesh et al. 2000), QUAL2K (Hemant 2014) are two type of water quality simulation tools. Availability of sufficient data for the model development, calibration and validation process is the most important factor to ensure the accuracy (Himesh et al. 2000) for all these tools. The QUAL2K is an Excel-based software, and according to the literature, this has been used in number of water quality studies in rivers.

End User measurements: The studies on end user consumption are first-hand experience in Sri Lanka. The literature shows three major methods for predicting domestic water consumptions on end usage.

1. The direct measurements from outlet of the household,
2. Estimations through interviews and determination of each micro-component,
3. Collect time series data of total residential water consumption and calculate by the water flow pattern.

The first two options are identified as suitable for developing countries as time series measurement will not possible without highly built-up facilities. Adaptation of correct methods has been considered to be significant for the end user consumption (Otaki et al. 2008). Another study carried out in UK noted that "only way to record an individuals' water use behavior is through qualitative means (i.e., diary-based questionnaires and interviews) combined with micro-component data. This method can be subjective when it comes to identify actual versus perceived (or intended) user behavior" (Zadeh et al. 2014). Being a developing nation and due to constrain in water supply infrastructures in Sri Lanka, adaptation of more complex method is not useful. The direct measuring of individual household is the most useful method for estimation of consumption.

17.2 Materials and Methods

The area selected for the analysis starts 59 km chainage at Avissawella and ends 14 km chainage at Ambathale. Based on the water quality data, industrial influences and the river utilization as drinking water source are the most significant part of the river in this section. Figure 17.1 shows Kelani River Basin and the research area. The quality behavior was investigated through water quality modeling software (QUAL2k) and the consumer consumption, and the perceptions were studied through the direct household measurements and the questionnaire surveys.

Figure 17.2 shows overview of the methodology adopted for the data collections, model development, calibration, validation, simulations and analysis of the results. The river bathymetric and discharge data were collected from Department of Irrigation. The water quality data was used as secondary data and was collected from the Pavithra Ganga program conducted by National Water Supply and Drainage Board (NWSDB) and the Central Environmental Authority (CEA). Qual2K software was used for the development of river model. The intake volumes and outfall volumes from Avissawella to Ambathale are shown in Table 17.1.

The calibration and validation of the model were carried out referring to the water quality parameters obtained from the NWSDB for wet and dry season, respectively. The validated model was used for the analysis of three different flow conditions forecasted on the water usage point of view.

20-year dry weather flow with the present water demand (2020 year) and pollution loading
20-year dry weather flow with projected water demand as at year 2030 with same pollution loading
20-year dry weather flow with the present water demand (2020 year) and increased pollution loading of 50% from the existing.

The same scenarios were simulated under the wet flow conditions and compared variations between wet and dry flow conditions.

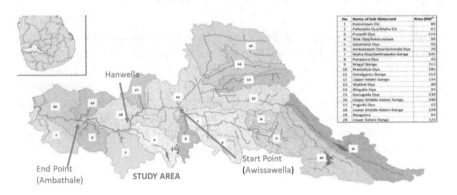

Fig. 17.1 Study areas in downstream of Kelani River

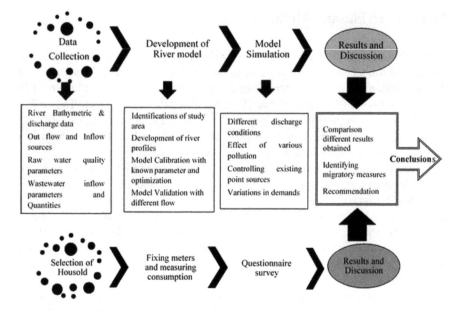

Fig. 17.2 Flow diagram of the methodology adopted

The end user household measurements were carried out through direct measurement of each consumer outlet of 25 houses located in the area. A questionnaire survey was carried out for 50 number of houses including end user consumption measured households.

17.3 Results and Discussion

17.3.1 Present Status Downstream of Kelani River

Figure 17.3 shows the average river discharge over one year and the tributary flow data during the wet and dry periods. As shown in the figure, the dry period of the year falls in December, January and February months of the year (Fig. 17.4).

Ruwanwella is the next intake located around 65 km chainage in upstream of Avissawella. The secondary data collected with referring to the Ruwanwella intake shows PH, Coliform values of 7.7, 6000/(100/ml) and 7.3, 1100 (100/ml) for dry and wet seasons, respectively. In addition, turbidity is around 2 (NTU) and Nitrate is less than 1 for both seasons.

Table 17.1 Intakes and outfall source to the Kelani river downstream (million meter cube per day)

Chainage (distance from the sea)/km	Name of the facility/location	Outfall/(MCM)	Intake/(MCM)
51.25	Seethawaka Industrial Zone WTP		0.1200
50.50	Seethawaka Industrial ZoneWWTP	0.1160	
41.34	Pugoda Oya (H5)	3.6900	
33.44	Kosgama Water Intake		0.0600
30.04	Wak Oya	1.5900	
29.90	Samanbedda (Water Intake)-Future		
29.71	Hanwella Water Intake-Future		
29.80	Varun Beverages	0.1200	
27.20	Amunugam Ela	1.7000	
25.89	Nawagamuwa Water Treatment Plant		0.0200
22.21	Chico Water Intake		
21.07	Fonterra Brand Lanka	0.1200	
20.01	Pahuru Oya	1.0400	
19.72	Maha Ela	1.9200	
19.34	Coca Cola Beverages	0.1000	
19.13	New Biyagama Village	0.2000	
18.92	Lion Brevery (Ceylon) PLC	0.1200	
17.6	Raggahawatta Ela	1.9400	
16.27	Biyagama Water Treatment Plant (KRBI)		2.3100
14.9	Ambathale Water Treatment Plant		6.3600

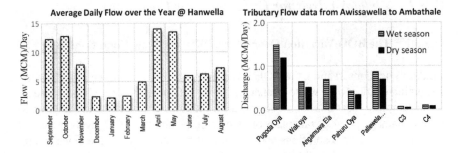

Fig. 17.3 Daily discharge variation and the tributary flow data over the year

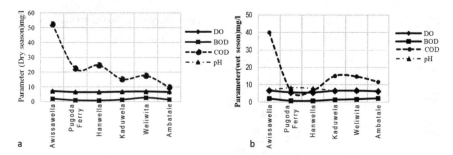

Fig. 17.4 Average water quality data obtained from the NWSDB and its variation across the river during the dry season (**a**) and wet season (**b**)

17.3.2 Calibration and Validation

The calibration was carried out for the month of February which is the driest month of the year, and validation was carried out for the month of July. The calibration and validation results are shown in Fig. 17.5.

Five parameters were considered for the model calibration and validation. Initially, the rate coefficients were defined based on the similar previous studies. Then, the coefficient was optimized to achieve the minimum difference between the model representation and the field measured figures. Table 17.2 shows the route mean squared errors of different parameters for calibration and validation.

According to the values shown in Table 17.2, the error percentages of all parameters are less than 25% both in the calibration and validation processes except BOD. The BOD has been further deviated from the observation where validation values having best approximation of 32% deviation. The finalized model was used to investigate various river flow conditions and the pollution loadings. The finalized rate parameter used in the validated model is shown in Table 17.3

17.3.3 Model Analysis for Different Flow Conditions

17.3.3.1 The BOD Variation Over Three Different Flow Conditions

Similar to field observation, the BOD levels are high at the beginning of the section predominantly due to the waste discharges from industrial zone located just upstream of the sampling location. Subsequently, the conditions got diluted and stable up to 21 km chainage point downstream of Hanwella. Again, the BOD levels increased with the increase of absorptions and addition of pollutant toward the downstream. There is risk of exceeding the pollutant level beyond the tolerance limits of drinking raw water quality just before Ambathale. Further, it is noted that worst condition will occur once the absorption is increased to cater the future demand (Fig. 17.6).

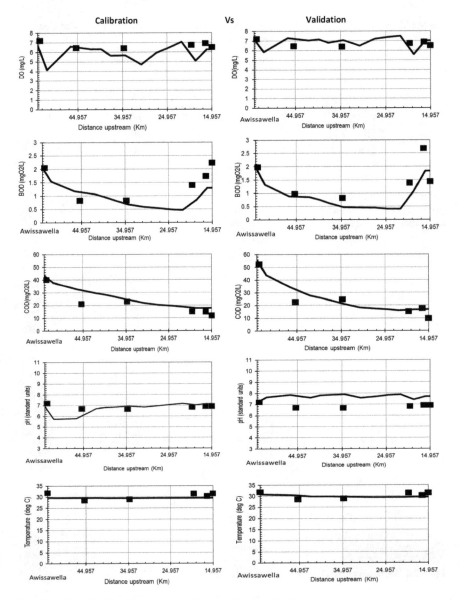

Fig. 17.5 Model calibration (dry season) versus validation wet season

17.3.3.2 The COD Variation Over Three Different Flow Conditions

Similar to characteristics shown in for BOD, the higher COD levels prevail at the beginning of the section. The conditions beyond the tolerance limits for drinking raw water are present at the upstream. Subsequently, the conditions get diluted into

Table 17.2 Route mean squared errors of different parameters for calibration and validation

S. No.	Parameter	RMSE			
		Calibration	%	Validation	%
1	Temperature	1.4	4.6	1.1	3.7
2	DO	1.3	25.3	0.8	11.5
3	BOD	0.5	41.6	0.4	32.0
4	pH	1.0	13.9	0.9	11.2
5	COD	3.9	15.1	6.0	22.4

Table 17.3 Major model parameters finalized at end of calibration and validation

Parameter	Used value/day
Reaeration rate, k_a	0–75
Hydrolysis rate, k_{hc}	1.8
BOD oxidation rate, k_{dc}	3.5
Pathogens, decay rate, k_{dx}	0.05
Decay rate	0.75

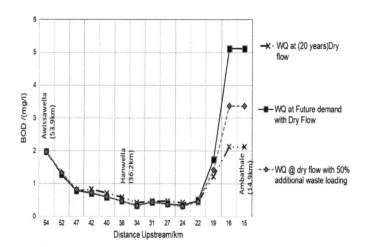

Fig. 17.6 Variation of BOD of three different flow conditions

stable level. However, marginal increase can be observed from 21 km chainage point at downstream of Hanwella. The influence of COD levels is not seemed to be severe compared to BOD, but possibility is there to increase with absorptions and mixing of additional pollutants (Fig. 17.7).

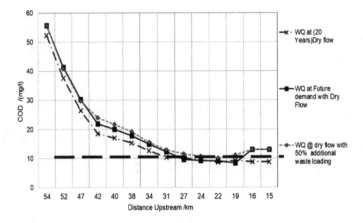

Fig. 17.7 Variation of COD of three different flow conditions

17.3.3.3 The BOD/COD Variation Over Three Different Flow Conditions

As shown in the figure, the BOD/COD ratio is increasing with growth of demand and pollution loading. The highest change was occurred at highest demand. As per the Figs. 17.6 and 17.8 biological pollutions creased toward the downstream and the pollution level was increased with increased demand. The overall results shows that increase in demand and the biological pollution are the key aspects to be addressed to mitigate the future risk of water supply.

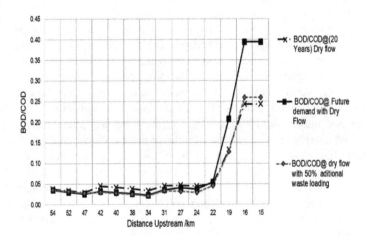

Fig. 17.8 Variation of BOD/COD of three different flow conditions

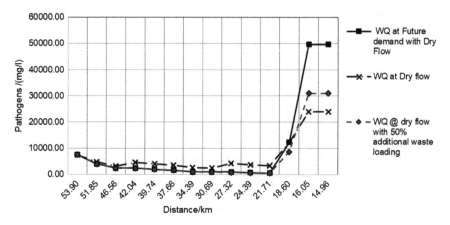

Fig. 17.9 Variation of pathogen for three different flow conditions

17.3.3.4 The Pathogen Variation Over Three Different Flow Conditions

Similar to characteristics shown in for BOD and COD, the pathogen concentration
has significant influence toward the downstream. The worst conditions were predicted
at the future demand with 50% additional pollutions. The results indicate that
extraction of the river causes to increase the concentration of pathogen and conditions
aggravated with the addition of more pollution (Fig. 17.9).

17.3.3.5 The River Characteristics at the Wet Season

During the wet season, the average river discharge volumes increase substantially
than dry flow volumes. Therefore, river carries high volume of water, and potential of
diluting pollutant loading is high. However during the wet flow conditions, the rate
of pollution entering the river through the industries and the extraction from the river
has no changes. Accordingly, the four scenarios considered for dry flow conditions
were modeled at wet flow situation. Fig. 17.10 show the results of the modeling.

As per the results shown from Fig. 17.10a–d for wet season, the influence
extraction of water from the river and addition of more pollutant to the river have no
significant effect over the overall pollution in river. The BOD levels vary between
2.0 mg/l and 0.5 mg/l which is well above the raw water quality requirements. The
COD variation was similar to the field observations and shows continuous reduction
from the high levels of COD at the start point, but the value is above the 15 mg/l
throughout the river indicating less oxidation characteristics. The BOD/COD ratios
have only minor increments compared to dry conditions. The pathogen concentration
has similar status both in wet and dry flows up to 22 km; thereafter only slight increase
could be observed in wet flows compared to rapid increasing shown in the dry flow.

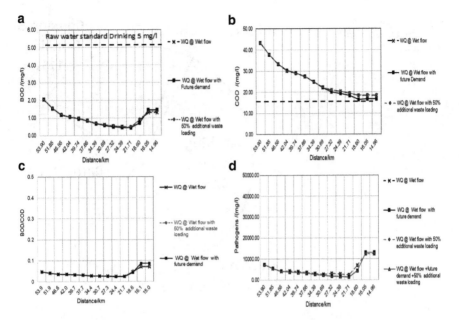

Fig. 17.10 **a** BOD variation at wet flow. **b** COD variation at wet flow. **c** BOD/COD variation at wet flow. **d** Pathegen variation at wet flow

The pathogen values of 50,000–20,000 in dry season have changed to 10,000–15,000 at wet season in the downstream areas.

According to overall observation made for the wet flow molding, it is emphasized that most critical condition of the river occurred only in the dry flow condition.

17.3.4 End User Measurements

As per the results shown in Figs. 17.11 and 17.12, the average per capita water consumption is 161 l/p/day. The water requirement for drinking and cooking is 28 l/d/(17%) which is the highest water quality requirement (WQR). Water requirement for washing clothes, washbasin, bathing, bidet and outdoor is 110 l/p/d and accounts for consumption 68% which needs average WQR. Toilet flushing and others are 23 l/p/d and account for 14% of total consumption and could be utilized with lowest WQR.

Accordingly, the healthy purified water requirement is 17% of total consumption, and water needs to match with all drinking water requirements. Out of the total consumption, 14% is for toilet flushing, and alternative method could be easily adopted by means of rainwater harvesting and partially treated recycled water replacing 30–70% of the consumption. The other catenary has the highest consumption but needs to be utilized with average quality water. This too could be

Fig. 17.11 End user consumption versus different applications

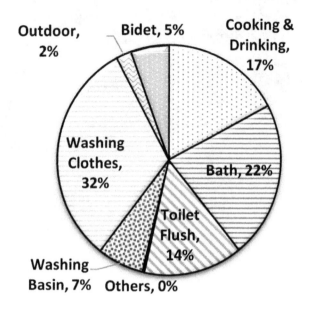

Fig. 17.12 Water quality requirements versus utilization

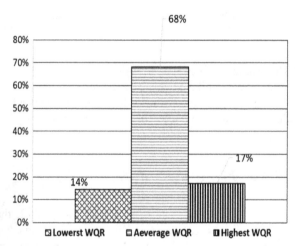

utilized with alternative sources such as groundwater or treated rainwater. However, finding average quality water is not possible as low quality water. Therefore, replacement of major percentage of average quality water has constrains, and around 20–30% of the consumption will be within manageable level.

17.4 Conclusion and Future Directions

17.4.1 Water Quality Variations in Kelani River Downstream

BOD levels have significant increase in river with high industrial and urbanization effects. Increased BOD to COD ratios show comparatively higher biological degradations than the industrial pollution. Mitigatory measure are needs to control biological pollution stimulated from uncontrolled wastewater discharges, solid waste discharges and other human activities. Future demand at dry flow increases BOD level above tolerance limits (5 mg/l) for raw water. Risk in drinking water supply systems at future occurs if pollution and extraction are not managed.

17.4.2 End User Measurement

As per the outcome of the end user measurement, the alternative water sources could be adopted to reduce the water demand. As shown in Table 17.4, 30% alternative consumption for toilet flushing could save 600–1300 million rupees per year, and water demand can be reduced by 0.043 MCM/Day as minimum. In addition, alternative use of groundwater and water with limited treatment to cover 30% of washing, bathing and bidet will reduce water demand by 0.377 MCM/Day in year 2030 and millions of investment required for the development of maintenance of

Table 17.4 Alternative water usage and respective projected saving

Type of usage	Per capita consumption (L/P/day)	Proposed percentage to use alternatives (%)	Saving volume/(MCM)		Project cost saving			
			Present demand (0.85)	Future demand (1.85)	Present demand (million rupees/year)		Future demand (million rupees/year)	
					O&M	O&M + Capital	O&M	O&M + Capital
Toilet flushing and out door	28	75	0.09	0.19	1499	5180	3261	11,273
		50	0.06	0.13	999	3453	2174	7516
		30	0.04	0.08	599	2072	1305	4509
Washing bathing and bidet	110	30	0.17	0.38	2911	10,063	6337	21,902
		20	0.12	0.25	1941	5709	4224	14,602
		10	0.06	0.13	970	3351	2112	7301
Total project saving	84	34	0.26	0.57	4410	15,243	9598	33,176
	102	22	0.18	0.38	2940	10,162	6399	22,117
	118.6	12	0.09	0.20	1570	5426	3417	11,810

water infrastructure facilities. It will also reduce water absorption from the river and risk of adverse pollution in Kelani River downstream in future.

Acknowledgements The work was supported by the Dr. Thushara Chaminda, Dr. Otaki. Ms. Kaori. NWSDB official, officials of the Department of Irrigation, CEA officials, BOI officials, officials of Department of Meteorological and the individual house owners.

References

Abeysinghe NMDeA, Samarakoon MB (2017) Analysis of variation of water quality in Kelani River, Sri Lanka. Int J Environ Agric Biotechnol (IJEAB) 2(6):2772. http://dx.doi.org/10.22161/ijeab/2.6.1

Alcamo J, Flörke M, Märker M (2007) Future long-term changes in global water resources driven by socio-economic and climatic changes. Center for Environmental Systems Research (USF), University of Kassel, p 247

Das N, Sarma KP, Patel AK, Deka JP, Das A, Kumar A, Shea PJ, Kumar M (2017) Seasonal disparity in the co-occurrence of arsenic and fluoride in the aquifers of the Brahmaputra flood plains, Northeast India. Environ Earth Sci 76(4):183

Herath G, Amarasekara T (2004) Assessment of urban and Industrial pollution on water quality Kelani River. In: The second international symposium on Southeast Asian water environment, Hanoi, Vietnam, pp 92

Himesh S, Rao CVC, Mahajan AU (2000) Application of steady state stream water quality model. J Indian Assoc Environ Manag 27:33–37

Kim J, Kuwahara Y, Kumar M (2011) A DEM-based evaluation of potential flood risk to enhance decision support system for safe evacuation. Natural Haz 59:1561–1572

Kumar M., AL Ramanathan, S. Chidambram, R. Goswami (2015) Criterion, indices, and classification of water quality and water reuse options. In: Eslamian S (ed) Urban water reuse handbook (UWRH), 1st ed, chap 6. CRC Press, pp 163–176. ISBN: 13-978-1-4822-2915-8

Kumar M, Ram B, Honda R, Poopipattana C, Canh VD, Chaminda T, Furumai H (2019a) Concurrence of antibiotic resistant bacteria (ARB), viruses, pharmaceuticals and personal care products (PPCPs) in ambient waters of Guwahati, India: urban vulnerability and resilience perspective. Sci Total Environ 693:133640. https://doi.org/10.1016/j.scitotenv.2019.133640

Kumar M, Chaminda T, Honda R, Furumai H (2019b). Vulnerability of urban waters to emerging contaminants in India and Sri Lanka: resilience framework and strategy. APN Sci Bull 9(1). https://doi.org/10.30852/sb.2019.799

Mallawatantri A, Rodrigo A, De Silva K (2016) Medium to long-term multi-stakeholder strategy and action plan for management and conservation of the Kelani River Basin 2016–2020. Natural Resource Management and Monitoring Unit, Central Environment Authority, pp 2–3

Otaki Y, Otaki M, Pengchai P, Ohta Y, Aramaki T (2008) Micro-components survey of residential indoor water consumption in Chiang Mai. Drinking Water Eng Sci 1:17–22

Patel AK, Das N, Goswami R, Kumar M (2019) Arsenic mobility and potential co-leaching of fluoride from the sediments of three tributaries of the Upper Brahmaputra floodplain, Lakhimpur, Assam, India. J Geochem Explor 203:45–58

Rajapaksha RWCN, Subasinghe SMCK, Halgahawatte HRLW, Weerasinghe WLK, Manoranjan M, Nandalal KDW, Raveenthiran K, Cassim M (2016) Water resources modelling study to update Water supply master plan for western Province metropolitan area. In: 20th congress of the Asia Pacific Division of the International Association for Hydro Environment Engineering & Research, Colombo, Sri Lanka, pp 5, 6

Water Quality in Kelani River Introduction Last Updated on Wednesday, 26 March 2014, Children's Fund UNICEF, and Cited on 19 Nov 2017 from. http://www.cea.lk/web/en/water?id=160 https://www.ck12.org/book/CK-12-Earth-Science-Concepts-For-Middle-School/section/6.1/

Zadeh SM, DVL Hunt, CDF Rogers (2014) School of Civil Engineering, College of Engineering and Physical Sciences, University of Birmingham, Birmingham B152TT, UK, pp 1961–1965

Chapter 18
Environmental Assessment and Implementation of Mitigation Plan to Protect the Environment: A Case Study of New Nizamuddin Bridge Project, Delhi (India)

Akansha Bhatia, Ankur Rajpal, Veerendra Sahoo, Vinay Kumar Tyagi, and A. A. Kazmi

Abbreviations

EIA   Environmental impact assessment
EMP   Environmental monitoring plan
MoEF Ministry of environment and forests
GoI   Government of India
MPN   Most probable number

18.1 Introduction

India is the fastest growing country with high rate of urbanization and industrial growth. Development through the use of technologies is required to improve the standard of living. In view of the fact that development projects are interfering with the environment, environmental concerns have to be addressed rationally before any project is grounded.

Environmental Impact Assessment (EIA) is an effort to anticipate measure and weigh the socio-economic and biophysical changes that may result from a proposed project. The objectives of the EIA include prediction of environmental impact of projects, finding ways and means to reduce adverse impacts, shaping the project to

A. Bhatia · A. Rajpal · V. Sahoo · V. K. Tyagi · A. A. Kazmi (✉)
Department of Civil Engineering, Indian Institute of Technology, Roorkee, Roorkee 247667, India
e-mail: kazmifce@iitr.ac.in

© Springer Nature Singapore Pte Ltd. 2020   339
M. Kumar et al. (eds.), *Resilience, Response, and Risk in Water Systems*,
Springer Transactions in Civil and Environmental Engineering,
https://doi.org/10.1007/978-981-15-4668-6_18

Table 18.1 MoEF proposed activities to conduct an EIA

Steps	Activities
i	Describe the proposed project as well as the options
ii	Describe the existing environment
iii	Select the impact indicators to be used
iv	Predict the nature and the extent of the environmental effects
v	Identify the relevant human concerns
vi	Assess the significance of the impact
vii	Incorporate appropriate mitigating and abatement measures into the project plan
viii	Identify the environmental costs and benefits of the project to the community
ix	Report on the assessment

suit local environment and presenting the predictions and options to the decision-makers. Ministry of Environment and Forests (MoEF), Government of India (GoI), provided the EIA rules (2006) as a major tool to minimize adverse impacts of developmental projects on the environment and to achieve sustainable development through timely, adequate, corrective and protective mitigation measures. MoEF has made it mandatory to conduct EIA for all the developmental projects, where tabulated (Table 18.1) steps should be taken in sequence to complete an EIA in an efficient manner.

Potential environmental impact and assessment, i.e., physical-chemical, biological and socio-economical evaluation are likely to affect the environmental components. The quality of water, air and noise pollution will be an environmental concern in the operational stage. Soil erosion and sedimentation will mainly be significant during the construction period, and with proper management, the impact can be minimized. Solid waste management will be an important component to be considered. Biologically, the impact on the surrounding flora and fauna needs to be accounted. Disturbance to the flora and fauna can be controlled to the minimum level especially in the constructional phase. Socio-economic aspects involve the land use, demographic variable (number of individuals and households), literacy rate of the residing populations, availability of social institutions and health care centers. Also, the Nizammudin bridge is used for the greater accessibility and improved communication facilities that are expected to be the part of benefit to the surrounding area.

An environmental monitoring plan has been prepared to check the efficacy of mitigation measures during construction and operation phases. The study plan includes the monitoring of Nizamuddin bridge project and surrounding areas for air quality, water quality of surface (Yamuna River) and ground water, soil quality, noise level measurements and Floral ecology. An EMP is included as part of this mitigation plan, which includes (i) mitigation measures for environmental impacts during implementation; (ii) an environmental monitoring program and the responsible entities for mitigating, monitoring and reporting.

- To evaluate the adequacy of environmental assessment.

- To suggest ongoing improvements in management plan based on the monitoring and to devise fresh monitoring on the basis of the improved EMP.
- To enhance environmental quality through prior implementation of suggested mitigation measures.
- Noise pollution analysis.
- Floral biodiversity identification.

18.2 Materials and Methods

18.2.1 Study Sites

The study sites were located near the eight lane new Nizamuddin bridge project over river Yamuna, which was the part of "development of 6/8 lane highway (NH-24) from km 0.000 to km 49.34 (Nizamuddin to U.P. border) in the state of Delhi."

Two new bridges, each of four lane configuration was constructed on both sides of existing Nizamuddin bridge. Two sites located near the construction zone (28°37'1.76" N, 77°19'32.97" E, Site-1) and the camp zone (28°6'0.38" N, 77°15'50.26" E, Site-2) were studied (Fig. 18.1).The site camp zone was the place to store construction materials where the office of the contractor was located about 5 km far from the construction site.

18.2.2 Sampling

In the present study, samples of river and ground water were collected at two different locations near the construction site of bridge (Fig. 18.2). Grab water samples were collected in 1 L bottles for physicochemical analysis. Samples for microbiological analysis were collected in sterilized 250 mL borosilicate glass bottles. Samples were preserved in an ice container at 4 °C prior to the analysis. Ample air space was left in the bottle (at least 2.5 cm) to facilitate mixing by shaking, before examination. Soil samples were collected at two different locations in the flood plain of the river Yamuna (downstream and upstream) to analyze the effect of construction on soil quality. Particulate matter$_{2.5}$ and PM_{10} in the air around the site were monitored. Floral vegetation including herbs, grass and weeds were collected from the site and identified in the laboratory.

The study period was divided in two phase, construction phase (August 2017–May 2018) and operation phase (June 2018–November 2018).

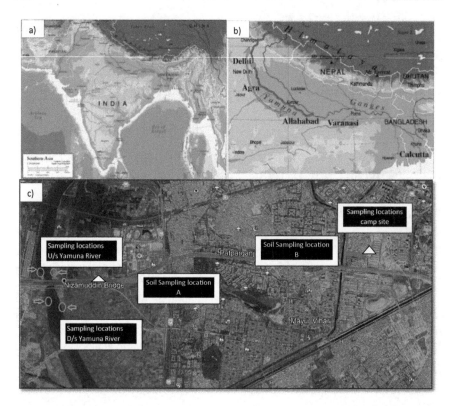

Fig. 18.1 **a** River map of India. **b** River Yamuna (Blue line) with red spot highlighted location at Delhi. **c** Location map of Nizamuddin bridge over river Yamuna at Delhi (*source* **a**, **b**: Harada 2008)

18.2.3 Methodology

The collected samples were transported to the Environmental Engineering laboratory at IIT, Roorkee, for further analysis. Standard operating procedures/methods were adopted to analyze the samples.

18.2.3.1 Water Quality Analysis

Physicochemical and microbiological analysis of the river and ground water samples were carried out for the parameters such as pH, turbidity, alkalinity, total solids, ammonia, nitrate, phosphorus, boron, heavy metals (Cr, Cd, Cu, Zn, Pb, Fe) and most probable numbers (MPN) for coliforms according to Standard Methods (APHA 2005).

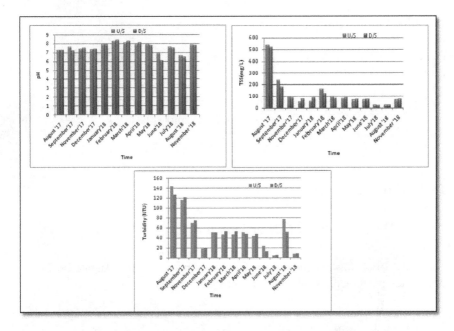

Fig. 18.2 pH, TSS and turbidity in upstream and downstream of river Yamuna water

18.2.3.2   Air Monitoring

Particulate matter$_{2.5}$ and PM$_{10}$ in the air around the site were monitored. Particulate matter was tested using the Respirable Dust Sampler (SLE RDS 103, Spectro, India).

18.2.3.3   Soil Testing

Soil samples were collected from two different locations at construction site near the river Yamuna (downstream and upstream) to analyze the effect on soil quality. Parameters such as pH, conductivity, organic matter, ammonia nitrogen, nitrate, phosphorus, calcium, magnesium, sodium and potassium were analyzed according to Standard Methods (APHA 2005).

18.2.3.4   Noise Pollution Analysis

Sound level meter (SLM 100, Envirotech, India) was used to measure the equivalent continuous linear weighted sound pressure level (LEQ) and sound pressure level (SPL) of noise near the sites.

18.2.3.5 Flora Identification

Plants were identified, and the density was calculated as plants/m^2 by using Quadrate method (Aneja 1996).

$$\text{Density} = \frac{\text{Number of individual of species in all sample plots}}{\text{total number of sample plots}}$$

18.3 Findings

18.3.1 Water Quality of River Yamuna

The study period was divided in two phase, construction phase (August 2017–May 2018) and operation phase (June 2018–November 2018).

18.3.1.1 PH, Solids and Turbidity

pH values varied from 6.6 to 7.9 during the construction and operation phases. TSS values ranged from 25 to 186 mg/L, except for the months of August and September 2017. Turbidity values reduced from 144 to 5 NTU and from 127 to 5 NTU in upstream and in downstream river water, respectively. pH values were slightly lower (<7) during the month of July and August attributing to the rainy season. TSS and turbidity were higher during the constructional phase. Variations in TSS values during the study period especially construction phase may be due to waste demolition near the banks of the Yamuna. However, the values of TSS significantly reduced during the operational phase of the project.

18.3.1.2 Dissolved Oxygen (DO) and Biochemical Oxygen Demand (BOD)

The dissolved oxygen (DO) concentration in river water was observed at the lower side (0.11–0.29 mg/L) during the construction period and varied from 1.5 to 2.22 mg/L during the operation phase. DO in river water should be at least 5 mg/L according to CPCB water quality standards. Values of BOD in the river water were found higher (12–45 mg/L) than the standard limits of CPCB (30 mg/L) during the construction and operational period. Low DO values the river water was mainly due to the high load of waste dumped in the river near the construction site. Demolition wastes from the construction sites include the heterogeneous mixtures of building materials such as concrete, wood, paper, metal and glass, contaminated with paints, fasteners, adhesive and dirt. These types of non-biodegradable wastes increase the toxins in the river water and become the homesite for the growth of unwanted weeds in the river that directly or indirectly compete for the dissolved oxygen present in the river. Simultaneously, the Yamuna River was not clean in the construction sites, and the biodegradable wastes consume the dissolved oxygen. However, the values of DO enhanced during the operational phase of the project owing to the rainy months (August to October). Similarly, values of BOD were higher during the constructional phase owing to the waste dumping in the river site due to construction activities, which decreased in the operational phase due to minimized disturbance to the river water, and reduced BOD values were depicted in rainy season attributing to the higher dilution of the river water.

18.3.2 Fecal Coliforms (FC)

FC values were higher in both upstream and downstream samples and ranged from 21,000 MPN/100 mL to 93,000 MPN/100 mL during the study period. This level of contamination is not allowed in the river water samples according to CPCB river water quality standards (500 MPN/100 mL). Fecal coliforms values were estimated to be higher owing to the discharge of untreated and/or partially treated sewage into the river during its course from the Delhi regions (DPCC 2018). The coliforms values were found to be the least affected due to construction activities. Dumping of fecal matters into the river from the hamlet settlement of the workers leads to the much higher values in upstream as compared to the downstream. More number of worker population was colonized near the upstream construction sites relating to higher coliforms contamination (Fig. 18.3).

18.3.2.1 Heavy Metals in the River Yamuna

Heavy metals in river Yamuna varied for the different elements such as lead, nickel, manganese, iron, copper, cobalt, chromium and zinc. All the tested heavy metals were in standard limits except, zinc, iron and lead in the monthly samples monitoring (Fig. 18.4). The higher values of iron and zinc were not considered as a threat to river flora and fauna, but the values of lead were higher in the downstream river water, which could be detrimental to biological activities. Lead (Pb) was incorporated into the use electrical conduits, water tank pipe coating and gutters and cladding. Construction materials disposal into the river slightly increased the lead concentration in river water especially in the rainy season (July to September'2017) due to leaching of the heavy metals (Tables 18.2 and 18.3).

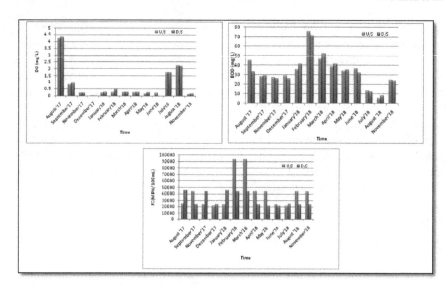

Fig. 18.3 DO, BOD and coliforms tested in Yamuna River water (upstream and downstream)

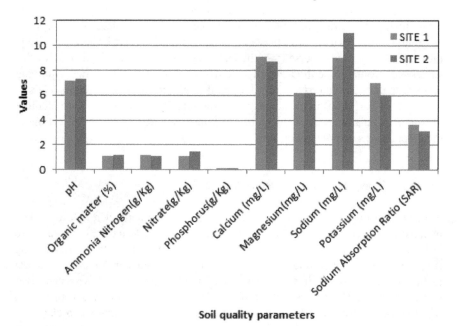

Fig. 18.4 Soil quality parameters at Site 1 (Construction site) and Site 2 (Camp site)

Table 18.2 Heavy metal results of river Yamuna upstreams during the construction and operational phase

Heavy metals	August'17	September'17	November'17	December'17	January'18	February'18	March'18	April'18	June'18	July'18	August'18	November'18	IS 10500:2012
Pb	0.02	0.01	0.03	0.03	0.02	0.02	0.02	0.01	0.01	0.01	0.01	0.02	0.01
Cr	0.07	0.03	0.06	0.07	0.04	0.04	0.04	0.03	0.04	0.04	0.03	0.07	0.05
Co	0.01	0.00	0.01	0.01	0.02	0.02	0.02	0.02	0.00	0.01	0.01	0.01	0.05
Mn	0.42	0.16	0.38	0.51	0.01	0.01	0.00	0.00	0.66	0.25	0.15	0.01	0.1
Ni	0.02	0.02	0.02	0.01	0.01	0.01	0.01	0.01	0.03	0.02	0.02	0.02	0.02
Cu	0.04	0.02	0.04	0.05	0.03	0.03	0.02	0.01	0.03	0.02	0.01	0.03	0.05
Zn	0.80	0.81	1.05	1.90	4.90	7.90	4.91	3.92	4.68	1.64	1.24	1.78	5
Fe	12.29	8.85	13.74	11.12	10.12	15.12	11.11	10.19	12.61	3.75	3.91	10.80	0.3

Results of heavy metals in River Yamuna (upstream)

Table 18.3 Heavy metal results of river Yamuna downstream during the construction and operational phase

Results of heavy metals in River Yamuna downstream

Heavy metals	August'17	September'17	November'17	December'17	January'18	February'18	March'18	April'18	June'18	July'18	August'18	November'18	IS 10500:2012
Pb	0.03	0.02	0.03	0.03	0.01	0.01	0.02	0.02	0.02	0.01	0.01	0.01	0.01
Cr	0.06	0.03	0.06	0.08	0.03	0.02	0.02	0.01	0.05	0.04	0.02	0.03	0.05
Co	0.01	0.00	0.01	0.01	0.02	0.03	0.02	0.01	0.00	0.00	0.01	0.01	0.05
Mn	0.43	0.11	0.35	0.63	0.03	0.02	0.02	0.01	0.61	0.20	0.21	0.01	0.1
Ni	0.02	0.02	0.03	0.02	0.01	0.01	0.01	0.01	0.04	0.02	0.02	0.01	0.02
Cu	0.04	0.02	0.04	0.05	0.04	0.04	0.03	0.02	0.04	0.02	0.01	0.02	0.05
Zn	1.29	0.81	1.36	2.29	3.29	5.29	5.12	5.12	4.19	1.24	1.37	1.28	5
Fe	13.67	12.06	16.73	12.66	12.36	13.46	12.43	11.21	12.5	3.91	3.75	9.80	0.3

18.3.3   Quality of Ground Water

Ground water samples were collected from upstream and downstream sites (near the construction site) of river Yamuna and analyzed for physicochemical and microbiological quality. The water quality of all the analyzed samples was found satisfactory and within the limit as per Indian Standards (Table 18.4).

Table 18.4 Physicochemical and microbiological quality of groundwater samples collected from upstream and downstream sites of river Yamuna

Parameters	Ground water (U/S)	Ground water (D/S)	IS 10500:2012 (drinking water)
pH	6.78	6.9	6.5–8.5
Alkalinity (mg/L)	370	390	200
Total acidity (mg/L)	60	80	–
EC (μS/cm)	1220	1182	–
Turbidity (NTU)	1.21	1.13	5
Total dissolved solids (mg/L)	732	709	500
Total suspended solids (mg/L)	Nil	Nil	–
COD (mg/L)	Nil	Nil	–
BOD (mg/L)	Nil	Nil	Nil
Ammonia nitrogen (mg/L)	Nil	Nil	–
Nitrate (mg/L)	2.2	2.3	45
Phosphorus(mg/L)	0.02	0.02	–
Total Hardness as $CaCO_3$ (mg/L)	330	350	300
Calcium as Ca^{2+} (mg/L)	128.6	132.7	75
Magnesium as Mg^{2+} (mg/L)	4.7	7.1	30
Sodium (mg/L)	120	110	–
Potassium (mg/L)	17	16	–
Chlorides (mg/L)	210	215	250
Sulfate (mg/L)	56	51	200
Iron (mg/L)	0.46	0.41	0.30
Boron (mg/L)	Nil	Nil	1
Fecal coliforms (MPN/100 mL)	Nil	Nil	Absent
SAR (meq/L)	1.49	1.69	0–9

The pH concentrations ranged from 6.78 to 6.86 and found within the desirable limits as per Indian Standards for drinking water. BOD values were tested to be negligible in the ground water samples. TSS were absent in the all the samples collected and analyzed. Fecal coliforms were absent in the ground water samples. The analyzed samples are suitable for the drinking purpose as per the IS 10500:2012 standards (BIS 2012). Nutrients like ammonia, nitrate and phosphorus were analyzed. Ammonia–nitrogen was absent, whereas the nitrate ranges around 2–2.5 mg/L. Phosphorus limits were low, i.e., 0.2–0.5 mg/L. Chlorides values were at higher side ranging around 210–250 mg/L. Sulfates values were around 55–60 mg/L.

18.3.3.1 Sodium Absorption Ratio (SAR)

SAR is a factor in determining the suitability of water for irrigation. SAR values between 0 and 3 meq/L reflect toward excellent water quality, whereas if greater than 9 meq/L, soil is not suitable for the irrigation purpose. However, the river water values <3 meq/L depicted that the groundwater was not affected by the construction activity and quality was satisfactory throughout the study period.

18.3.4 Sediment Quality Analysis of River Yamuna

Heavy sediment deposits affect river water quality as it can fill in drains, lakes, rivers and sewage blockage. Sediment loss from the construction site includes plastic, cement, etc., that can also cause flooding, which can carry toxins into the river water. Table 18.5 summarized the sediments quality of Yamuna River near the project sites. Organic matter content of sediment ranged from 2 to 3%. Sandy–clayey type of sediments was observed. Ammonia nitrogen varied from 2 to 3 g/kg for both the upstream and downstream sediment samples. Nitrate nitrogen and phosphorus were

Table 18.5 Sediment samples quality parameters near the sites

Parameters	Yamuna (U/S)	Yamuna (D/S)
pH	8.2	8.3
EC (μS/cm)	768	762
Organic matter (%)	2.8	2.9
Ammonia Nitrogen (g/kg)	2.34	2.32
Nitrate (g/kg)	1.9	1.8
Phosphorus (g/kg)	0.21	0.24
Calcium (mg/L)	8.5	8.6
Magnesium (mg/L)	6.12	7.22
Sodium (mg/L)	8	6.9
Potassium (mg/L)	7.8	6.9

Table 18.6 Noise level monitoring near the sites	Parameters	Site 1 (construction site)	Site 2 (camp site)
	SPL (dB)	64.3 ± 12.4	65.2 ± 13.6
	LEQ (dB)	66.4 ± 12.8	67.2 ± 16.4

1.9 g/kg and around 0.2 g/kg, respectively, in both U/S and D/S samples. Sodium and calcium values ranged about 7–8 mg/L for both the samples.

18.3.5 Noise Level Measurement

Noise level was checked by using portable noise meter and measured in sound pressure level (SPL) and equivalent continuous sound pressure level (LEQ) at the sites. Average values for the study period are shown in Table 18.6. The noise pollution measured was observed to be slightly affected during the construction phase owing to the construction machines activity and vehicular movement nearby the monitoring sites. However, it does not surpass the standard limits of 80–90 dB during the entire study period that can adversely affect human audible range

18.3.6 Soil Quality Analysis

Sandy–silt type soil was collected having organic matter (1–2%) was analyzed in the samples. However, the amount of nitrate-nitrogen was sufficient for the plant growth. Phosphorus values were low in the analyzed soil samples on an average, and the soil quality was satisfactory during the sampling at both the sites (Fig. 18.4).The findings revealed that soil quality was not affected by the construction activities as the quality of soil was sandy–silt type in nearby regions. However, the plantation activity was recommended at the project sites to increase the afforestation in the area.

18.3.7 Air Pollution Monitoring

The results depicted the higher values of particulate matter in reference to the National Ambient Air Quality (NAAQ) standards, 2009 (100 $\mu g/m^3$ for PM_{10} and 60 $\mu g/m^3$ for $PM_{2.5}$ @ 24 h). Values of PM_{10} and $PM_{2.5}$ were higher near the construction site, while the values considerably reduced at the camp site in the post construction phase as compared to the construction phase. This may be attributed to the high vehicular pollution in the winter season and the construction work carried near the sampling locations (Fig. 18.5a, b). Particulate matter values were depicted to be higher throughout the constructional phase attributing to the construction activities.

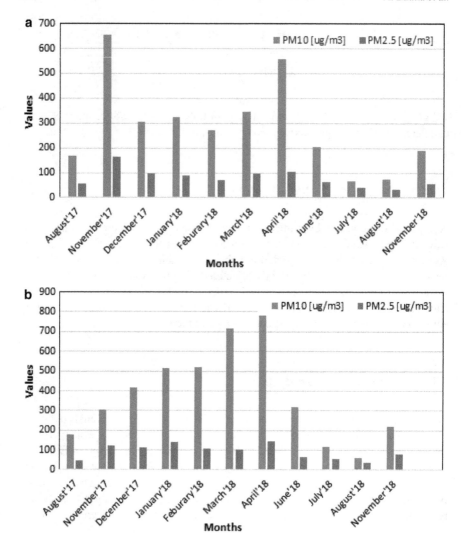

Fig. 18.5 **a** Particulate matter at Construction (Site-1). **b** Particulate matter at camp office (Site-2)

The values were found to be higher in the winter season due to the accumulation of pollutants in the lower atmosphere. Although the values reduced during the operational phase attributing to lesser constructional and vehicular activities at both the sites.

Table 18.7 Floral identification and density

Scientific name	Common name (local name)	Density (plants/m^2)
Ipomoea alba	White morning glory	0.01
Dactyloctanium sindicum	Crow foot grass	0.01
Ipomea aquatica	Water morning glory	0.01
Alternanthera philoxeroides	Alligator weed	0.04
Amaranthus viridis	Cholai	0.05
Cynodon grass	Dhoob	0.00
Datura stramonium	Datura	0.05
Phyllanthus niruri	Amla	0.01
Portulaca oleracea	Purslane	0.21
Persicaria decipiens	Swamp willow weed	0.11

18.3.8   Flora Identification

Floral vegetation nearby the study sites was identified, and density for each of the found species was calculated (Table 18.7). Species found were locally known as *White Morning Glory, Crow foot grass, Alligator Weed, Dhoob grass, Datura* and *Swamp Willow Weed.* These species were observed to be located in the study region for whole of the period without any disturbance from the construction of the bridge.

18.4   Socio-economic Analysis

Socio-economic survey near the Delhi–Meerut Expressway (DME) construction site revealed that the project causes the least damage to the heritage sites of the national capital. Social infrastructure such as schools and hospitals was established near the sites. There was one Government hospital and two private hospitals in nearby area. In addition, vaccination and health checkups of the workers were carried out through the efficient team of doctors, hired by the contractor after every three to four months. The contractor provided basic first aid facilities to workers.

18.5 Mitigation Measures

Various mitigation measures and recommendations were given by the experts from IIT, Roorkee. This includes water sprinkling should be carried out to suppress the plant dust emissions. The site should be clean and tidy. The contractor must conduct safety audit for site improvement and safety of workers. Safety barriers at construction site must be installed. Silt bearing water pumped from the cofferdam should be diverted through an effective silt trap prior to discharge into the watercourse. Contractor has to make more efforts to prevent any kind of construction waste spill in Yamuna River. Demolishing waste dumped upon Yamuna River Bank side should be disposed properly. It is suggested to provide regular training sessions to the workers about correct working techniques and on preventing injuries. Mitigative measures such as wet suppression should be taken by the contractor to minimize exposure for the workers to the particulate matters especially respirable dust matter. Planting of eight trees for every single tree deforested was recommended. The floodplains around the bridge required restoration as per the committee's recommendation accepted by NGT. The issue of trained and adequate manpower for enforcement of EIA recommendation should be given due consideration.

18.6 Conclusions

The result of river and ground water testing near the construction site indicates the satisfactory values. There is no improvement in the measured values of coliforms in the river water samples. The results indicate the higher values of particulate matter in reference to the National Ambient Air Quality (NAAQ) standards, 2009, at both the sites during the winter season. Values for the PM_{10} and $PM_{2.5}$ were higher near the construction site, while the values considerably reduced at the camp site in the operational phase as compared to the earlier month reports. Organic matter content of sediment ranged from 2 to 3%. Sandy–clayey quality of sediment was observed. No records were observed for the zooplanktons and phytoplanktons in the river water. In conclusion, EIA for any project will depend on the production of high quality impact statements as well as proper reviewing and the subsequent policing to ensure that the prediction of impacts is fair and mitigation measures are effective.

References

Aneja KR (1996) Experiments in microbiology. Plant Pathology and Biotechnology

APHA (2005) Standard methods for water and wastewater, 21st ed

Bureau of Indian Standard (2012) Drinking water standards 10500:2012 (Second Revision), 1–8

Delhi Pollution Control Commitee (2018). https://www.dpcc.delhigovt.nic.in

Environment Impact Notification (2006) Gazette of India, 14th September

Harada H (2008) India-Japan international collaboration for an innovative Sewage treatment technology with cost-effective and minimum energy requirement. Asian Science and Technology Seminar (ASTS). Thailand, 9–11 March
https://www.designingbuildings.co.uk/wiki/Lead_in_construction
National Ambient Air Quality Series (2009) Guidelines for the measurement of ambient air pollutants volume-I (guidelines for manual sampling & analyses). NAAQMS/36/2012-13

Chapter 19
Assessment of the Significance of Water-Energy-Food Nexus for Kuwait

Dr Amjad Aliewi and Dr Husam Alomirah

19.1 Introduction

The water, energy and food (WEF) nexus means that the securities of the water, energy and food sectors are linked in a complex manner such that actions in one sector do have impacts on the other two sectors. Hence, cross-sectoral and dynamic perspective between the WEF sectors is important to identify so that the impacts of a sector-related decisions on the other sectors should be considered. This effectively helps determine the potential trade-offs and synergies between the sectors. Therefore, it is important to assess and manage the water- energy-food nexus collectively to suggest to the decision-makers suitable interventions, management options and policy statements regarding the nexus resources. Managing natural resources for water-food-energy security is a global challenge. Their sustainable management is more profound in the hyper-arid areas which face harsh environment, infertile soil and little rainfall. The WEF nexus is about keeping a balance among the needs of the society, people, industry, agriculture and the environment for the WEF resources. The nexus will help ease sectorial tension and encourage regulations and policies that benefit water, energy and food security sectors equally. The nexus inter-linkages should be used and managed to take into consideration changes in population, industry, agriculture, regional trades, technological development and changes in climate and diets (Fig. 19.1). As Kuwait is moving to achieve sustainable development goals, it faces some stresses on the elevated demands on water, energy and food resources in

Dr A. Aliewi (✉)
Water Research Center, Kuwait Institute for Scientific Research, Kuwait City, Kuwait
e-mail: aaliewi@kisr.edu.kw

Dr H. Alomirah
Food and Nutrition Program, Environment and Life Sciences Center, Kuwait Institute for Scientific Research, Kuwait City, Kuwait

© Springer Nature Singapore Pte Ltd. 2020
M. Kumar et al. (eds.), *Resilience, Response, and Risk in Water Systems*,
Springer Transactions in Civil and Environmental Engineering,
https://doi.org/10.1007/978-981-15-4668-6_19

357

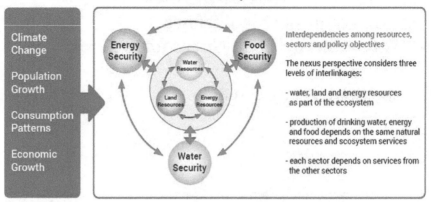

Fig. 19.1 Interconnectivity between the water, energy and food nexus (Ref. Al-Zubari et al. (2018)

an already water-energy-food-stressed country. Assessing the integration and inter-connectivity among water, energy and food sectors to achieve sustainable development goals for Kuwait is becoming essential because of the challenges this country is facing. These challenges include climate change, population growth, economic development, urbanization, extended agriculture and the increase in water demand. In a study about the water-energy nexus, Siddiqi and Anadon (2011) assess the importance of the water-energy nexus approach in the Middle East and North Africa (MENA countries). Their results show a weak reliance of energy systems on freshwater but a strong reliance of water abstraction and production systems on energy. They reported that 9% of the total annual electrical energy consumption in Saudi Arabia is attributed to water generation. They also reported that other Arabian Gulf countries consume 5–12% of total electricity production for desalination. They concluded that the policy makers in Saudi Arabia have to take into consideration energy implications during development of the agricultural sector and expensive desalination systems. Abderrahman (2001) carried out a nexus study for Saudi Arabia. He showed that the energy plants at the Red Sea produce about 20% of the total national electricity production in Saudi Arabia and that the energy requirements to pump 1 m^3 of groundwater from wells range between 0.4 and kWh. In his analysis, he suggested that there is a need to reduce power consumption and water demand to conserve energy and water and to minimize negative impacts on the environment.

In 2015, the World Resources Institute placed Kuwait among the seven highest-ranked countries that will be at 'extremely high water stressed' by 2040 (Luo et al. 2015) The only natural water resource in Kuwait is the groundwater as there are no permanent rivers, dams or lakes and the long-term average rainfall is around 121 mm per annum; high evaporation rates and deficient soil moisture mean only a small percentage of this infiltrates aquifers. The country has almost no internally renewable sources of groundwater except for some lateral underflow from Saudi Arabia. The groundwater withdrawal rate is 255 million m^3 per year which is more than 10 times the annual recharge rates. In addition to this limited resource, Kuwait relies on desalination, and recently on treated wastewater, to provide water for all sectors including agriculture, which consumes 54% of this freshwater (FAO 2009; Ismail 2015). The nexus is about improved coordination and integration among the three sectors to provide wise ideas to support decision makers for better strategic planning to achieve sustainable development goals. The nexus approach is normally understood to produce a resource recovery through improving resource use efficiency. However, Majgl et al. (2015) illustrated that the interactions among nexus sectors are ought to be dynamic. The basis for the idea of the nexus is to try to establish balance between different uses of ecosystem resources (such as energy, water, land and soil) with socioeconomic factors (FAO 2014). This balance helps achieve sustainable development (Biggs et al. 2015; Andrews-speed et al. 2015). One more major challenge in the context of this nexus is governance. What regulations and governance arrangements can help address the growing industry, agriculture, economy and the securities in terms of water, energy and food? Toward its sustainable development, Kuwait makes sure that the ecosystem is the center of the WEF nexus (through environmental protection) which will assist in the development of policy and management tools. In this context, it is always important to involve many relevant sectors in meetings and dialogues about the proposed nexus in order to reduce risks. Therefore, proper interconnectivity with an increasing number of trade-offs among the resources of the nexus should be clearly identified for Kuwait, although it is somehow complex because of the growing demands on these resources. The use of renewable energy (as power sources) for water desalination plants (that produce freshwater) and the use of virtual water concept for food production in Kuwait are examples of integrated interconnectivity between the three sources. A number of the GCC countries acquire farmlands from foreign countries (rich with water and agricultural lands) for purchase or lease to grow essential crops. As an example, UAE and KSA utilize 380,000 and 500,000 hectares from Sudan and Kenya, respectively, for agricultural activities (Saif et al. 2014).

The aim of this chapter is to assess the interactions between water, energy and food systems in Kuwait in order to develop possible interventions and policy statements with suggested effective governance against overarching national challenges such as food security and the sustainable use/supply of energy and water. Kuwait is developing in a way to have sustainable and stable availability of water, energy and food sources at all levels: individual, household and national (Loring et al. 2013). Domestic production of food in Kuwait can only supply a small fraction of the current dsssemand (http://www.phenometricsinc.com/; http://www.algaeindustrymagazine.com/algae-wastewater-cleaning-technology-scores-wiscons). Unfortunately, Kuwait's arid climate, low soil fertility and scarcity of natural water resources make sustainable food cultivation very challenging due to the associated water and cooling energy requirements. Kuwait aims to become self-sufficient in food by 2040 (Ismail 2015). However, the sustainability of this action is questionable as Kuwait cannot be a country with self-sufficiency in food because of water resources scarcity and the poor suitability of its lands for agriculture. Such strategy risks depleting all freshwater resources and increases the need for more desalination with all the related environmental and economic consequences. Therefore, optimization of resources among sectors is mandatory for environment and production sustainability rather than aiming to achieve food self- sufficiency. Climatic changes of witnessing more droughts and floods events with changes in rainfall patterns over Kuwait will make the nexus concept more challenging. The nexus requires identification of the inter-linkages between water-energy-food sectors and how the inter-linkages can be used to suggest policy statement principles about sustainable development in the sectors (Endo et al. 2015). In this concept of WEF nexus, cities in Kuwait should be developed to use water, energy and food resources toward the optimal use and reuse as much as possible. There is a need to improve water quality management, optimal allocation of water and energy resources, in addition to sustainability of food supplies. Energy should be generated from waste, less water should be used for energy generation, renewable energy should be used for desalination, waste must be reduced and the supply management should be improved. The integrated concept and planning of the nexus require suggesting a set of policy statements that help to understand the inter-linkages between nexus sources, trade-offs and risks in order to develop integrated solutions with minimum associated risks. There is a need to illustrate that Kuwait will benefit on the national level from the nexus approach. The WEF nexus for Kuwait is linked to several SDGs by directly addressing SDG2 (zero hunger), SDG6 (clean water and sanitation), SDG7 (affordable and clean energy) and SDG12 (responsible consumption and production). The impacts of this study will be very useful to serve SDG13 (climate action) and SDG17 (partnerships for the goals).

19.2   Identifying Potential Inter-Linkages Between Water-Energy-Food Sources in Kuwait

The authors conducted a field survey and met several decision makers in the three sectors of water, energy and food supplies in Kuwait. They managed to visualize some potential inter-linkages that can be used to identify synergies and trade-offs between sustainable water, energy and food security. It was realized that electricity generated from solar power and geothermal (or renewables) can be used for pumping groundwater for irrigation, to boil and sterilize water for drinking and cooking purposes and for water desalination. Growth of some crops with smaller amount of water can be used for biofuel production. Marginal water (e.g., brackish water, produced water in oil production) can be used in energy production.

Energy can be saved and used more efficiently in adopting technologies that can also make efficient use of water resources. In these discussions, the decision makers of WEF sources in Kuwait emphasized that irrigation systems can use energy efficiently. In the same direction, they stressed that energy can be recovered from biomass, organic waste and wastewater. Some crops that can be used for biofuel production may be grown with smaller amount of water. There was an understanding at the decision-makers' level that the use of brackish water should be maximized to produce oil. As far as food and water linkages ar concerned, the improvements in water use efficiency should support agricultural activities in a way to use less water for more produced food mainly through irrigation. There is a need to use non-renewable energy sources to increase food yields. Food imports and storage are big issues in the context of the WEF nexus for Kuwait. The decision makers agree that there must be reasonable investments in food imports and storage. The use of treated wastewater for the development of inland fisheries and aquaculture in agricultural activities should be encouraged in Kuwait. Hence, the investment in reuse of drainage water and treated municipal and industrial wastewater to maintain stable food production should be encouraged too. Kuwait cannot continue pumping groundwater at unsustainable rates leading to its salinization and deterioration in its quality. It can be illustrated that using virtual water concept may be relevant to the overall economy of Kuwait. The cost of utilized water for agriculture and industry will be significantly reduced if virtual water policy is implemented widely in Kuwait. The preliminary discussions with Kuwait dairy received warm attention and interest. Perhaps government cost share to offset the initial investment of virtual water and desalination from renewable energy sources concepts could be worthwhile, as done in the western countries. The inter-linkages between water and agriculture require investigation about the efficiency to use phytoremediation for the plants to remove chemical contaminants from water and to reduce salinity levels for efficient irrigation to generate food. Hence, Kuwait may focus on selected food production and promote water-efficient and drought-resistant crops and increase the reuse of treated wastewater in agriculture, while keeping virtual water trade through some food imports (especially high-water demanding foods) to alleviate the water stress, achieve sustainability and build a food secure nation. Importing food and embedded 'virtual water' releases groundwater for more strategic use (Aliewi et al. 2017; Zubari 2014). In fact, sustainable growth of food crops in Kuwait requires (i) selection of suitable and sustainable cultivation strategies, (ii) sustainable sources of irrigation water and (iii) conditioned growth climates.

In the Arab region, there are few studies which were carried out to illustrate the viability for conducting nexus concept. These studies were cited in (Al-Zubari et al. 2018). An integration of good agricultural practices that combine technology such as smart irrigation, green energy, optimization through modeling, crop selection for better water productivity, improved efficiency, alternative water resources, land management and food trade is needed to address the major challenges to sustainable agricultural development in Kuwait. This integration and the modeling optimization require understanding of the inter-linkages between the water-energy-food (WEF) resources which allows the quantification of the trade-offs between water, energy and food (Amy et al. 2017; Daher and Mohtar 2015; Degirmencioglu et al. 2019; Lee et al. 2019; Mohtar 2017; Mohtar and Daher 2016; Mohtar et al. 2014; Mortada et al. 2018). The WEF nexus has become the center of global policy, development and research in order to meet the ever-increasing demand on water, energy and food against strong resource limits (Al-Zubari et al. 2018). Another challenge that requires attention in the development of the WEF nexus approach for Kuwait is to model and manage the availability of the WEF sources, waste generation and cost in order to evaluate their development and management options.

19.3 Discussion on Suitable Strategic Solutions for Kuwait that Maximize the Benefits from the WEF Nexus

The strategic solutions that this study realizes to be so important to Kuwait to maximize the benefits from the call of the WEF nexus approach are (1) virtual water, (2) the use of treated wastewater in irrigation, (3) the use of renewable energy in pumping water, wastewater treatment and desalination plants and (4) the reuse of waste from livestock manure and organic waste These strategic solutions cannot be interconnected and integrated in practical terms without developing suitable policy principles, regulations and governance that increase efficiency and security in the fields of water, energy and agriculture for Kuwait. In other words, some of the above integrated solutions are vital for sustainable water reuse and efficient cultivation of crops in conditioned greenhouses in Kuwait. These integrated strategic solutions will quantify trade-offs and interactions between water, energy and food. These solutions will be implemented while observing and assessing the environmental impacts as well as the socioeconomic feasibility of food imports to Kuwait and how it should be addressed. Some more details are provided below about these strategic solutions.

1. The virtual water as a strategic solution.

The concept of virtual water is about importing embedded water in food instead of generating water supplies from scarce indigenous water resources for food (Aliewi et al. 2017; Al-Rashed and Aliewi 2018). Allan and Lant (2003) illustrate that countries (like Kuwait) can save their scarce water resources for more critical needs by relying more on imported food because 95% of water needs by a person is embedded (hidden) as virtual water in food. In addition, if Kuwait develops as a food self-sufficiency country, then it has to consume 90% of its national water resources (including a good portion of the desalinated water) to produce its food. The concept of the virtual water in Kuwait is very much connected with water security. Kuwait is still behind with regards to implementing the concept of virtual water because its benefits are not well recognized. Importers of essential food to Kuwait are in real terms practicing the non-self-sufficiency policy in food. It is important to note that Kuwait is already importing most of its food from abroad, but the concept of virtual water that this paper is calling for Kuwait is special. This paper calls for a special connection between trade in food with Kuwait shortages of freshwater through purchasing large farms in countries rich with fertile lands for agriculture and rich with freshwater resources. This purchase of these farms will benefit the economy of Kuwait in two directions: from the ownership of the farms and from the embedded water in the food produced from these farms especially if the labor forces belong to Kuwait. As a result, Kuwait economy will be boosted by this special formula of trade and from bridging the gap of freshwater needs for agriculture.

This chapter analyzes that there is a lack of realizing the extent to which the food trade can alleviate the worsening of water deficits on the national economies of Kuwait. The link between livelihoods of Kuwait people (who live in an arid environment) and water security can be made through this special concept of virtual water as explained above. It should be noted that food self-sufficiency policy for Kuwait is a dangerous hope as it strengthens the high water using sectors. It encourages farmers to overpump brackish groundwater resources and overuse of freshwater resources from desalination plants which affect the overall economy of Kuwait. Kuwait must make up its food deficit by importing food from its own farms from abroad. Japan, Hong Kong and Singapore have enough water to meet domestic and municipal demands, but they do not have enough water resources to generate their foods. Because of implementing the virtual water concept wisely, they are now water secure and food secure (Allan and Lant 2003). It should be emphasized that the contribution of agriculture in Kuwait to employment, exports and national economy is weak. Therefore, food imports from its own farms are much better for Kuwait economy than generating food locally to export. In real terms, Kuwait economy is a non-farm economy, and as a result, it should not be diverted to depend on farming inside Kuwait itself. Kuwait strongest income comes from the oil industry and that is far away from farm industry which cannot be sustainable due to poor lands for agriculture and scarce freshwater resources to irrigate in Kuwait.

2. The link between using brackish groundwater and treated wastewater for irrigation.

Kuwait uses at the moment about 90 million cubic meter per year (Mm^3/yr) as brackish groundwater for agriculture and oil industry. The official statements of the Ministry of Electricity Water (MEW) indicate that this figure will increase in the near future to 250 Mm^3/yr for irrigation purposes. This is a large increase knowing that agriculture consumes 60–90% of groundwater and provides a low contribution (less than 1%) to Kuwait national economy (Aliewi et al. 2017; Al-Rashed and Aliewi 2018). Also, Kuwait at the moment uses 110 Mm^3/yr treated wastewaters for greenery purposes and 120 Mm^3/yr as reverse osmosis-treated wastewater for crop agriculture (Aliewi et al. 2017; Al-Rashed and Aliewi 2018). This total of 230 Mm^3/yr (36% of the generated raw wastewater) of the reuse of treated wastewater in irrigation is still low (Aliewi et al. 2017; Al-Rashed and Aliewi 2018). Our target is to improve the sustainability of the existing brackish groundwater resources through recycling (reuse) treated wastewater. This is a wise reuse of the waste that should be maximized in Kuwait. Therefore, Kuwait should increase its reuse of treated wastewater in irrigation to much higher than the current percentage of 36%. The concept of meeting irrigation water needs from brackish groundwater or from desalinated water is not a strategic solution. The utilization of waste (in this case wastewater) to a suitable level of treatment, for irrigation and aquifer recharge is a strategic solution in this nexus approach we are addressing.

3. The use of renewable energy in pumping water for irrigation, in desalination plants and in treatment of wastewater.

The renewable energy is a strategic solution in the call for the nexus approach for Kuwait. Although its use in Kuwait is still humble, its use in the Arab region requires special attention. Al-Zubari (Al-Zubari et al. 2018) showed that the improved efficiency in the WEF sectors can be achieved by integrating hydropower and renewable energy. (a) He discussed that increasing the efficiency of irrigation through coupling hydropower with renewable energy has a potential to enhance food security in Egypt. The plans for using renewable energy sources in Egypt include wind and solar energy. Al-Zubari (Al-Zubari et al. 2018) explained that the farm powers 30 kW submersible pump that has an average flow rate of 120 m^3/h. The well which is powered by solar energy serves a pivot irrigation area of around 120 acres. In our analysis, this is a major achievement in this discipline as operating a well with 120 m^3/h pumping rate is considered suitable to meet irrigation needs in Kuwait. (b) Also, Al-Zubari (Al-Zubari et al. 2018) showed that the use of renewable energy for wastewater treatment in Jordan is very feasible. He showed that the annual average energy consumption of the activated sludge system at a plant there was around 61.58 GWh. Improving the efficiency of the operation of this plant by utilizing hydraulic renewable energy and biogas produced through anaerobic digestion has resulted in better results for this nexus approach. The results indicate that the biogas production generates thermal and electrical power of 5.4 MW, and hydraulic energy accounts for 3.45 MW. In addition to that, the system had reduced CO_2 emission by around 300,000 tons/yr, and the effluent of the treatment plant (100 Mm^3/yr) is used for agricultural production. (c) The use of renewable energy to operate desalination plants. Al-Zubari et al. (2018) shows that Saudi Arabia pumps 300,000 barrels of oil per day to operate desalination plants. Using renewable energy according to the Saudi Arabia plans will reduce production costs of desalinated water from 2.2 to 5.5 SR/m^3 to 1–1.5 Saudi Riyals/m^3. This is a saving in operational cost of 55–73%. Any increase in oil consumption due to reason of natural development will definitely mean less oil export and subsequently reduced revenues of the national economy. Saudi Arabia is planning complete use of solar energy to all water desalination plants in the near future. The results will be extremely beneficial to Saudi Arabia economy and environment as the operational costs, and GHGs emissions will be reduced significantly.

4   The reuse of waste from livestock manure and agricultural activities.

Most of Kuwaiti population lives in urban areas which increases the amount of biowaste. One of the main challenges here is to ensure economic growth in Kuwait without exponential growth in the volume of waste. Kuwait will benefit from this waste as a valuable resource. Kuwait will develop to minimize the hazardousness of urban wastes through separation of organic waste, efficient collection and transport of organic waste that will be composted to produce organic fertilizers and gases. The organic fertilizers are good for the soil life, and for the cultivation of crops. The reuse of waste from livestock manure and organic waste through an anaerobic digester to produce effluents rich with nutrients and high carbon solids will eventually be used to generate power (as renewable energy), improve soil health, reduce GHG emissions, control odor, control fly and improve livestock health. All these benefits will have economic and environmental returns to Kuwait once it is utilizing them.

19.4 Conclusions

The water-energy-food approach in Kuwait is needed because there is an urgent requirement to address interconnectivity among WEF sectors in Kuwait so that sustainable development goals can be optimized and achieved. This necessitates policies, regulations and governance arrangements to support the nexus approach. Kuwait cannot be a self-sufficiency food country because water resources are scarce and its lands for agriculture are poor. Instead, Kuwait should seek the improvements in water use efficiency and productivity in support of agricultural activities. Kuwait development plans should be developed to use less water for more produced food mainly through irrigation. Also, the use of modern energy should be utilized to increase food yields with partial investments in food imports and storage.

In order to apply the WEF nexus approach in Kuwait successfully, strategic solutions should be developed such as:
Use of wastewater from inland fisheries and aquaculture in agriculture activities.
Investment in reuse of drainage water and treated municipal and industrial wastewater to maintain stable food production.
Use of plants to improve water quality.
The use of virtual water concept for food production.

Because the above solutions are not developed and integrated yet in order to benefit Kuwait from them on the national level, it is believed that Kuwait should develop a local technology for sustainable, cost-effective food production based on the needs of Kuwaiti nationals and residents, including advanced crop production systems and advanced water use. The option of using desalinated water in irrigation and testing the use and efficiency of an integrated system driven by renewable solar energy should be assessed for Kuwait. In addition, the options of using treated wastewater in irrigation should also be assessed. Based on the quality of the treated water and the use of biowaste as organic fertilizers, assessment of crops that can be irrigated should be conducted. The assessment of the mixing of the treated wastewater and desalinated water with freshwater for irrigation should also be assessed for the best water use efficiency option. This assessment will take into consideration the social acceptability by the farmers and end users of wastewater use in agriculture. Finally, the critical components and design elements of the system that supplies water and cooling energy to greenhouses in a local farm in Kuwait should be field tested.

19.5 Acknowledgements

The authors would like to extend their appreciation to the support of the management of the Kuwait Institute for Scientific Research (KISR) in carrying out this research. The authors would also like to express their thanks to several research centers and department with KISR that made a number of relevant documents available for this study.

References

Abderrahman W (2001) Energy and water in arid developing countries: Saudi Arabia, a case study. Int J Water Resour Dev 17(2):247–255. https://doi.org/10.1080/07900620120031306

Aliewi A, El-Sayed E, Akbar A, Hadi K, Al-Rashed M (2017) Evaluation of desalination and other strategic management options using multi-criteria decision analysis in Kuwait. Desalination 413:40–51

Allan T, Lant C (2003) Virtual water- the water, food and trade nexus. Useful concept of misleading metaphor. Water Int 28(1):106–113

Al-Rashed M, Aliewi A (2018) Water resources sustainability in Kuwait against United Nations sustainable development goals. In: Infrastructure management and urban solutions in sustainability of life in the Arabian gulf countries, Cambridge Research Centre, Cambridge

Al-Zubari W, ElSadek A, Mohamed A (2018) The water-energy-food nexus in the Arab region, Arabian gulf university. A policy paper published by the League of Arab States (LAS), with technical and financial support from the Deutsche Gesellschaft fur Internationale Zusammenarbeit (GIZ)

Amy G et al (2017) Membrane-based seawater desalination: present and future prospects. Desalination 401:16–21

Andrews-speed P, Bleschwitz R, Boersma T, Johnson C, Kemp G, Van Deveer SD (2015) Want, waste or war: the global resource nexus and the struggle for land, energy, food, water and minerals. Earthscan from Routledge. ISBN: 978-1-138-78446-8 (hbk)

Biggs EM, Bruce E, Boruff B, Duncan JMA, Horsley J, Pauli N, McNei K, Neef A, Van Ogtrop F, Curnow J, Haworth B, Duce S (2015) Sustainable development and the water-energy-food nexus: a perspective on livelihoods. Environ Sci Policy 54:389–397

Daher B, Mohtar RH (2015) Water-energy-food (WEF) Nexus Tool 2.0: guiding integrative resource planning and decision-making. Wate Int. /02508060.2015.1074148

Degirmencioglu A, Mohtar RH, Daher B, Ozgunaltay-Ertugrul G, Ertugrul O (2019) Assessing the sustainability of crop production in the Gediz Basin, Turkey: a water, energy, and food nexus approach. Fresen Environ Bull. 28(4/2019):2511–2522

Endo A, Burnett K, Orencio P, Kumazawa T, Wada C, Ishii A, Tsurita I, Taniguchi M (2015) Methods of the water-energy-food nexus. Water 7:5806–5830. https://doi.org/10.3390/w7105806)

FAO (2009) Kuwait report in Irrigation in the middle east region in figures—AQUASTAT Survey 2008. Water Report 34

FAO (2014) The water–energy–food nexus: a new approach in support of food security and sustainable agriculture. Food and Agriculture Organisation of the United Nations, Rome

Ismail H (2015) Kuwait: food and water security. Strateg Anal Pap Future Dir Int

Lee SH, Mohtar RH, Yoo S (2019) Assessment of food trade impacts on water, food, and land security in the MENA region. Hydrol Earth Syst Sci (HESS) 23, 557–572. Copernicus Publications EGU. https://doi.org/10.5194/hess-23-557-2019

Loring P, Gerlach S, Huntington H (2013) The new environmental security: Linking food, water and energy for integrative and diagnostic social-ecological research. J Agric Food Syst Community Dev 3:55–61

Luo T, Young R, Reig P (2015) Aqueduct projected water stress country rankings. Technical note. World Resources Institute, Washington, DC. Available online at: www.wri.org/publication/aqueduct-projected-water-stresscountry-rankings

Majgl A, Eard J, Pluschke L (2015) The water-food-energy nexus-realising a new paradigm. J. Hydrol. 530–540. https://doi.org/10.1016/j.jhydrol.2015.12.033

Mohtar RH (2017) Opportunities in the food-energy-water nexus. In: White paper for the ASABE global initiative—special session. ASABE Conference, Spokane, WA, 17 July 2017

Mohtar RH, Daher B (2016) Water-energy-food nexus framework for facilitating multi-stakeholder dialogue. Water Int. https://doi.org/10.1080/02508060.2016.1149759

Mohtar RH, Daher B, Mekki I, Chaibi T, Chebbi RZ, Salaymeh A (2014) The water, energy and food (WEF) nexus project: a basis for strategic planning for natural resources sustainability-Challenges for application in the MENA region. Geophys Res Abs. 16, EGU2014-15330

Mortada S, Abou Najm M, Yassine A, El Fadel M, Alamiddine I (2018) Towards sustainable water-food nexus: an optimization approach. J Clean Prod 178(2018):408–418

Saif O, Mezher T, Arafat HA (2014) Water security in the GCC countries: challenges and opportunities. J Environ Stud Sci 4:329–346

Siddiqi A, Anadon A (2011) The water-energy nexus in middle east and North Africa. Energy Policy 39(8):4529–4540. https://doi.org/10.1016/j.enpol.2011.04.023

Zubari W (2014) Virtual water trade as a policy instrument contributing to the achievement of food security in the GCC Countries. In: Sadik A, El-Solh M, Saab N (eds) Annual report of the Arab forum for environment and development. Technical Publications, Beirut, Lebanon, pp 30–34

Chapter 20
Energy Recovery from Anaerobic Digestion of Wastewater

Savita Ahlawat and Meenakshi Nandal

20.1 Introduction

The population of world is increasing day by day and directed toward urban centers. According to a report by United Nations, 2012, by 2030, an additional 2.1 billion population is predictable to be living in metropolitan cities. Enormous amount of waste, i.e., billions of tons of waste is emitted every year by these cities, including sludge and wastewater. These trashes/wastes can be end up by many ways depending on the local framework: Firstly, they can be accumulated or not. Secondly, treated or not and thirdly used directly, indirectly or end devoid of beneficial use.

The whole world is facing severe menace due to water pollution problems. Access to safe drinking water is not guaranteed to a majority of the population. Mostly, in developing countries like India, it is very vital to maintain the quality of surface water sources (Banu et al. 2007). According to the Census 2011, the demand for freshwater for increasing population will become unmanageable. As per Indian Infrastructure report 2011, due to no other source for irrigation, millions of small-scale farmers depend on wastewater or polluted water sources to irrigate high-value edible crops in urban and peri-urban areas. In 2015, it was computed that the sewage treatment capability was only 22,963 MLD in contrast to municipal wastewater generation of 61,754 MLD in the India itself (CPCB Bulletin, July 2016). So, it can be concluded that there is a hefty void among generation and treatment capacity. As per CPCB Bulletin, July 2016, around 38,791 MLD of unprocessed sewage (62% of the total sewage) is dumped openly into close-by water bodies. The demand for water supply is incomparably raising due to increase in industrialization and urbanization. The total number of cities and towns in India has swelled from 2250 to 5161 and 7936 in 1991, 2001 and 2011, respectively. Discharges of inadequately treated municipal wastewater may have a great impact on natural water sources, i.e., surface and

S. Ahlawat (✉) · M. Nandal
Department of Environmental Science, Maharshi Dayanand University, Rohtak, India
e-mail: ahlawatsavi09@gmail.com

© Springer Nature Singapore Pte Ltd. 2020
M. Kumar et al. (eds.), *Resilience, Response, and Risk in Water Systems*,
Springer Transactions in Civil and Environmental Engineering,
https://doi.org/10.1007/978-981-15-4668-6_20

underground water bodies or land. When it is directly discharged into rivers, canals and ponds, it can damage the aquatic life and quality of water sources. Due to hike in BOD and COD, eutrophication happens due to heavy amount of organic material and nutrients. So, due to all these reasons, there is a great need for treatment of municipal wastewater in order to save our natural water resources from contamination or depletion and also for the fulfillment of the water demand. So with the help of wastewater treatment system, this problem can be figured out, and the treated water can be reused for various purposes like industrial, agricultural, aquacultural and municipal purposes. The treatment of wastewater has been an issue of high antecedence in most of the developing countries, and they have therefore reached a very satisfactory quality of their wastewater discharges.

This chapter basically focused on the basic principles of anaerobic digestion, basic anaerobic technologies to establish their prospective for biogas generation. Anaerobic digestion basic principles are structured in Sect. 20.2. Various technologies for anaerobic digestion of wastewater are depicted in the Sect. 20.3. The utilization of biogas as a source renewable energy is highlighted in Sect. 20.4. Conclusions and recommendations are discussed in the Sect. 20.5.

20.2   Key Principles of Anaerobic Digestion

20.2.1   Theory of the Process

Anaerobic digestion is a convoluted progression which transforms organic matter into methane with the help of tons of microbial populations allied by their individual substrate and product specificities, and this process also illustrates the direct and indirect symbiotic association between different groups of bacteria. A balance is formed with the help of a chain mechanism in which the product of one bacterium is substrate for other, and in this way, the substrate concentration is maintained. However, this biological conversion takes place in four steps, i.e., hydrolysis phase, acidogenesis phase, acetogenesis phase and methanogenesis phase. All the chemical reactions are described below in a flow diagram (Figs. 20.1 and 20.2).

20.2.2   Atmospheric Dynamics Impinging Anaerobic Treatment

As anaerobic digestion is organic process, many environmental factors upset this process such as temperature, pH, alkalinity and toxicity.

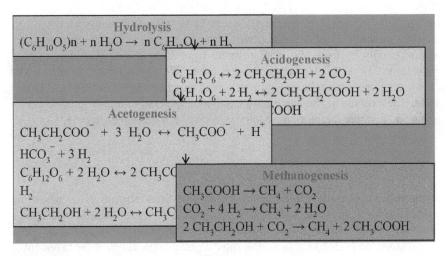

Fig. 20.1 Chemical reactions involved in anaerobic digestion process

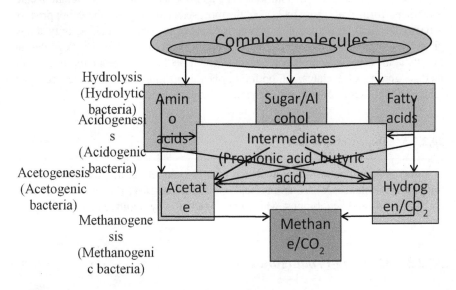

Fig. 20.2 Anaerobic microorganism digestion procedure

20.2.2.1   Functioning Temperature

Controlled digestion is splitted in three temperature ranges, i.e., psychrophilic (10–20 °C), mesophilic (20–40 °C) or thermophilic (50–60 °C). Due to low temperature conditions, bacterial growth and conversion processes are slower. Hence, a long retention time is required by psychrophilic digestion which may result in hefty reactor volumes in contrast to mesophilic digestion which requires less reactor volume. For thermophilic process, high loading rates can be beneficial. At temperatures as low as 0 °C, anaerobic digestion can occur; however, the pace of methane production augments only by fostering of temperature from 35 to 37 °C. (Lettinga and van Haandel 1993). At this temperature range, mesophilic organisms are involved. The choice of temperature relies upon the relativity among energy constraint and biogas yield. The extreme methanogenic bacterial activity befall at higher temperature of about 55 °C where thermophilic bacteria supersedes mesophilic bacteria.

20.2.2.2 pH

In anaerobic digestion, the first step can transpire at a broad range of pH values, while methanogenesis only ensues when pH is neutral (Lettinga and van Haandel 1993). pH range of 6.3–7.8 seems to be most affirmative condition for the methanogenesis process (Mirron et al. 2000). In an anaerobic digester while carrying the treatment of domestic wastewater owing to buffering capability of the acid base system, the pH lingers in this range devoid of addition of any additional chemical. The methanogenic bacteria perform at pH close to 7.0, but the most selected range for the entire process is 6.0 to 8.0. Acid fermentation may preponderate over methanogenic activity at low pH because acidogenic bacteria are less susceptible to pH variations. The degradation of fatty acids especially propionate takes place at pH less than 6 due repressions of methanogenesis of acetates. An optimum pH value should be maintained between 7.5 and 8, in order to obtain soaring yield of biogas.

20.2.2.3 Toxic Compounds

A numerous compounds show evidence of a noxious effect at excessive concentrations, for instance, VFAs, ammonia, cations such as Na+, K+ and Ca++, heavy metals, sulfide and xenobiotics, which unsympathetically impinge methanogenesis.

20.2.3 Methane Production Potential

For the prediction of potential for biogas generation, the COD is exploited to quantify the total organic contents present in wastewater. The oxygen correspondent of organic matter that can be oxidized is appraised using a strong chemical oxidizing agent in an acidic medium. The conclusive products of anaerobic degradation of organic matter are gas in form of methane and fresh bacterial biomass.

Biochemical oxygen demand (BOD) is another extensively used stricture of organic pollution. It is the scheme used for the measurement of dissolved oxygen availed by oxygen-consuming microorganisms in biochemical oxidation of organic matter during 5 days at 20 °C. By retaining time of 50 days at least, the total anaerobic biodegradability is measured.

The gas yield relies on various dynamics like digestibility of the organic matter, digestion kinetics, the retention time and the digestion temperature. The progression can be optimized by controlling conditions, for instance, temperature, humidity, microbial activity.

20.2.4 Necessities for Anaerobic Treatment

In disparity to the aerobic treatment systems, the loading rate of anaerobic reactors is not halted by the delivery of a reagent, except by the processing competency of the microorganisms. Consequently, it is imperative that a satisfactory hefty bacterial mass is retained in the reactor. The retention of biomass should be enlarged in contrast with the retention of liquid in high-rate treatment systems. The following conditions are vital for high-rate anaerobic reactors (Letting et al. 1999):

- It is exceedingly vital to keep hefty bacterial sludge under soaring organic (>10 kg/m^3/day) and high hydraulic (>10 m^3/m^3/day) loading conditions.
- Maximum contact among the influent wastewater and the hefty bacterial mass.
- Also negligible transport dilemma should be practiced with respect to substrate compounds, intermediate and end stuffs.

Sludge Retention Time (SRT) is a vital parameter. The process will not crop up at very low SRT, and as a result, it will acidify the reactor. To ensure the methanogenesis process, ample hydrolysis and acidification of lipids at 25 °C, SRT of not less than 15 days is required (Miron et al. 2000). The SRT should be longen at lesser temperatures since the growth rate of methanogens and the hydrolysis constant fades with temperature. The SRT should be increased to ensure same effluent standards. In completely mixed systems, the SRT is equivalent to the HRT, while in systems with inbuilt sludge retention; the SRT is elevated than the HRT.

Hydraulic Retention Time (HRT): This is the most important parameter that usually distress the performance of UASB reactor during the treatment of municipal sewage (Vieira and Garcia 1992). HRT is that time in which wastewater remains in the reactor and is calculated as:

$$HRT = \frac{\text{Reactor volume}}{\text{Wastewater flow}}$$

HRT is the factor which affects the COD reduction rate and important parameter with respect to the aimed degradation rate. The HRT should not be less than 2 h. According to the study by Trnovec and Britz (1998), during the treatment of a carbohydrate-rich effluent of the canning industry with UASB reactor, it was reported that COD removal performance was higher than 90% at an HRT of 10 h. As per an investigation done by Fang at 37 °C, the impinge of HRT on acidogenesis of dairy wastewater and HRT stretching from 4 to 24 h and by escalating HRT from 4 to 12 h, it was reported a boost in the acidification, i.e., from 28 to 54% [15]. But however, it is also accounted by number of scientists that there is no distinct impression of HRT on the treatment effectiveness of UASB reactor (Halalsheh 2002), and this disparity of outlook in scientific community is perhaps due to the discrepancy in the reactor design, operating procedures and range of HRT.

20.2.5 Pros and Cons of Anaerobic Digestion

The pros and cons of the anaerobic digestion progression are formulated in the figures given below (Lettinga 1999) (Figs. 20.3 and 20.4).

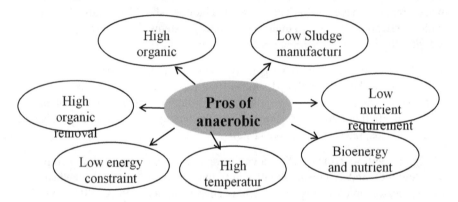

Fig. 20.3 Pros of anaerobic digestion process

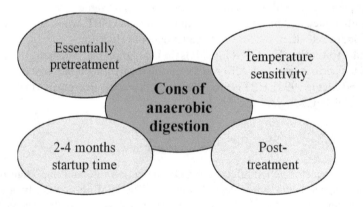

Fig. 20.4 Cons of anaerobic digestion process

20.3   Tools and Techniques for Anaerobic Digestion of Wastewater

There are various techniques for the treatment of various wastewaters, but high-rate anaerobic treatment systems are highly suitable for the diluted and concentrated wastewater in spite of concentrated slurries. In these type of systems, SRT is much soaring than the HRT. The sludge is protected in the reactor with the help of internal settler systems or external settlers with sludge salvaging or fixation of biomass on support material. These types of systems are highly recommended for the wastewaters containing low SS contents in them. The paramount examples used worldwide are contact process, upflow anaerobic sludge bed, anaerobic fixed film reactor, fixed film fluidized bed system, expanded granular sludge bed, hybrid systems and anaerobic filter (Lettinga et al. 1999). The above are conversed briefly:

Upflow Anaerobic Sludge Bed Reactor (UASBR): Till the advancement of the upflow bed reactor, the anaerobic digestion was very atypical in the main treatment system. Lettinga and co-workers in the late 1970s brought up the idea of this process (Lettinga et al. 1980). In original, this reactor was blueprinted to treat the concerted effluents of preliminary industries, but as with the extension of its relevance, its use expanded to sewage treatment. In these days and age, the UASB reactor is comprehensively exercised for the treatment of several types of wastewater, shaping part of the high-rate anaerobic technology (Kavitha and Murugesan 2007) (Fig. 20.5).

The UASBR assembly chiefly consisted of influent tank, peristaltic pump, cylindrical UASB reactor, gas/liquid/solid separator, effluent outlet, gas collection system. As stated by Simpson (1971) and Pretorius (1971), predecessor of UASBR can be observed in alleged anaerobic contact process. An analogous system to UASBR was

Fig. 20.5 Upflow anaerobic sludge blanket

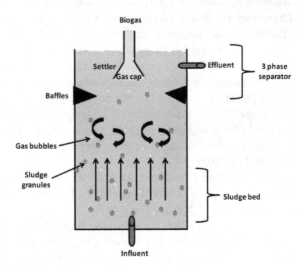

inspected in the name of 'Biolytic tank' somewhere in 1910. So, in view of above, UASBR technology can be adopted for a variety of wastewater treatments (Lettinga 1995, 1996a, b).

Development of dense granular sludge bed at lowest part of the UASBR where all microbial digestion primarily comes to pass illustrates the success of the UASB process. Incoming suspended solids and bacterial growth get accumulated for the configuration of sludge bed. According to some studies, flocs and granules get naturally aggregated by bacteria under convinced conditions in upflow technologies (Hulshoff Pol 1989). Under practical reactor conditions, these granules have excellent settling property and hence non-vulnerable to get out of the arrangement. A UASB arrangement facilitates good treatment at higher OLR rates due to retention of active sludge, either the cylindrical reactor may contain granular or flocculent. A pleasant contact between biomass and the wastewater is imparted by the natural turbulence facilitated by the influent flow and the biogas production. As per Kato et al. (1994), lofty organic loads can be applied in the UASB arrangement due to all the above positives which cannot be supplied by the aerobic process.

As a result, a reduced amount of the reactor space and volume is required, and additionally, paramount power is generated as biogas. Numerous alterations can be adopted in UASB equipment with sand trap, screens for coarse material, drying beds for the sludge, and this attributes to advancement of wastewater treatment system. The primary settler, the anaerobic sludge digester, the aerobic step (activated sludge, trickling filter, etc.) and the secondary settler of a conventional aerobic treatment plant can be redeemed by UASB reactor.

Several treatments approximating stabilization ponds and activated sludge plants are required for the UASBR bilge water to remove remnant organic matter, nutrients and pathogens due to all these leverages. Lettinga discussed various economics of anaerobic treatment in UASB arrangements.

Expanded Granular Sludge Bed (EGSB): The expanded granular sludge bed (EGSB) process integrates the sludge granulation concept of UASBs. The focal improvement of the EGSB system, trademarked 'Biobed,' compared to other types of anaerobic fluidized or expanded bed technologies is the abolition of carrier material as a mechanism for biomass retention contained the reactor. Hence, this technology can be picked out as mutated conventional fluidized bed or ultra-high-rate UASB. It can be used for the wastewaters of breweries, chemical plants, fermentation industries and pharmaceutical industries. This system is invented to operate at high COD loading; it is very space proficient, entailing a smaller footprint size than a UASB system (Figs. 20.6 and 20.7).

Anaerobic Fixed Film Reactors (AFFR) Effluents of high strength are treated by AFFR with the help of mixed populance of bacteria immobilized on the surfaces of support medium (Chua and Fung 1996).

To conquer the glitch between UASB and AF systems, the hybrid system was invented. In anaerobic filter reactor (AF), there is a presence of channeling and deceased precincts in the subordinate part of the filter. In UASB, the washout of the sludge may be a dilemma. So, to cope with these problems, the hybrid systems may

Fig. 20.6 Anaerobic filter
process

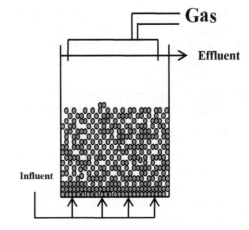

Fig. 20.7 Anaerobic contact
process

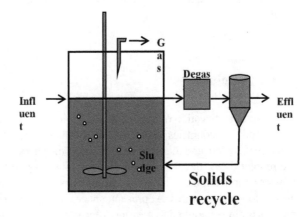

be helpful. Elmitwalli (2000) states that for COD diminution, the filter area plays a physical job for biomass preservation and biological activity.

20.4 Recovery of Biogas as an Energy Source

After CO_2, methane being the second most man-made greenhouse gas which attributes 14% of the global GHG emissions. Although its natural life is shorter than the CO_2, its global warming potential (GWP) is 21 times extra than the CO_2 which means it has more knack to block the heat. There are quite a lot of approaches to wastewater methane mitigation and recovery and also several options for availing of recovered methane. There are numerous recovery approaches given below:

- Fixation of anaerobic sludge digestion (novel construction or retrofit of the existing aerobic treatment systems).
- The new confine biogas systems for the existing open air anaerobic lagoons.
- Installing simple degassing devices at the effluent release of anaerobic municipal reactors.
- Systems which are not operated correctly should be optimized by implementing proper operation and maintenance (O&M).

Uses of Methane gas:

1. Methane can be used as fuel for the production of electricity and heat within the treatment systems.
2. After proper treatment, the methane gas can be a best option for local gas supply.
3. Methane gas can be used to sell to the local industrial user or power producer for the production of heat or power.
4. For a suitable fleet vehicle fuel, methane gas can be treated and compressed on the site.

20.5 Conclusion

The anaerobic digestion of the municipal sewage is very limited to the tropical countries mainly. In countries like Asia and South America, large UASB treatment systems are installed, and there are limitations of the temperature disparity. To overcome the problem, new developments in high-rate anaerobic treatment systems can lead to wider application of anaerobic treatment of conventionally collected wastewater even at low temperatures. At present, anaerobic treatment is a proven technique. The process is almost applied to a variety of wastewater streams excluding the fact that it is still limited. But owing to this, a hefty potential energy source is being abandoned. Moreover some potential sources, which are now treated otherwise, are an excellent substrate for anaerobic treatment. This could contribute to renewable energy production rather than consuming energy during treatment. Anaerobic digestion merges aspects of treatment and utilization, in form of treatment of wastewater with production of methane with less greenhouse emissions in comparison with the use of traditional fuel combustion. Millions of people in the world do not have effective sanitation at their disposal. So, in the near future, more studies should be done on the concept of anaerobic treatment of sewage water of isolated buildings so that each and every person gets the facility of sewage disposal, and use of methane energy can be also availed.

References

Banu JR, Kaliappan S, Yeom IT (2007) Treatment of domestic wastewater using upflow anaerobic sludge blanket reactor. Int J Environ Sci Technol 4(3):363–370

Chua H, Fung JPC (1996) Hydrodynamics in the packed bed of anaerobic fixed film reactor. Water Sci Technol 33(8):1–6

CPCB Bulletin, vol 1 July 2016. (http://cpcb.nic.in/openpdffile.php?id= TGF0ZXN0RmlsZS9MYXRlc3RfMTIzX1NVTU1BUllfQk9PS19GUy5wZGY)

Elmitwalli TA (2000) Anaerobic treatment of domestic sewage at low temperature, Ph.D.-thesis. Wageningen University, Wageningen, The Netherlands

Halalsheh M (2002) Anaerobic pre-treatment of strong sewage: a proper solution for Jordan. Ph.D. thesis. Department of Environmental Technology, Agricultural University, Wageningen

Hulshoff Pol LW (1989) The phenomenon of granulation of anaerobic sludge. Ph.D. thesis. Wageningen Agricultural University, Wageningen, The Netherlands

Kato MT, Field JA, Kleerebezem R, Lettinga G (1994) Treatment of low strength soluble wastewater in UASB reactors. J Ferment Bioeng 77(6):679–686

Kavitha K, Murugesan AG (2007) Efficiency of upflow anaerobic granulated sludge blanket reactor in treating fish processing effluent. J Ind Pollut Control 23(1):77–92

Lettinga G (1995) Anaerobic digestion and wastewater treatment systems. Antonie Van Leeuwenhoek 67:3–28

Lettinga G (1996a) Sustainable integrated biological wastewater treatment. Water Sci Technol 33(3):85–98

Lettinga G (1996b) Advanced anaerobic wastewater treatment in the near future. Aquatech 1996. In: IAWQNVA Conference on Advanced Wastewater Treatment, Amsterdam, The Netherlands, pp 24–32

Lettinga GAFM, Van Velsen AFM, Hobma SD, De Zeeuw W, Klapwijk A (1980) Use of the upflow sludge blanket (USB) reactor concept for biological wastewater treatment, especially for anaerobic treatment. Biotechnol Bioeng 22(4):699–734

Lettinga G, van Haandel AC (1993) Anaerobic digestion for energy production and environmental protection. Anaerobic digestion energy production environmental protection 817–839

Lettinga G, Hulshoff-Pol LW, Zeeman G (1999) Lecture notes: biological wastewater treatment; part i anaerobic wastewater treatment. Wageningen University and Research, Wageningen, The Netherlands

Miron Y, Zeeman G, Van Lier JB, Lettinga G (2000) The role of sludge retention time in the hydrolysis and acidification of lipids, carbohydrates and proteins during digestion of primary sludge in CSTR systems. Water Res 34(5):1705–1713

Pretorius WA (1971) Anaerobic digestion of raw sewage. Water Res 5:681–687

Simpson DE (1971) Investigations on a pilot-plant contact digester for the treatment of a dilute urban waste. Water Res 5:523–532

Trnovec W, Britz TJ (1998) Influence of organic loading rate and hydraulic retention time on the efficiency of a UASB bioreactor treating a canning factory effluent. Water S. A. 24(2):147–152

Vieira SMM, García AD (1992) Sewage treatment by UASB-reactor. Operation results and recommendations for design and utilization. Water Sci Technol 25(7):143–157

Chapter 21
Dark Fermentative Hydrogen Production from Lignocellulosic Agro-waste by a Newly Isolated Bacteria *Staphylococcus Epidermidis* B-6

Payal Mazumder, Dhrubajyoti Nath, Ajay Kumar Manhar, Kuldeep Gupta, Devabrata Saikia, and Manabendra Mandal

21.1 Introduction

Air pollution, exhaustion of fossil fuel reservoir and global warming have led to the relentless investigation for sustainable alternative fuels (Nagarajan et al. 2017). So, the production and use of non-carbonaceous fuels are getting much more attention nowadays. To elaborate, bio-based energy is a tenable and propitious replacement for unsustainable sources of energy; this can substitute and fortify against a catastrophe in the energy supply and the ever-increasing demand. Lately, hydrogen (H_2) gas has gained worldwide heed and is a promising future fuel (Anwar et al. 2019; Kumar et al. 2019b). H_2 is a conceivable multifaceted energy currency that could convert the utilization of non-renewable fossil fuels due to its an elevated yield per unit mass of energy (~122 kJ/g), which is 2.75 times substantial than the yield from hydrocarbon fuels (Mohan and Pandey 2019). Furthermore, upon ignition H_2 produces water (H_2O) after combining with O_2, the only externality of the process. It is an evidently commendatory denouement for greenhouse gases (GHG) emissions. Discretely, H_2 is more alike electrical energy and hence, an eminent solution (Onaran and Argun 2019). Currently, molecular H_2 is fundamentally generated from fossil fuels via the process of steam reforming of methane (CH_4) and/or natural gas. The global produce of H_2 presently surpasses at the rate of 1 billion m^3 per day, out of which around 48% is manufactured from 30% from oil, natural gas, 18% from coal

P. Mazumder (✉)
Centre for the Environment, Indian Institute of Technology Guwahati, North Guwahati, Assam 791039, India
e-mail: payal93@iitg.ac.in

D. Nath · A. K. Manhar · K. Gupta · D. Saikia · M. Mandal
Department of Molecular Biology and Biotechnology, Tezpur University, Napaam, Tezpur, Assam 784028, India

© Springer Nature Singapore Pte Ltd. 2020
M. Kumar et al. (eds.), *Resilience, Response, and Risk in Water Systems*,
Springer Transactions in Civil and Environmental Engineering,
https://doi.org/10.1007/978-981-15-4668-6_21

and rest 4% is generated from splitting water through electrolysis (vikaspedia). When amalgamated with the steam reforming process, refined H_2 is produced attained via water gas shift reaction process, a crucial industrial process utilized peculiarly for manufacturing ammonia (Singh et al. 2019). There are many more thermochemical procedures/techniques obtainable for the manufacturing of H_2 that comprise autothermal reforming, catalytic oxidation, thermal decomposition, pyrolysis and steam gasification, etc. (Devasahayam and Strezov 2018; Bu et al. 2019; Waheed et al. 2016; Kouhi and Shams 2019). Nonetheless, the generation of H_2 established on natural fossil fuel reservoirs augment the discharge of harmful GHGs. In this context, biologically produced hydrogen which is a carbon-neutral, sustainable and eco-friendly technique is contemplated as the most propitious candidate. It possesses high energy, is clean in nature and can be obtained from a broad range of inexhaustible feedstock (Kotay and Das 2008; Winter 2009; Chong et al. 2009).

Considering the aspect of biomass conversion to energy, dark fermentative method of H_2 generation has been delineated to have huge prospective and more economic over other physicochemical methods (Bundhoo 2019). Lignocellulosic waste biomasses are promising raw materials for biofuel production due to high carbohydrate percentage (Taherzadeh and Karimi 2008; Nissilä et al. 2014). However, the sugar polymer cellulose and hemicelluloses in lignocellulosic biomass remain bound in compact form with lignin, which restrict them from easy microbial degradation. A few fungi are capable of solubilizing lignin, and some bacterial species can degrade cellulose, but the process takes a very long time (Lee 1997; Lynd et al. 2002; Kumar et al. 2008, Kumar et al. 2019a). Thus, these substances require proper pre-treatment prior to fermentative hydrogen production. Xylose shares a major fraction (35–45%) of total sugar yield from hydrolysis of lignocellulosic materials (Lavarack et al. 2002). Extensive research has been done on H_2 production via fermentation from substrates such as glucose and sucrose. However, due to inefficiency of microbes for xylose utilization, there are a very few reports (Li et al. 2010; Cheng et al. 2012; Chenxi et al. 2013; Wu et al. 2014; An et al. 2014; Poladyan et al. 2018; López-Aguado et al. 2018; Zhao et al. 2019; Kongjan et al. 2019) on H_2 production from xylose using a pure culture of microbes. Hence, for complete utilization of sugar released from lignocellulosic material, it is important to isolate bacterial species with efficiency to convert xylose to H_2. Thus, our current work is based on the isolation and identification of new microbial (bacteria) species having the potential to utilize xylose for fermentative H_2 production. Parameters that are crucial for fermentation was also studied to ascertain the optimum parameters for maximal H_2 yield and high production rate. This study also investigates the feasibility of the isolated bacterium for H_2 production from acid hydrolysed rice straw under batch culture condition.

21.2 Materials and Methods

21.2.1 Isolation of H_2-Producing Bacteria

The bacterial strain for H_2 production was isolated from the soil sample collected from the outskirt of Kaziranga national park of North East India. For isolation of bacterial strain, 1 g of the soil sample was homogenized with 0.85% NaCl (w/v) followed by several-fold dilution. 100 uL of the diluted sample was then spread on nutrient agar plates and kept at 37 °C for 24 h. Several bacterial colonies differing by colony morphology were obtained and were subsequently maintained as a pure culture for screening of H_2 production ability from xylose.

21.2.2 Screening for H_2 Production from Xylose and Culture Condition

The preliminary screening of bacterial strains for H_2 production from xylose was done under batch culture condition. The medium used was GM-2 (Yeast extract—1.0 g/L, K_2HPO_4 —1.0 g/L, $MgSO_4.7H_2O$—0.5 g/L and $FeSO_4.7H_2O$—1 mg/L) with slight modification (Patel et al. 2014). The experiments were conducted in 125 mL BOD bottle under anaerobic condition. 100 mL of the medium supplemented with 5 g/L xylose was inoculated with 2% (v/v) of culture (1 O.D at 600 nm). After bacterial inoculation, the bottles were made airtight and initial anaerobic condition was established by flushing N_2 gas. The bottles were then kept at 37 °C and evolved biogas was collected by water displacement method under acidic water. The potential strain with maximum biogas production ability was selected for this study and was named as strain B-6.

21.2.3 Identification of the Bacterial Strain B-6

For identification of the bacterial strain B-6, 16S rRNA gene sequencing was done by using universal primers, 27F (5′AGAGTTTGATCCTGGCTCAG3′) and 1492R (5′GGTTACCTTGTTACGACTT3′). The amplified product was purified by using PCR purification kit (Qiagen) and sent for sequencing (Eurofins Genomics India Pvt. Ltd., India). The obtained sequence was subjected to BLAST in National Centre of Biotechnology information (NCBI) BLAST search tool and the phylogenetic tree was constructed using the neighbour-joining method with MEGA 5.2 software.

21.2.4   Optimization of Culture Condition

The parameters crucial for fermentative H_2 production including initial pH, nitrogen source and substrate concentration were optimized under batch culture condition. The effect of pH was studied by adjusting the initial pH of the fermentation medium with the pH range of 5–9 with incremental step of 1. The fermentation was carried out at 37 °C with xylose concentration 10 g/L. The effect of nitrogen source on H_2 production was studied by fermentation medium GM-2 amended with inorganic (ammonium sulphate, ammonium chloride) and organic (yeast extract, peptone) nitrogen sources at a concentration of 1.0 g/L. The initial pH was 7.0, fermentation temperature 37 °C and xylose concentration 10 g/L. The effect of xylose concentration on fermentation was studied by varying the initial xylose load from 5–50 g/L under optimum pH, and nitrogen source at 37 °C.

21.2.5 H_2 Production with Different Carbon Source

The ability of the selected strain for utilization of other carbon sources were studied at optimum conditions (pH–7, N_2 source–yeast extract and incubation temperature –37 °C). The different carbon sources used were glucose, fructose, sucrose, mannose, lactose (10 g/L) and glycerol (10 mL/L).

21.2.6   Acid Hydrolysis of Rice Straw

Acid hydrolysis of the hemicellulose fraction of the rice straw was conducted by treating the dry rice straw (1%, 2%, 3%, 5% and 7%, w/v) with diluted H_2SO_4 (0.5%, v/v) at 121 °C for one hour in autoclave. After hydrolysis, the hydrolysate was filtered through a thin cloth to remove the solid fraction. Over-liming of the hydrolysate was done by adding $Ca(OH)_2$ with frequent stirring and the final pH adjusted to 10. The resulting precipitate was removed by centrifugation at 1500 rpm for 15 min. The supernatant was then re-acidified by lowering the pH to 7 and again centrifuged. The final supernatant thus obtained was then used for fermentative H_2 production (Nigam 2000).

21.2.7   Analytical Methods

The amount of biogas evolved during fermentation was measured by the water displacement method. The gas components were analysed by gas chromatograph (Nucon GC5765, India) equipped with Porapak-Q and molecular sieve columns using a thermal conductivity detector (Nath et al. 2015). Argon was used as carrier gas with a flow rate of 20 mL/min, and temperature of oven, injector and detector was set to 60, 80 and 110 °C, respectively. The xylose concentration was measured by DNS method (Miller 1959).

21.3 Results and Discussion

21.3.1 Strain Identification and Phylogenetic Analysis

The sequenced 16S rRNA gene of the strain B-6 was aligned with gene bank NCBI (http://blast.ncbi.nlm.nih.gov) using BLAST program. A phylogenetic tree was constructed by neighbour-joining method with MEGA 5.2 software (Fig. 21.1). The tree indicates that the strain B-6 belonged to the genus *Staphylococcus* and showed maximum similarity with *Staphylococcus epidermidis* strain NBRC100911. Thus, the strain was identified and named as *Staphylococcus epidermidis* B-6. The gene sequence was also submitted to the NCBI gene bank with an accession number KT072716.

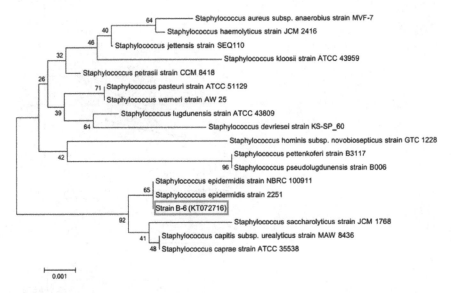

Fig. 21.1 Neighbour-joining tree showing the phylogenetic relationship of the isolated strain B-6 and related species based on 16S rRNA gene

21.3.2 *Effect of Various Nitrogen Sources on H₂ Production*

The effect of various nitrogen sources on H_2 production by *Staphylococcus epidermidis* B-6 is shown in Fig. 21.2a–c. A significant change in H_2 production was observed by changing the source of nitrogen in the fermentation medium. The maximum H_2 yield of 1.55 mol H_2/mol xylose was observed using yeast extract with the highest bacterial growth and 98% of xylose consumption (Fig. 21.2a, b). This is due to the fact that yeast extract is a complex nitrogen source comprising of peptides and amino acids, which can be easily taken up by the bacterium during fermentation and directly incorporated into proteins or transformed into other cellular nitrogenous constituents (Large 1986; Ferchichi et al. 2005). On the other hand, when inorganic

Fig. 21.2 Effect of N_2 source on H_2 production performance of *Staphylococcus epidermidis* B-6, **a** volume and rate of H_2 production, **b** H_2 yield and bacterial growth and **c** xylose degradation rate and final pH value

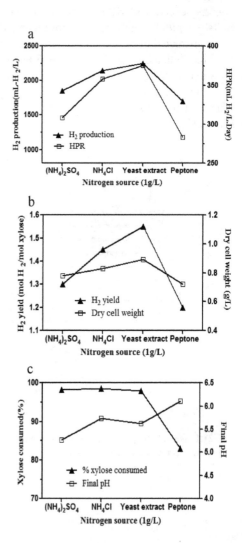

nitrogen sources are used, the cells have to spend more energy in synthesizing amino acids, and as a result, they spend a longer period of lag phase and decrease in H_2 yield (An et al. 2014). The results suggest that yeast extract as a better nitrogen source for maximum H_2 production with high production rate and bacterial growth. However, the other cheap nitrogen sources like ammonium chloride and ammonium sulphate also showed relatively higher production of H_2 (Fig. 21.2a). Thus, this can be beneficial for industrial-scale H_2 production by using cheap inorganic nitrogen sources instead of expensive organic nitrogen sources.

21.3.3   Effect of Initial pH on H₂ Production

The effect of initial pH on H_2 production by *Staphylococcus epidermidis* B-6 was investigated at pH 5–9 (with an interval of pH 1). The result (Fig. 21.3a–c) showed that the initial pH of the medium is an important factor in the bacterial growth and H_2 production process. The yield of H_2 and cell growth was increased significantly by increasing the initial pH from 5 to 7 and then decreased by further increasing the pH (Fig. 21.3b). The rate of hydrogen production (Fig. 21.3a) and xylose degradation (Fig. 21.3c) also showed a similar trend with cell growth and H_2 yield as shown in Fig. 21.3b. The maximum H_2 yield is 1.6 mol H_2/mol xylose with hydrogen production rate of 342 mL H_2/L. Day was observed at pH 7. The cell growth and xylose consumption were also observed maximum at this pH. The H_2 yield and cell growth were very low at pH 5 and bellow that no growth and H_2 evolution was observed. This can be the fact that, at high concentration of H^+ ion environment, the cell's ability to maintain internal pH get destabilized, consequently intracellular ATP level drops and inhibiting xylose uptake (Nigam et al. 1985; Xu et al. 2010). However, the strain B-6 was found to produce H_2 above 1 mol/mol xylose within the pH range of 6–9 and very little or no gas production was observed by further increasing the pH. This can be the fact that at higher pH range, the activity of the key enzyme [Fe − Fe] hydrogenase gets decreased and the changing direction of metabolic pathway from acidogenesis to solventogenesis results into low H_2 production (Zhu and Yang 2004; Gadhe et al. 2013).

21.3.4   Effect of Initial Xylose Concentration on H₂ Production

The initial substrate load usually plays a crucial role in cell growth and H_2 production. Figure 21.4a–c shows the various effects of initial xylose concentration on fermentation. An increase in the H_2 yield from 1.4 to 1.6 mol H_2/mol xylose was observed by increasing the xylose concentration from 5 to 10 g/L and was then decreased by further increasing the concentration. Though the volume of H_2 production (Fig. 21.4a)

Fig. 21.3 Effect of initial pH on H_2 production performance of *Staphylococcus epidermidis* B-6, **a** volume and rate of H_2 production, **b** H_2 yield and bacterial growth and **c** xylose degradation rate and final pH value

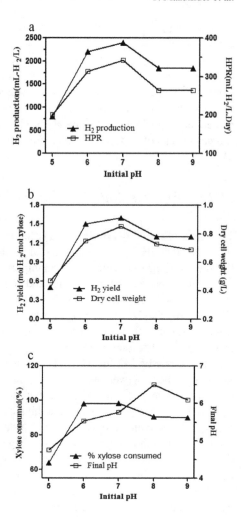

was relatively higher by increasing xylose concentration above 10 g/L, but the rate of production was found gradually decreases with H_2 yield and xylose consumption rate (Fig. 21.4b, c). However, cell growth was observed maximum at 20 g/L xylose concentrations and was then decreased above this concentration. This is due to the fact that at higher substrate concentration, the yield is decreased by inhibitory effect of substrates (Lin and Cheng 2006). Another reason may be that at higher substrate concentration, the carbon flux is directed more towards the production of reduced by-products like organic acids and alcohols (Chittibabu et al. 2006). The undissoci-ated organic acids get accumulated in the fermentation broth with higher substrate concentration, which would leak into the cell and decrease the pH of the intracellular environment. As a consequence, the cell growth and H_2 yield get inhibited (Akutsu et al. 2009).

Fig. 21.4 Effect of xylose concentration on H_2 production performance of *Staphylococcus epidermidis* B-6, **a** volume and rate of H_2 production, **b** H_2 yield and bacterial growth and **c** xylose degradation rate and final pH value

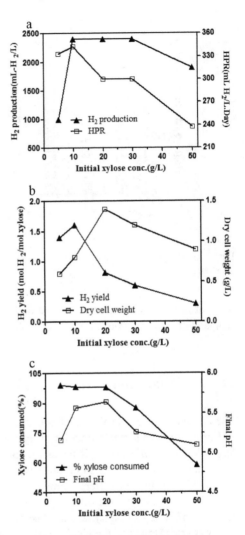

21.3.5 Utilization of Different Carbon Sources

It is important for H_2 producing bacterial strain to have the ability to use various carbon sources for better utilization of complex waste biomass. A variety of carbon sources have been reported for fermentative H_2 production. Therefore, different carbon sources were fed to evaluate their effect on H_2 production by *Staphylococcus epidermidis* B-6. The hydrogen production data (Fig. 21.5) showed that the bacterium can utilize diverse carbon sources. H_2 production was observed with lactose (305 mL H_2/g), maltose (280 mL H_2/g), xylose (240 mL H_2/g), fructose (180 mL H_2/g), glycerol (120 mL H_2/g) and glucose (70 mL H_2/g). Thus, the ability of the strain B-6 to utilize a wide variety of carbon sources for H_2 production would be

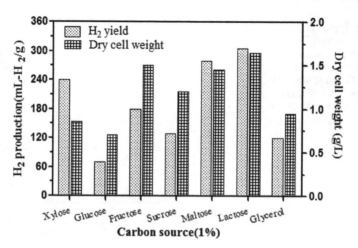

Fig. 21.5 H_2 production performance of *Staphylococcus epidermidis* B-6 by using different substrates

of great impact on waste management and converting the waste biomass to energy (Kapdan and Kargi 2006; Chong et al. 2009).

21.3.6 Production of H_2 from Rice Straw Hydrolysate

The lignocellulosic hydrolysate obtained from wood, agricultural waste by-product and crop contains a major fraction of xylose (Lavarack et al. 2002). As the strain *Staphylococcus epidermidis* B-6 was found potential in utilizing xylose for H_2 production, thus its feasibility to produce H_2 from acid hydrolysate of rice straw was examined. For this, the different concentration of dried rice straw was treated with diluted H_2SO_4 and the hydrolysate was used for batch fermentation. The maximum H_2 yield of 30 L/kg rice straw was observed with hydrolysate prepared by treating 1% (w/v) rice straw and the yield was decreased by further increasing the rice straw concentration (Fig. 21.6). Diluted acid was used for the hydrolysis process (0.5% v/v). Increasing the acid concentration in acid hydrolysis could provide a strong or complete reaction for hydrolysis, yielding more hydrolysed product (Chong et al. 2004). However, at higher acid concentration, the conversion of sugars to various inhibitory compounds takes place, which retard the cell growth. Furfural is one such compound, which is generated as a degradation product from xylose at higher H_2SO_4 concentration (Aguilar et al. 2002). The results suggest that the strain can be used for large scale H_2 production and can help on the way of complete utilization of lignocellulosic waste hydrolysate. This will be a great benefit for the conversion of waste into energy and reduce the waste generation (Fig. 21.7; Table 21.1).

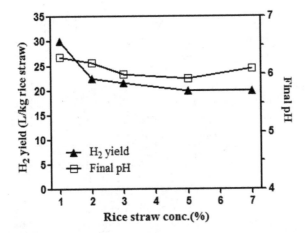

Fig. 21.6 Yield of H₂ and final pH value at different concentration of rice straw hydrolysate by *Staphylococcus epidermidis* B-6

Fig. 21.7 Food–Energy–Water ecosystem services nexus web

Table 21.1 Comparison of H_2 yield from various process utilizing different substrate

Substrate	Treatment	H_2 yield	References
Food waste	Taihu Algae (Ultrasonic pre-treatment)	31.42 mLH_2/g-VS	Xu et al. (2019)
Digested sludge	*Laminaria japonica* (Microwave irradiation)	15.8 mLH_2/g-VS	Yin et al.(2019)
Corn stover, wheat straw, rice straw, Corncob, sorghum stalk	Simultaneous saccharification and fermentation	80.09, 62.49, 95.21, 102.62 and 81.94 mL/g TS	Li et al. (2018)
	Asynchronous hydrolysis and fermentation	66.44, 62.86, 86.31, 90.03 and 77.36 mL/g TS	
Rice husk	Anaerobic granular sludge	320.6 mL/g biomass	Gonzales and Kim (2017)
Sugarcane top	White rot fungus *Pleurotus pulmonarium* MTCC 1805	77.2 mL/g-VS	Kumari and Das (2016)
Cassava residue	Microwave-heated acid pre-treatment (1% v/v H2SO4, 135 C, 15 min) þ Enzymatic hydrolysis	106.2 mL/g-VS	Cheng et al. (2015)
Anaerobic sludge	*Chlorella vulgaris* (pre-treatment with Onozuka R-10 Enzyme)	39 mLH_2/g-VS	Wieczorek et al. (2014)
Oil palm empty fruit bunch hydrolysate	Acid/heat pre-treatment (6% w/v H2SO4, 120 C, 15 min)	1.98 moL H_2/mol xylose	Chong et al. (2013)
Rice straw hydrolysate	*Clostridium butyricum* CGS5	0.76 moL H_2/mol xylose	Lo et al. (2010)
Corn stover	*Thermoanaerobacterium thermosaccharolyticum* W16	2.24 moL H_2/mol sugar	Cao et al. (2009)
Wheat straw hydrolysate	Thermophilic mixed culture	178.0 mL/g sugars	Kaparaju et al. (2009)

21.4 Conclusions

Biohydrogen production by dark fermentative microbes is a sustainable solution to manage lignocellulosic wastes which are hard to solubilize. *Staphylococcus epidermidis* B-6 was used for the first time to produce bio-H_2 from xylose and rice straw. The highest H_2 yield obtained was 1.6 mol H_2/mol and 30 L/kg from xylose and rice straw respectively. The results suggested that the strain could be used for complete utilization of lignocellulosic waste hydrolysate and production of H_2 on a large scale. This will be a great benefit for the conversion of waste into energy and provide an economic technology to manage generated waste.

Acknowledgements This research work was financially supported by Department of Biotechnology (Ref. No. BT/212/NE/TBP/2011 dated December 14, 2011), New Delhi, India. Mr. D. Nath was the recipient of Junior Research Fellowship (File No: 09/796(0060)/2015-EMR-I) from the Council of Scientific and Industrial Research (CSIR), New Delhi. The authors also acknowledge the help extended by Tezpur University, Assam, India, by providing infrastructure and other necessary facilities to carry out the research work successfully.

References

Aguilar R, Ramírez JA, Garrote G, Vázquez M (2002) Kinetic study of the acid hydrolysis of sugarcane bagasse. J Food Eng 55:309–318

Akutsu Y, Li YY, Harada H, Yu HQ (2009) Effect of temperature and substrate concentration on biological hydrogen production from starch. Int J Hydrogen Energy 34:2558–2566

An D, Li Q, Wang X, Yang H, Guo L (2014) Characterization on hydrogen production performance of a newly isolated Clostridium beijerinckii YA001 using xylose. Int J Hydrogen Energy 39:19928–19936

Anwar M, Lou S, Chen L, Li H, Hu Z (2019) Recent advancement and strategy on bio-hydrogen production from photosynthetic microalgae. Bioresour Technol. https://doi.org/10.1016/j.biortech.2019.121972

Bu E, Chen Y, Wang C, Cheng Z, Luo X, Shu R, Zhang J, Liao M, Jiang Z, Song Q (2019) Hydrogen production from bio-derived biphasic photoreforming over a raspberry-like amphiphilic Ag2O-TiO2/SiO2 catalyst. Chem Eng J 370:646–657

Bundhoo ZMA (2019) Potential of bio-hydrogen production from dark fermentation of crop residues: a review. Int J Hydrogen Energy 44:17346–17362

Cao G, Ren N, Wang A, Lee DJ, Guo W, Liu B (2009) Acid hydrolysis of corn stover for biohydrogen production using Thermoanaerobacterium thermosaccharolyticum W16. Int J Hydrogen Energy 34:7182–7188. https://doi.org/10.1016/j.ijhydene.2009.07.009

Cheng J, Song W, Xia A, Su H, Zhou J, Cen K (2012) Sequential generation of hydrogen and methane from xylose by two-stage anaerobic fermentation. Int J Hydrogen Energy 37:13323–13329

Cheng J, Lin R, Ding L, Song W, Li Y, Zhou J (2015) Fermentative hydrogen and methane cogeneration from cassava residues: effect of pre-treatment on structural characterization and fermentation performance. Bioresour Technol 179:407–413. https://doi.org/10.1016/j.biortech.2014.12.050

Chenxi Z, Wenjing L, Hongtao W, Xiangliang P (2013) Simultaneous hydrogen and ethanol production from a mixture of glucose and xylose using extreme thermophiles I: effect of substrate and pH. Int J Hydrogen Energy 38:9701–9706

Chittibabu G, Nath K, Das D (2006) Feasibility studies on the fermentative hydrogen production by recombinant Escherichia coli BL-21. Process Biochem 41:682–688

Chong AR, Ramírez JA, Garrote G, Vázquez M (2004) Hydrolysis of sugarcane bagasse using nitric acid: a kinetic assessment. J Food Eng 61:143–152

Chong ML, Sabaratnam V, Shirai Y, Hassan MA (2009) Biohydrogen production from biomass and industrial wastes by dark fermentation. Int J Hydrogen Energy 34:3277–3287

Chong PS, Jahim JM, Harun S, Lim SS, Mutalib SA, Hassan O (2013) Enhancement of batch biohydrogen production from prehydrolysate of acid treated oil palm empty fruit bunch. Int J Hydrogen Energy 38, 9592–9599. https://doi.org/10.1016/j.ijhydene.2013.01.154

Devasahayam S, Strezov V (2018) Thermal decomposition of magnesium carbonate with biomass andplastic wastes for simultaneous production of hydrogen and carbonavoidance. J Clean Prod 174:1089–1095

Ferchichi M, Crabbe E, Hintz W, Gil GH, Almadidy A (2005) Influence of culture parameters on biological hydrogen production by Clostridium saccharoperbutylacetonicum ATCC 27021. World J Microbiol Biotechnol 21:855–862

Gadhe A, Sonawane SS, Varma MN (2013) Optimization of conditions for hydrogen production from complex dairy wastewater by anaerobic sludge using desirability function approach. Int J Hydrogen Energy 38:6607–6617

Gonzales RR, Kim SH (2017) Dark fermentative hydrogen production following the sequential dilute acid pretreatment and enzymatic saccharification of rice husk. Int J Hydrogen Energy 42:27577–27583

Kongjan P, Inchan, S, Chanthong S, Jariyaboon R, Reungsang A, O-Thong S (2019) Hydrogen production from xylose by moderate thermophilic mixed cultures using granules and biofilm up-flow anaerobic reactors. Int J Hydrogen Energy 44, 3317–3324

Kaparaju P, Serrano M, Thomsen AB, Kongjan P, Angelidaki I (2009) Bioethanol, biohydrogen and biogas production from wheat straw in a biorefinery concept. Bioresour Technol 100:2562–2568. https://doi.org/10.1016/j.biortech.2008.11.011

Kapdan KI, Kargi F (2006) Bio-hydrogen production from waste materials. Enzyme Microb Technol 38:569–582

Kotay SM, Das D (2008) Biohydrogen as a renewable energy resource-prospects and potentials. Int J Hydrogen Energy 33:258–263

Kouhi M, Shams K (2019) Bulk features of catalytic co-pyrolysis of sugarcane bagasse and ahydrogen-rich waste: the case of waste heavy paraffin. Renew Energy 140:970–982

Kumar R, Singh S, Singh OV (2008) Bioconversion of lignocellulosic biomass: biochemical and molecular perspectives. J Ind Microbiol Biotechnol 35:377–391

Kumar M, Ram B, Honda R, Poopipattana C, Canh VD, Chaminda T, Furumai H (2019a) Concurrence of antibiotic resistant bacteria (ARB), viruses, pharmaceuticals and personal care products (PPCPs) in ambient waters of Guwahati, India: urban vulnerability and resilience perspective. Sci Total Environ 693:133640. https://doi.org/10.1016/j.scitotenv.2019.133640

Kumar M, Chaminda T, Honda R, Furumai H (2019c) Vulnerability of urban waters to emerging contaminants in India and Sri Lanka: resilience framework and strategy. APN Sci Bull 9(1). https://doi:10.30852/sb.2019.799

Kumari S, Das D (2016) Biologically pretreated sugarcane top as a potential raw material for the enhancement of gaseous energy recovery by two stage biohythane process. Bioresour Technol 218:1090–1097. https://doi.org/10.1016/j.biortech.2016.07.070

Large PJ (1986) Degradation of organic nitrogen compounds by yeasts. Yeast 2:1–34

Lavarack BP, Griffin GJ, Rodman D (2002) The acid hydrolysis of sugarcane bagasse hemicellulose to produce xylose, arabinose, glucose and other products. Biomass Bioenergy 23:367–380

Lee J (1997) Biological conversion of lignocellulosic biomass to ethanol. J Biotechnol 56:1–24

Li S, Lai C, Cai Y, Yang X, Yang S, Zhu M, Wang J, Wang X (2010) High efficiency hydrogen production from glucose/xylose by the ldh-deleted thermoanaerobacterium strain. Biores Technol 101:8718–8724

Li Y, Zhang Z, Zhu S, Zhang H, Zhang Y, Zhang T, Zhang Q (2018) Comparison of bio-hydrogen production yield capacity betweenasynchronous and simultaneous saccharification and fermentation processesfrom agricultural residue by mixed anaerobic cultures. Biores Technol 247:1210–1214

Lin CY, Cheng CH (2006) Fermentative hydrogen production from xylose using anaerobic mixed microflora. Int J Hydrogen Energy 31:832–840

Lo YC, Lu WC, Chen CY, Chang JS (2010) Dark fermentative hydrogen production from enzymatic hydrolysate of xylan and pretreated rice straw by Clostridium butyricum CGS5. Bioresour. Technol 101, 5885–5891. https://doi.org/10.1016/j.biortech.2010.02.085

López-Aguado C, Paniagua M, Iglesias J, Morales G, García-Fierro JL, Melero JA (2018) Zr-USY zeolite: efficient catalyst for the transformation of xylose into bioproducts. Catal Today 304:80–88

Lynd LR, Weimer PJ, Van ZWH, Pretorius IS (2002) Microbial cellulose utilization: fundamentals and biotechnology. Microbiol Mol Biol Rev 66:506–577

Miller GL (1959) Use of dinitrosalicylic acid reagent for determination of reducing sugar. Anal Chem 31:426–428

Mohan SV, Pandey A (2019) Chapter 1—Sustainable hydrogen production: an introduction. Biomass, Biofuels, Biochemicals, Biohydrogen (second edition) 1-23

Nagarajan D, Lee DJ, Kondo A, Chang JS (2017) Recent insights into biohydrogen production by microalgae–From biophotolysis to dark fermentation. Biores Technol 227:373–387

Nath D, Manhar Ak, Gupta K, Saikia D, Das SK, Mandal M (2015. Phytosynthesized iron nanoparticles: effects on fermentative hydrogen production by Enterobacter cloacae DH-89. Bull Mater Sci 38:1533–1538

Nigam JN (2000) Cultivation of Candida langeronii in sugarcane bagasse hemicellulosic hydrolysate for the production of single cell protein. World J Microbiol Biotechnol 16:367–372

Nigam JN, Margaritis A, Lachance MA (1985) Aerobic fermentation of D-xylose to ethanol by Clavispora sp. Appl Environ Microbiol 50:763–766

Nissilä ME, Lay CH, Puhakka JA (2014) Dark fermentative hydrogen production from lignocellulosic hydrolyzates -a review. Biomass Bioenergy 67:145–159

Onaran G, Argun H (2019) Direct current assisted bio-hydrogen production from acid hydrolyzed waste paper. Int J Hydrogen Energy 44:18792–18800

Patel SKS, Kumar P, Mehariya S, Purohit HJ, Lee JK, Kalia VC (2014) Enhancement in hydrogen production by co-cultures of Bacillus and Enterobacter. Int J Hydrogen Energy 39:14663–14668

Poladyan A, Baghdasaryan L, Trchounian A (2018) Escherichia coli wild type and hydrogenase mutant cells growth and hydrogen production upon xylose and glycerol co-fermentation in media with different buffer capacities. Int J Hydrogen Energy 43:15870–15879

Singh A, Patel AK, Deka JP, Das A, Kumar A, Kumar M (2019) Prediction of arsenic vulnerable zones in groundwater environment of rapidly urbanizing setup, Guwahati, India. Geochemistry 125590. https://doi.org/10.1016/j.chemer.2019.125590

Taherzadeh MJ, Karimi K (2008) Pretreatment of lignocellulosic wastes to improve ethanol and biogas production: a review. Int J Mol Sci 9:1621–1651

Waheed QMK, Wu C, Williams PT (2016) Hydrogen production from high temperature steam catalytic gasification of bio-char. J Energy Inst 89(2):222–230

Wieczorek N, Kucuker MA, Kuchta K (2014) Fermentative hydrogen and methane production from microalgal biomass (Chlorella vulgaris) in a two-stage combined process. Appl Energy 132:108–117

Winter CJ (2009) Hydrogen energy—abundant, efficient, clean: a debate over the energy-system-of-change. Int J Hydrogen Energy 34:S1–S52

Wu XB, Huang GF, Bai LP, Long MN, Chen QX (2014) Enhanced hydrogen production from xylose and bamboo stalk hydrolysate by overexpression of xylulokinase and xylose isomerase in Klebsiella oxytoca HP1. Int J Hydrogen Energy 39:221–230

Xu JF, Ren NQ, Wang AJ, Qiu J, Zhao QL, Feng YJ, Liu BF (2010) Cell growth and hydrogen production on the mixture of xylose and glucose using a novel strain of Clostridium sp. HR-1 isolated from cow dung compost. Int J Hydrogen Energy 35:13467–13474

Xu J, Upcraft T, Tang Q, Guo M, Huang Z, Zhao M, Ruan W (2019) Hydrogen generation performance from Taihu Algae and food waste by Anaerobic Codigestion. Energy Fuels 33(2):1279–1289

Yin Y, Hu J, Wang J (2019) Fermentative hydrogen production from macroalgae Laminaria japonica pretreated by microwave irradiation. Int J Hydrogen Energy 44(21):10398–10406

Zhao L, Wang ZH, Wu JT, Ren HY, Yang SS, Nan J, Cao GL, Sheng YC, Wang AJ, Ren NQ (2019) Co-fermentation of a mixture of glucose and xylose to hydrogen by Thermoanaerobacter thermosaccharolyticum W16: characteristics and kinetics. Int J Hydrogen Energy 44:9248–9255

Zhu Y, Yang ST (2004) Effect of pH on metabolic pathway shift in fermentation of xylose by Clostridium tyrobutyricum. J Biotechnol 110:143–157

CPSIA information can be obtained
at www.ICGtesting.com
Printed in the USA
LVHW010204030820
662190LV00002B/66